This is a study of the nature and role of science in the exploration of the Canadian Arctic, beginning with the Northwest Passage expeditions of the Royal Navy in 1818, and ending with the Canadian Arctic Expedition of 1913–18. The numerous expeditions launched during these years not only made important contributions to the sciences, but also formed significant parts of the national and imperial histories of the nineteenth and early twentieth centuries.

Many of these expeditions began as British, Russian, American, and Swedish goals in the North came gradually into competition with the development of Canada's emergent nationhood, and with the growth of internationalism in science. Also in competition were notions of the Arctic as a source of knowledge, and as a resource base.

Arctic science today, in itself and in its political, social, military, and economic contexts, is built on the pioneering work discussed in this book. Today's concerns about the Arctic resonate with those of earlier generations of explorer-scientists and the aboriginal people from whom they learned.

# Science and the Canadian Arctic

# Science and the Canadian Arctic

## A Century of Exploration 1818–1918

TREVOR H. LEVERE

*Institute for the History and Philosophy of Science and Technology
University of Toronto*

CAMBRIDGE
UNIVERSITY PRESS

PUBLISHED BY THE PRESS SYNDICATE OF THE UNIVERSITY OF CAMBRIDGE
The Pitt Building, Trumpington Street, Cambridge, United Kingdom

CAMBRIDGE UNIVERSITY PRESS
The Edinburgh Building, Cambridge CB2 2RU, UK
40 West 20th Street, New York NY 10011–4211, USA
477 Williamstown Road, Port Melbourne, VIC 3207, Australia
Ruiz de Alarcón 13, 28014 Madrid, Spain
Dock House, The Waterfront, Cape Town 8001, South Africa

http://www.cambridge.org

First published 1993
First paperback edition 2003

*A catalogue record for this book is available from the British Library*

*Library of Congress Cataloguing in Publication data*
Levere, Trevor Harvey.
Science and the Canadian Arctic: a century of exploration,
1818–1918 / Trevor H. Levere.
p.    cm.
Includes index.
ISBN 0 521 41933 6
1. Scientific expeditions – Arctic regions – History.    2. Scientific
expeditions – Canada – History.    3. Arctic regions – Discovery and
exploration – Canada – History.    I. Title.
Q115.L554    1993
508.719–dc20    92-12570

ISBN 0 521 41933 6 hardback
ISBN 0 521 52491 1 paperback

# Contents

# Illustrations

# Acknowledgments

━━━━━━━━━━━━━━━━━━━━━━━━━━━━━━━━━━━━━━━━━━━━━━━━━━━━━━━━━━

In preparing this book, I have been the fortunate recipient of support, advice, and criticism. I began the project in 1983 in the unrivaled collections of the Scott Polar Research Institute, Cambridge, where Clive Holland and Henry King were tirelessly and generously helpful, and Terence Armstrong was unfailingly encouraging; Robert Headland and Valerie Galpin made subsequent brief visits enjoyable and productive. That year was one of efficient academic luxury, thanks to sabbatical leave from the University of Toronto, the hospitality of the president and fellows of Clare Hall, a fellowship from the John Simon Guggenheim Memorial Foundation, and a grant from the Social Sciences and Humanities Research Council of Canada. That council provided further research support, as did Victoria University in the University of Toronto. I am also grateful to the staff of the Baker Library, Dartmouth College Library, and Philip Cronenwett; the Arctic Institute of North America; the Royal Geographical Society, and Christine Kelly; the Baldwin Room, Metropolitan Toronto Library; the British Library; the British Museum, Natural History, and Dorothy Norman; The Castle Museum, Norwich, and Tony Irwin; Edinburgh University Library; the Huntington Library; the Library of Congress; the Linnean Society, and Gina Douglas; the McCord Museum, Montreal, and Pamela Miller; McGill University Archives, and Robert Michel; the National Archives, Washington, D.C., and Alison Wilson; the Museum of the History of Science, Oxford, and Anthony Simcock; the National Maritime Museum, Greenwich; the National Museum of American History; the National Museum of Natural Science, Ottawa, and David Gray and Stewart MacDonald; the Naval Historical Library; the Nederlands Scheepvaart Museum and Willem Mörzer Bruyns; the Public Archives of Canada; the Public Record Office, Kew; the Redpath Museum, McGill University; the

Royal Botanical Gardens, Kew, and Miss L. E. Thompson; the Royal Scottish Museum, and Alison Morrison-Low; the Royal Society of London, and Sally Grover; the Science Museum, London; the Smithsonian Institution Archives and Susan Glenn; and the libraries of the University of Toronto, especially the Fisher Rare Book Library and Richard Landon, and the Map Library and Joan Winearls. David Knight gave valuable criticism; Suzanne Zeller and Heather Jackson read and commented on an entire draft; Rod Home, Gregory Good, and Anita McConnell helped me with geomagnetism; Jack Cranmer-Byng read and commented on Chapter 10; James Savelle, Elmer Harp, and William Taylor, Jr. helped me to approach anthropology; Gordon Smith and Donat Pharand told me about boundary disputes and freedom of passage; David Hoeniger generously shared with me his transcriptions of manuscripts now in Kew and Cambridge. Tom Symons and Trevor Lloyd encouraged me at the outset; Susan Sheets-Pyenson pointed me toward relevant archives in Montreal; Gordon McOuat helped with geographical distribution and geology. Julie Andrews and Suzanne Zeller were valued research assistants.

In addition to permission from institutions (acknowledged in notes throughout), I am grateful to Mrs. Dorothy A. Smith, Dr. Stuart E. Jenness, and Evelyn Stefansson Nef (Mrs. John Nef), for permission to quote from family papers. Without such generosity, historical research would grind to a halt.

Helen Wheeler and John Kim of Cambridge University Press saw this book through the press; they were helpful, tolerant, and supportive throughout, and I am much in their debt.

# Abbreviations

| | |
|---|---|
| AAAS | American Association for the Advancement of Science |
| AMNH | American Museum of Natural History |
| *ANH* | *Annals and Magazine of Natural History* |
| BAAS | British Association for the Advancement of Science |
| *BJHS* | *British Journal for the History of Science* |
| BL | British Library |
| CAE | Canadian Arctic Expedition 1913-16/18 |
| *DCB* | *Dictionary of Canadian Biography* |
| DCL | Dartmouth College Library |
| FLS | Fellow of the Linnean Society |
| FRGS | Fellow of the Royal Geographical Society |
| FRS | Fellow of the Royal Society of London |
| GSC | Geological Survey of Canada |
| HBC | Hudson's Bay Company |
| IGY | International Geophysical Year |
| IPY | International Polar Year |
| MBA | Mae Bell Allstrand/Anderson |
| NARS | National Archives and Records Service, Washington, D.C. |
| NHSM | Natural History Society of Montreal |
| NMNS | National Museum of Natural Science, Ottawa |
| PAC | Public Archives of Canada |
| *Phil. Trans* | *Philosophical Transactions of the Royal Society of London* |
| PRO | Public Record Office |
| *Proc. R. S. E.* | *Proceedings of the Royal Society of Edinburgh* |
| RA | Royal Artillery |

*Abbreviations*

| | |
|---|---|
| RBG | Royal Botanic Gardens, Kew |
| RGS | Royal Geographical Society |
| RLB | Robert Laird Borden |
| RMA | Rudolph Martin Anderson |
| RN | Royal Navy |
| RSL | Royal Society of London |
| SIA | Smithsonian Institution Archives |
| SPRI | Scott Polar Research Institute, University of Cambridge |
| VS | Vilhjalmur Stefansson |

# Introduction

This is a study of the nature and role of science in the exploration of the Canadian Arctic. It covers the century that began with the British Royal Naval expeditions of 1818 and ended with the Canadian Arctic Expedition of 1913–18, the first major Canadian government-sponsored scientific expedition to the Arctic. That century saw the high noon of the British Empire, and the first great stages in its decline. Canada was a colony in 1818; it achieved Dominion status within the empire in 1867, and was in many ways self-consciously a sovereign nation by 1918.[1] Integral to Canada's self-image as a sovereign nation was its northern status;[2] science in the Arctic was among the tools by which it sought to establish and extend that sovereignty. The century from 1818 encompasses the transition from colonial to national science, from a state of scientific dependence to one of nondependence and integration into an international scientific culture.[3] There are further reasons for ending this study around 1918. In 1920, oil was discovered at Norman Wells on Great Bear Lake in the Northwest Territories, precipitating prompt and major changes in the administration of the North. The ensuing decade marked the advent of the bush plane, transforming access to the Arctic; much work that might formerly have taken months or even years to complete because of problems of access could now be done in

[1] Intention precedes achievement; it was not for another decade that Ottawa believed that the question of its arctic sovereignty was fully resolved, and today that question is very much alive again, and its resolution is by no means assured.

[2] S. D. Grant, *Sovereignty or Security? Government Policy in the Canadian North, 1936–1950* (Vancouver, 1988) p. 3.

[3] R. W. Home and S. G. Kohlstedt, eds., *International Science and National Scientific Identity* (Dordrecht, Boston, and London, 1991) pp. 1–5, 32, 50. S. F. Cannon, *Science and Culture* (New York, 1978) p. 100; the essay by Cannon on Humboldtian science in that volume (pp. 73–110) provides context and background for many of the themes discussed here.

1

visits of a few days or weeks. The same ease of access facilitated the militarization of the Canadian Arctic, beginning in the 1920s. Taken together, these factors marked a radical transformation in the development of arctic science.[4]

Science is socially based, and has been incorporated in successive phases of national development.[5] The colonial phase was a manifestation of cultural hegemony,[6] with primarily British models, institutions, and even individuals dominating the scene. "Metropolitan science *was* science.... *Colonial science* . . . was . . . 'imperial science seen from below'. " Imperial science was carried on throughout the empire for economic and intellectual gain, and in such a case, "the central issue becomes no longer science *in* imperial history, but science *as* imperial history."[7] In these terms, a large part of this book is imperial history.

Science is more than its context: it has its inner dynamic, directed through its institutions and applied through instruments and concepts to an uncompromising natural world. Navigation and even survival in the Arctic demanded knowledge of landforms and ice conditions, currents, weather, the distribution of animals and birds that might serve as food, the limitations of the magnetic compass as one neared the pole, navigation by the sun and stars, ways of satisfying the physiological and psychological needs of men through arctic summers and winters – enough, indeed, to constitute an extensive and utilitarian program of scientific observation, in addition to whatever observations curiosity dictated and time allowed.

The sciences thus featured prominently in arctic exploration.[8] There were approximately two hundred expeditions to the Canadian Arctic be-

[4]  Home and Kohlstedt, *International Science*, pp. 14–15.
[5]  The notions of colonialism, imperialism, and nationalism have recently come under scrutiny. On colonialism, see N. Reingold and M. Rothenberg, eds., *Scientific Colonialism: A Cross-Cultural Comparison* (Washington, D.C. and London, 1987). For nationalism and internationalism, see Home and Kohlstedt, eds., *International Science*, and the broader account in E. J. Hobsbawm, *Nations and Nationalism Since 1780: Programme, Myth, Reality* (Cambridge, 1990). The most extended treatment of imperialism and science is L. Pyenson, *Cultural Imperialism and Exact Sciences: German Expansion Overseas, 1900–1930* (New York, 1985); *Empire of Reason: Exact Sciences in Indonesia* (Leiden and New York, 1989); see also his brief introduction, "Science and Imperialism," in R. C. Olby, G. N. Cantor, J. R. R. Christie, and M. J. S. Hodge, eds., *Companion to the History of Modern Science* (London and New York, 1990) pp. 920–33.
[6]  Reingold and Rothenberg, *Scientific Colonialism*, p. x.
[7]  R. Macleod in *Scientific Colonialism*, pp. 219, 220. Note also Macleod's statement (p. 232) that "where there was trade, there was the navy, and where the navy sailed, or the army rested, the natural sciences benefited."
[8]  Lamson in D. L. VanderZwaag and C. Lamson, eds., *The Challenge of Arctic Shipping: Science, Environmental Assessment, and Human Values* (Montreal and Kingston, London,

tween 1818 and 1918.[9] Roughly eighty of them, or forty percent, came back with results that led to scientific publication. In addition, most of the two hundred expeditions led to an extension of geographical knowledge; and in the nineteenth century, the sciences had not been narrowed to their present boundaries, so that geography was still recognized as a science.[10] Most of the officers commanding Royal Naval arctic expeditions in the nineteenth century were elected as Fellows of the Royal Society of London because of their contributions to geographical science. This aspect of northern exploration has been often and well described.[11] Discussions of geography here will, in contrast, mostly be either to indicate its scientific range, or to show the emerging tension between geographical discovery and competing forms of disciplined scientific observation. The heroism, suffering, and occasional dramatic disaster of arctic exploration have also been well described. Thus I have generally subordinated questions of temperament and personality to the scientific work of expeditions. Nonetheless, such questions may bear on scientific developments, and we shall then consider them: Adolphus Greely's limitations as a commander contributed to the fragmentation and failure of his scientific expedition, just as John Rae's robustness of character and physique underlay his successes.

More often, the stuff of myth and legend is only marginal to our story. John Franklin's first arctic expedition was a disaster, with its survivors being remembered as the men who ate their boots; his last expedition has entered British and Canadian mythology, transforming the "arctic sublime"[12]

Buffalo, 1990) p. 3 notes that some historians insist that early explorations, which made no systematic observations, "belong to a pre-scientific era of arctic operations," whereas others label early navigators "the 'founders' of arctic science." Here I take a social and institutional view: what was regarded then as science by scientists and the spokesmen for scientific institutions counts as science.

9   A. Cooke and C. Holland, *The Exploration of Northern Canada 500 to 1920: A Chronology* (Toronto, 1978).

10   M. B. Hall, *All Scientists Now: The Royal Society in the Nineteenth Century* (Cambridge, 1984) pp. 199–215 discusses the nineteenth-century Royal Society's encouragement of exploration, including (pp. 200–5) arctic exploration, and the scientific status of geography. For discussions of earlier scientific geography, see E. G. R. Taylor, *Late Tudor and Early Stuart Geography 1583–1650* (London, 1934), and L. Cormack, *Non Suffick Orbem: Geography as an Interactive Science at Oxford and Cambridge, 1580–1620*, Ph.D. thesis, University of Toronto (Toronto, 1988). Scientific geography as it emerged in the Renaissance and Scientific Revolution was virtually a precondition for the work discussed here. See also D. N. Livingstone, "The History of Science and the History of Geography: Interactions and Implications," *History of Science* 22 (1984) 271–302.

11   The most recent work is P. Berton's admirable *The Arctic Grail: The Quest for the North West Passage and the North Pole 1818–1909* (Toronto, 1988).

12   The phrase is Chauncey Loomis's, in his essay of that title in U. C. Knoepflmacher and G. B. Tennyson, eds., *Nature and the Victorian Imagination* (Berkeley, 1977) pp. 95–112.

into a threatening realm. Starvation, cannibalism, heroism, murder, execution, and ultimate disappearance without trace were part of Franklin's expeditions. The Admiralty's instructions, however, simply stressed geographical discovery and natural science. Franklin worked with the instrument maker Robert Were Fox to develop and promote an improved instrument for measuring magnetic dip and force; geomagnetic work formed a major motive for his last expedition. His second arctic expedition, less remembered than his first or last because it was successful, made major contributions to geography and natural history. Readers of this book will learn why one of the naval officers searching for Franklin described him and his crew as sacrifices in the cause of knowledge.

The framework for this study is set by history and geography. Until the second half of the last century, the history of what is now northern Canada and its arctic archipelago could be broadly split into two. South of the tree line, the fur trade dominated.[13] North of that line, the main motive was the search for the Northwest Passage.[14] The principal agents in the boreal forest were the Hudson's Bay Company and the rival North West Company, hunting and trading in Indian territory.[15] After the companies amalgamated in 1821, the Hudson's Bay Company enjoyed virtually exclusive sway. They provided logistical support for numerous expeditions, including Franklin's, gave selective help to collectors for museums, and sometimes incorporated scientific work in the expeditions they sent out. Their work was mostly on the mainland, stretching to the arctic coast, and much less in the islands to the north.

The northern sea and islands were the realm of whalers and sailors, and of the most northern aboriginal people, the Eskimos or Inuit, who lived and hunted on land and ice. Geographical discoveries in the Arctic were almost always discoveries only to the Europeans, Canadians, or Americans who made them: the explorers often depended for guidance and even survival on the Inuit, whose local knowledge was impeccable, who traveled widely, and who had a pretty fair idea of neighboring topography for many days' travel.[16] The search for the Northwest Passage took place within this context of native expertise, and was, in the aftermath of the Napoleonic Wars, primarily an enterprise of the Royal Navy: only gradually did other powers

---

[13]  The classic study is Harold Innis, *The Fur Trade in Canada: An Introduction to Canadian Economic History* (New Haven, Conn., 1930).

[14]  L. H. Neatby, *In Quest of the North West Passage* (Toronto, 1958).

[15]  Canada's northern peoples, Indian and Inuit, did not cede their land by treaty, purchase, or force of arms.

[16]  See, e.g., J. Ross and J. M. Savelle, "Round Lord Mayor's Bay with James Clark Ross: The Original Diary of 1830," *Arctic* 43 (1990) 66–79 at 67–9.

and other nations join the search. Those sciences important for navigation were inevitably a first concern for the navy; I shall in general not include in this study navigational practices merely dependent upon science, for example the use of quadrants and of the *Nautical Almanac,* although I shall have a little to say about them in Chapters 1 and 4.[17] What will concern us are observations, collections, and occasionally experiments directed to the confirmation or extension of knowledge.

It is striking that natural history and geology were regularly included in the Admiralty's instructions to its arctic commanders; indeed, these sciences featured prominently in the imperial mandate on land and sea, in Africa, India, Australia, and elsewhere during the nineteenth century.[18] Investigations in natural history might satisfy both scientific curiosity and economic opportunity. Geology was in like case.[19] The potential utility of geological investigations was a large part of their attraction: the discovery of valuable mineral ores might more than justify the expense of an arctic voyage. As coal burning steam vessels entered the Arctic, the search for coal was added to that for ores. Readily available fuel could extend the range and duration of an expedition, and might thereby facilitate a transit of the passage[20] through the archipelago.

[17] Accurate navigation was much improved thereby. See D. M. Knight, *The Nature of Science: The History of Science in Western Culture Since 1600* (London, 1976) p. 142.

[18] For the Australian case, see I. Inkster and J. Todd in R. W. Home, ed., *Australian Science in the Making* (Cambridge, 1988): "The early concentration on staple exploitation led to an emphasis on a widely-based 'natural history' programme within the scientific enterprise as a whole. Natural history was suited to both the economic needs of Britain and her colonies and the psychological requirements of the scientific workers concerned." See also D. Fleming, "Science in Australia, Canada, and the United States: Some Comparative Remarks," *Proceedings of the 10th International Congress of the History of Science, Ithaca 1962* (Paris, 1964) pp. 179–96, arguing that natural history was favored in colonial societies because it appeared to be the intellectual aspect of pioneering. The role of the "inventory sciences," encompassing natural history, in Canada is explored in S. Zeller, *Inventing Canada: Early Victorian Science and the Idea of a Transcontinental Nation* (Toronto, 1987). The British background is explored in D. A. Allen, *The Naturalist in Britain: A Social History* (London, 1976).

[19] The adoption of a stratigraphic order derived from Europe, and its application to remote realms, could serve conceptually as a form of territorial imperialism. J. A. Secord, *Controversy in Victorian Geology: The Cambrian-Silurian Dispute* (Princeton, 1986) pp. 30, 122, and "King of Siluria: Roderick Murchison and the Imperial Theme in Nineteenth-Century British Geology," *Victorian Studies* 25 (1982) 413–42. See also R. A. Stafford, "Geological Surveys, Mineral Discoveries, and British Expansion, 1835–71," *Journal of Commonwealth and Imperial History* 12 (1984) 5–32.

[20] As indicated here, the Northwest Passage is the entire northern route from Europe to Asia. I use passage here for that part of the route that lies through the Canadian Arctic archipelago, but in the ensuing chapters I shall follow the nineteenth-century usage of scientific navigators and explorers, and refer to the whole route and that part of it that lies within the archipelago as the Northwest Passage.

Until very recently, when the feasibility of building a class 8 icebreaker[21] appeared to enter the political as well as the technical realm in Canada,[22] the Northwest Passage from Europe to the Far East by way of the Canadian arctic archipelago was a seldom-realized dream. It dated back to the Renaissance and had endured in spite of repeated evidence that the passage, barely navigable at best, could not be a viable trade or military route. The problem was ice, which in the archipelago often took forms far harder to penetrate than the pack of the Arctic Ocean. Yet the Royal Navy and a host of successors kept sending ships and explorers to find and sail through the passage, or else to reach an economically and scientifically still more useless target, the North Pole. Dreams of glory, national pride, scientific curiosity, employment for the navy in peacetime, and aims at territorial and economic monopoly were all involved, in various combinations. But arching over them all was the continuing myth of the arctic sublime, with its vision of the passage. "The Northwest Passage is as much a cultural artifact as an Arctic navigation route. It is as much a metaphor for human perseverance and ingenuity as a physical reality."[23] Science too is a cultural artifact, and we shall be exploring the interaction of those two artifacts in investigating the sweep and the detail of the arctic environment.

"Nordicity" is an essential element in Canada's self-image.[24] The northern frontier has repeatedly served to focus national aspirations, in the Reform party's platform in 1857, in Diefenbaker's campaign for a Conservative majority in 1958, with it goal of a new Canada, "a Canada of the North," and at intervals before, between, and since. In such campaigns, and in the policies entailed by them, science has been co-opted or otherwise involved:[25] this has sometimes been to its advantage, for example in the in-

---

21   This is defined as an icebreaker capable of maintaining continuous headway of three knots through solid ice ten feet thick, without backing up and ramming. Such a vessel would, at least in theory, make possible year-round navigation through the archipelago. See K. R. Nossal, "Polar Icebreakers: The Politics of Inertia," in F. Griffiths, ed., *Politics of the Northwest Passage* (Kingston and Montreal, 1987) pp. 216–38.

22   J. Clark, statement on sovereignty, in Canada, House of Commons, *Debates,* 10 Sept. 1985, 6462–4; reprinted in Griffiths, *Politics of the Northwest Passage,* pp. 269–73. The Canadian government has since reneged on its commitment to build such an icebreaker. This and other developments have prompted inquiries of the kind informing John Honderich, *Arctic Imperative: Is Canada Losing the North?* (Toronto, 1987).

23   Griffiths, *Politics of the Northwest Passage,* p. 3.

24   An extended and thoughtful essay is L.-E. Hamelin, *Canadian Nordicity: It's Your North Too,* trans. W. Barr (Montreal, 1979).

25   Cf. Lamson in VanderZwaag and Lamson, *Arctic Shipping,* p. 5: "Debate about the origins of arctic science may be of interest to scholars; however, there is general consensus that political and economic circumstances were – and remain today – principal factors in

ception of Canada's Polar Continental Shelf project in 1958.[26] The Arctic has never before been so accessible.

Those politicians who over the years have directed attention northward have recognized what is for Canada a looming geographical fact. In a timely look at the circumpolar North, Armstrong, Rogers, and Rowley proposed a broad and comfortably elastic approach to the arctic and subarctic regions. There are, as they observed, many definitions of these regions, and we shall briefly note some of them in this book. The tree line, the Arctic Circle, the southern limit of permafrost, and other more complicated demarcations have been used as boundaries. Armstrong and his coauthors proposed, however, not to adopt any single definition, but rather to think of the regions as "a group of concepts and attributes, concerned with climate, vegetation, fauna, presence of ice and snow, sparseness of human habitation, ... and many other factors." In these terms, Canada "is a northern nation in a more important sense than the others; it is the second largest country in the world, and three-quarters of it lie within our area."[27] If one considers only land north of the tree line, Canada has more arctic territory than any of the other seven arctic nations.[28]

Our main geographical focus will be the Canadian archipelago, and the continental coast from Hudson Strait between Davis Strait and Baffin Island in the east to the Beaufort Sea and the Alaska boundary in the west. This was the maze of ice, sea, cold, reefs, and currents that confronted those who would navigate the passage in the few short months of arctic summer. Important also will be the sub-Arctic and Arctic down the Mackenzie River from Great Bear Lake, and northern regions reached from the Mackenzie Valley. We shall look at expeditions, at their mandates, and at their scientific work. Where this spills over, as it did into the Rocky Mountains in John Franklin's second expedition, we shall incorporate such work in our journey. Similarly, expeditions making preliminary stops on Greenland's west coast did scientific work there; except for such studies, Greenland will

promoting scientific enterprise." In fact, scholars have paid very little attention to "the origins of arctic science"; hence this book.

[26] There are many arguments about the deleterious effects of the combination of politics and bureaucracy on arctic science. See, e.g., chaps. 7 and 8 in VanderZwaag and Lamson, *Arctic Shipping*. For the Polar Shelf project, see M. Foster and C. Marino, *The Polar Shelf: The Saga of Canada's Arctic Scientists* (Toronto, 1986).

[27] T. Armstrong, G. Rogers, and G. Rowley, *The Circumpolar North: A Political and Economic Geography of the Arctic and Sub-Arctic* (London, 1978) pp. 1, 4. To arrive at this figure, they include the northern parts of the provinces (except Nova Scotia, New Brunswick, and Prince Edward Island) with the whole of Labrador, the Northwest Territories, the Yukon Territory, and the Arctic archipelago.

[28] Ibid., p. 73.

be outside our story.[29] The exclusion of Alaska in the west conforms largely though not entirely to political realities: Alaska was Russian before the United States purchased it, and did not welcome investigations by other nations. Where expeditions aiming at Canadian territory carried out work in Alaska, as happened with the Canadian Arctic Expedition of 1913–18, we shall note it. Baffin Bay was important for the earliest nineteenth-century naval expeditions: from the period of the Franklin searches, in mid-century, expeditions and explorers of several nationalities pushed northward between Greenland and the archipelago. These waters and their environing lands will assume increasing prominence during our discussion of the second half of the nineteenth century.

I refer interchangeably to the Canadian and the arctic archipelago. Britain assumed sovereignty over the archipelago during the nineteenth century, and some of the limits of that sovereignty were defined by treaty and diplomacy with Russia and with the United States. But there were serious potential conflicts, especially between Britain and the United States and in an attempt to head off such conflicts, Britain transferred the archipelago to Canada in 1880. Most but not all of the islands in the archipelago had been discovered by the British, but there remained whole areas of the Arctic Ocean that were unexplored even by 1880; the archipelago is difficult of access and it is large, with six of the thirty largest islands in the world. At the turn of the century, the Norwegian Otto Sverdrup discovered a major new group of islands west of Ellesmere Island. Canada's title under these circumstances was precarious, in spite of a plaque subsequently placed on Melville Island claiming the archipelago from the mainland to the North Pole.[30] Questions of sovereignty underlay much of the deployment of science in these years.

If science and colonialism, imperialism, and nationalism provide recurring themes, so too does the interplay of science and internationalism. The Arctic is circumpolar, shared by nations with similar geographical conditions and with continental shelves sloping into the same all but landlocked ocean. Ocean currents and ice movements, the migration of animals and birds, the distribution of plants, the variation of geophysical phenomena,

[29]   Greenland needs a separate study. The best early account is H. Rink, *Danish Greenland: its People and Products* (London, 1877; reprinted London, 1974); T. R. Jones edited a valuable collection of scientific papers, *Manual of the Natural History, Geology, and Physics of Greenland and the Neighbouring Regions . . .* (London, 1875) and there is an important journal, *Meddelelser om Grønland.*

[30]   M. Zaslow, *The Opening of the Canadian North 1870–1914* (Toronto and Montreal, 1971) p. 267. The plaque was placed there by Captain J. E. Bernier in the winter of 1908–9; see chap. 9.

and (with some contested restrictions of recent date) the movement of native peoples are no respecters of political frontiers. The need for international cooperative research was recognized in the nineteenth century. The Arctic was an important region for geomagnetic research, containing as it does the north magnetic dip pole: the international magnetic study instituted by the explorer-scientist Alexander von Humboldt and the mathematician Carl Friedrich Gauss of Göttingen led to the particular prominence of magnetic work in explorations of the Canadian Arctic. The Imperial German Council was the first government to endorse the principle of coordinated scientific work, in an enterprise that, drawing on the vision of one man, the Austro-Hungarian officer Weyprecht, led to the International Polar Year of 1882–3. And, to take just one other example, Charles Darwin's ideas about geographical distribution led to the incorporation of botanical studies in the Canadian Arctic with those of other arctic regions. As Weyprecht observed over a century ago, the earth "should be studied as a planet. National boundaries, and the North Pole itself, have no more or less significance than any other point on the planet, according to the opportunity they offer for the phenomena to be observed."[31] In these terms, the Arctic was a global laboratory, in contradistinction to a strategic resource depot.[32]

These issues will be pursued in the following chapters, sometimes thematically, more often within the chronological framework of the history of arctic expeditions and exploration. Since the development of a naval tradition of scientific work was a necessary prelude to the scientific mandate of Royal Naval arctic expeditions, a preliminary chapter discusses the emergence of that tradition, and its first deployment in the Arctic.[33] The choice of expeditions discussed is based on their scientific significance, and the degree to which they help to elucidate informing themes. This means that we shall often ignore expeditions that are prominent in the annals of exploration, but of only trivial scientific significance, such as Peary's attempts at reaching the pole. Where several well-documented expeditions could serve to make essentially the same points in a given period, I have discussed the expeditions that make those points with the greatest clarity, using economy, drama, and the extent of documentation as additional selection criteria. The most numerous scientific expeditions in the Canadian Arctic from 1818

---

[31] Weyprecht, as quoted in VanderZwaag and Lamson, *Arctic Shipping*, p. 16.

[32] Lamson in *Arctic Shipping*, pp. 16–18.

[33] This was Constantine Phipps's expedition, which got as far as Svalbard (Spitsbergen), and it is the only expedition to the east of Greenland that we shall notice. After the Napoleonic Wars, British Arctic exploration concentrated on the Canadian Arctic, the subject of chapters 2 to 10 in this book.

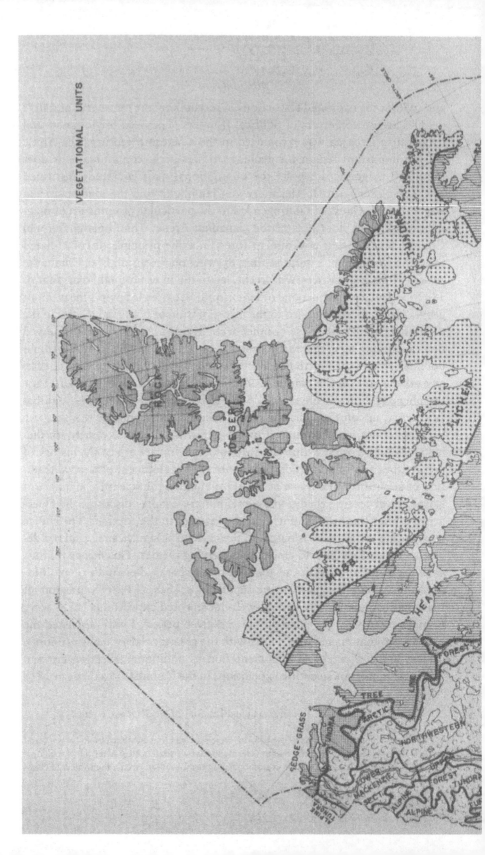

VEGETATIONAL UNITS

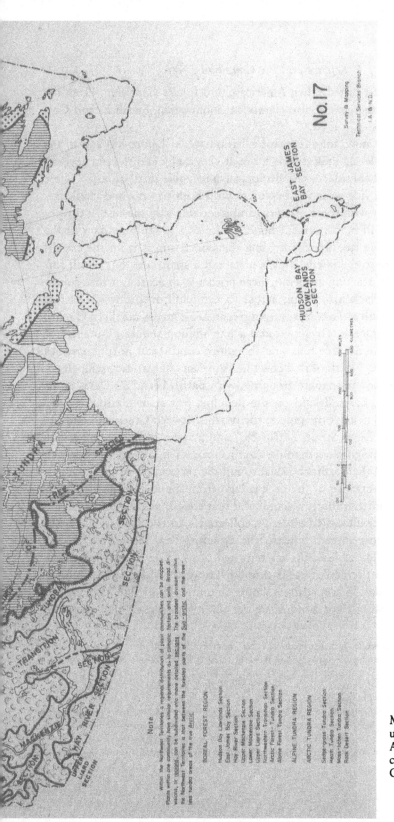

Map 1. Vegetational units of the Canadian Arctic, National Archives of Canada NMC C-138091.

to 1918 were first British, then American, and finally Canadian. From the 1880s onward, there were also significant Norwegian, Swedish, and German expeditions.

So much, for now, for the historical framework. Before moving to the main narrative, let us look a little more at geography and the physical environment, and consider some ecological and zoological issues.[34] These combined to provide scientific explorers with both context and challenge. Readers familiar with these matters can reasonably skip the rest of this introduction. It will be convenient here to take as a rough boundary of the arctic regions the tree line,[35] dividing the boreal forest from the barren lands or tundra to the north. This is often not a sharp line, but it will suffice. In Canada, starting at the northern border with Alaska, it runs southeast from the Mackenzie Delta, across to Churchill on the west coast of Hudson Bay, south to James Bay, around it and east across northern Quebec to the Atlantic Ocean. Over two and a half million square kilometers of Canada lie above the tree line, and the entire continental north coast lies within the Arctic. North of that coast lies the Canadian arctic archipelago, a vast and roughly triangular grouping, with Baffin Island on Davis Strait at its southeast, Banks Island on the Beaufort Sea at its southwest, and Ellesmere Island nearest the pole to the north. Between Greenland and the archipelago are Davis Strait, Baffin Bay, and a series of narrowing straits and sounds as one goes north to the Arctic Ocean. The passage lies between Davis Strait and the Beaufort Sea, through the sea and ice of the archipelago. The chief entrance to the archipelago from the east is through Lancaster Sound, north of Baffin Island; and that sound, passing into Melville Sound and thence into M'Clure Strait, collectively known as Parry Channel, provides the most direct route to the Beaufort Sea. Ice movements in that sea mean that at least the western end of M'Clure Strait is regularly blocked, and so the best route is through Prince of Wales Strait between Banks Island and Victoria Island, into Amundsen Gulf and then to the Beaufort Sea. Strictly speaking, there are several passages[36] that enter the

---

[34]    There are numerous specialist studies of these matters, but good introductory texts will give us the necessary background and framework here. In the following pages I am particularly indebted to Armstrong, Rogers, and Rowley, *The Circumpolar North;* B. Stonehouse, *Polar Ecology* (London and New York, 1989); B. Sage, *The Arctic and its Wildlife* (New York and Oxford, 1986).

[35]    This boundary corresponds fairly closely to the 10° July isotherm, and more closely to the "Nordenskiöld line," connecting places where the mean temperature for the warmest month is (9 − 0.1k), where k is the mean temperature for the coldest month.

[36]    For the possible routes, see D. Pharand, *Northwest Passage: Arctic Straits* (Dordrecht, 1984).

archipelago either through Lancaster Sound or through Hudson Strait, between northern Quebec and Baffin Island, and then go south to the continental coast to the east of Victoria Island, and proceed onward to Amundsen Gulf and the Beaufort Sea. These routes are likely to be less passable, because of ice, than the main one via Parry Channel and Prince of Wales Strait. The passage was found piecemeal in the course of the searches for John Franklin in the mid-nineteenth century, but its difficulty was such that it was not completed until 1906 by Roald Amundsen in the *Gjoa*, who began his voyage across the Atlantic three years earlier.

Sea ice was the greatest challenge. It is protean, but can be considered in three main divisions: polar pack ice over the deep ocean, beyond the continental shelf, and consisting of multiyear ice and ridges; landfast ice, growing out from the shore and generally newly formed each year, although containing some multiyear ice and ridges; and ice of the transition zone, a shear zone between the fixed landfast ice and the moving polar pack ice. The motion of the pack and the shear or transition zone arise from ocean currents, winds, and Coriolis forces produced by the earth's rotation. The main influx of sea water into the Arctic is through the North Atlantic drift, on the Norwegian side of the ocean; the main outflow is down the east coast of Greenland. There is, however, a lesser but important net influx from the Pacific Ocean through Bering Strait, and a net efflux eastward through the Canadian archipelago. This pattern is complicated by a surface current flowing northward along the west coast of Greenland, which turns west and south where it is met by the flow down Smith Sound. It continues down the west side of Baffin Bay, reinforced by the flow from Lancaster Sound, and bearing with it ice floes and icebergs. The latter, calved from the glaciers of west Greenland, can be huge, seventy meters high and more above water, five times as much below. Dodging icebergs in Baffin Bay and even in the North Atlantic was one of the first but not the least hazards for northern navigators coming from Europe, as the *Titanic* amply and tragically demonstrated. Within the Canadian Arctic, the main circulation of pack ice and floes is the Beaufort Gyre,[37] a clockwise circulation that is largely wind-driven, and that produces the buildup of ice at the western end of M'Clure Strait – hence the impassibility of that end of Parry Channel.[38]

---

[37] R. H. Herlinveaux and R. R. de Lange Boom, *Physical Oceanography of the Southeastern Beaufort Sea*, Beaufort Sea Technical Report No. 18, Dept. of the Environment (Victoria, B.C., 1975). The other principal circulation of ice in the Arctic begins north of Bering Strait and Siberia, crosses the middle of the Arctic Ocean, and heads for the Atlantic via the east coast of Greenland.

[38] L. K. Coachman, "Physical Oceanography in the Arctic Ocean," *Arctic* 22 (1969) 214–24.

Surface current patterns and therefore ice movements within the archipelago are decidedly complicated. Sea ice is always present in the northernmost channels; elsewhere it grows through the winter into June, until it is up to eight feet thick. Some of it melts in situ in July and August, some of it is carried into the Atlantic and melts there, and some of it is piled up, especially in early winter storms, to form thick rafts and rough ridges. Pack ice containing old ice floes is accordingly "much more difficult for a ship to penetrate and can cause more severe damage."[39] Within the archipelago, the prevailing winds are from the north and northwest, and pack ice is driven south and southeast, blocking channels to the south of Barrow Strait and Lancaster Sound. Some areas of the sea are free from ice or covered only with thin and early vanishing ice year after year, while all around them the sea is ice-covered; these open areas, called polynyas, are produced either by an upwelling of warmer water, for example from the Atlantic drift, or by currents removing ice, or by winds from the land blowing ice away from the coast as it is formed, or by some combination of these factors. The largest of them, the North Water[40] at the north of Baffin Bay, became important for explorers and whalers. Whalers in Baffin Bay learned early on about sea ice, but it is little wonder that the passage was so long impenetrable. At the beginning of the nineteenth century's spate of expeditions, the archipelago was largely terra incognita for explorers, although not for the Inuit. Only after the second world war were the last discoveries of new land made, when aerial photography revealed the existence of hitherto unknown islands in Foxe Basin, between Baffin Island and Hudson Bay.

Since access to the archipelago was by sea and ice, its landforms extending inland from the coasts long remained unknown. The main geological formation is the Canadian shield, the worn roots of Precambrian mountains, lying in an arc around Hudson Bay, and constituting well over half of the Canadian North; it is a region of ice-scoured granite and gneiss and countless lakes. Low in the east, it is tilted up in the west to form a line of cliffs and mountains along the coasts of Davis Strait, Baffin Bay, and the north Labrador Atlantic coast. The shield, dipping down and covered with the Paleozoic strata of the coastal plain in the west, surfaces to occupy most of the continental Canadian Arctic, as well as most of the central and eastern islands of the archipelago south of Parry channel, and parts of Devon

---

[39]  G. Rowley, "Bringing the Outside Inside: Towards Development of the Passage," in Griffiths, ed., *Politics of the Northwest Passage*, pp. 25–45 at p. 30.

[40]  Wind off the land is believed to be the main agent in keeping this free. See M. and M. J. Dunbar, "The History of the North Water," *Proc. Royal Society of Edinburgh* (B) 72 (1971/72) 231–241 at 240.

Island and Ellesmere Island. North of the shield are the Queen Elizabeth Islands, along the eastern edge of the Beaufort Sea, and most of the islands north of Parry Channel. There, a sequence of Paleozoic, Mesozoic, and Tertiary deposits has been folded into rich complexity, and constitutes enormously thick sedimentary strata. As in the shield, the highest mountains are in the east, on Ellesmere Island. There are ice fields on the higher mountains, and most of the archipelago north of Lancaster Sound is either ice field, desert, or semidesert.

One tends to think of deserts as hot, but the high Arctic (bounded to the south roughly by the 5° July isotherm) is mostly desert, and the main reason for this is the low precipitation it experiences. Through the long winter, a strong anticyclone sits over the central pack ice, producing clear skies and light centrifugal winds from November to May. Winters are cold, although not as cold as in the Antarctic; the mean minimum winter temperature at Devon Island, where total darkness lasts three months, is just below −40°. The thaw begins in March or April, but mean temperatures rise above freezing in the high Arctic for no more than three months. As the winter anticyclone weakens, depressions come in from the south, bringing precipitation with them. Cold limits the amount of water vapor that the air can hold, so that the arctic atmosphere in winter is especially dry, but the same principle leads to high relative atmospheric humidity, thereby limiting evaporation. Permafrost, lying within a meter of the surface of the soil, inhibits the absorption and run off of melt water. Thus desert and semidesert regions can still have enough available water to support suitably adapted vegetation, even though precipitation is light, from 70 mm to somewhat over 200 mm.

Cold polar regions, such as we have today, are in fact the exception rather than the norm in the earth's history. Without ice caps, solar radiation and the circulation of sea and air carry temperate climates into high latitudes. Global cooling, the upheaval of land, and continental drift have combined to give us the present frigid poles. In spite of the currently observed trend of global warming, we are still in an ice age: Antarctica, Greenland, and the ice caps along the east of the Canadian archipelago have not emerged from glaciation. The previous ice age was 250 million years ago. The present one began some 20 to 30 million years ago, and within it, the arctic climate changed from warm to cool 5 to 10 million years ago. The ice hardened 3 to 5 million years ago, and spread from the uplands to the plains. As the cold intensified and ice accumulated on land, the sea level dropped, and a land bridge, Beringia, appeared across Bering Strait between America and Asia. This happened repeatedly, and allowed Asian species of

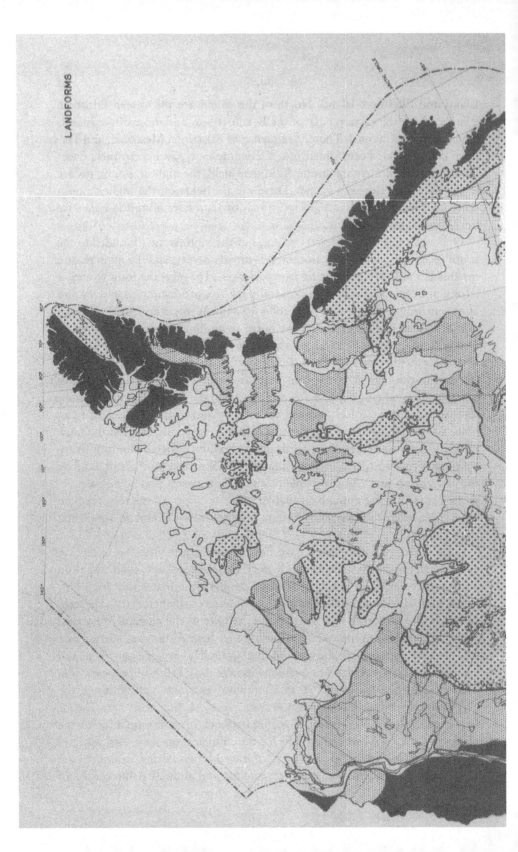

No.6

Surveys & Mapping
Technical Services Branch
I.A.B.N.D.

LEGEND

Mountains

Uplands

Plateaus

Lowlands

Only major landforms (mountains, uplands, plateaus and lowlands) are shown on this map

Map 2. Landforms of the Canadian Arctic. National Archives of Canada NMC C-137050.

animals to migrate to the American continent. This migration is the source of the circumpolar distribution of many species. The final glacial period began about 100,000 years ago, and included three cold spells, of which the most recent ended 20,000 years ago. Glaciation stripped the soil from many rocks, contributing to desertification.

Since in geological terms, and even with reference to the history of life on earth, this ice age is a recent phenomenon, the arctic regions are ecologically young. The last big glacial period ended 3,000 to 15,000 years ago, with the retreat of the Pleistocene ice sheets; tundra is the most recently evolved of the world's major ecosystems. Only a few pre-Pleistocene species survived the periods of glaciation, and for other species, there has not been much time to adapt to polar conditions: "most polar species are still currently invading and adapting from neighbouring temperate regions; the paucity of polar species, in comparison with temperate or tropical species, is due partly to lack of time."[41] There are around 4,000 species of land mammals in the world, but only 50 or so have moved into the circumpolar Arctic, including 8 found only on the polar fringe, and more than another dozen never found in the northernmost zones. There are 182 species of birds in the circumpolar Arctic, 105 of which are found in the Canadian Arctic, but only a handful of which remain there year-round. Strategies of survival[42] for animals and birds include migration, insulation, and, for small mammals like lemmings, taking advantage of microclimates; their tunnels beneath the snow may enjoy temperatures many degrees warmer than the atmosphere. Plants keep low, may take several seasons to complete their reproductive cycle, are resistant to freezing, and are able to cope with dramatic changes of temperature. Leaves and catkins that have started to shoot in spring sunshine and are then frozen solid in a sudden cold spell continue their growth unharmed as soon as the temperature rises again. In general, coping with severe fluctuations in temperature, both seasonal and short-term, is more of a challenge than coping with cold: "polar habitats are harsh and highly variable, unbuffered and unpredictable."[43] The combination of these environmental variables with ecological systems where species are few, food webs simple, and food sources generally limited, leads to boom and bust cycles of animal population. The lemming cycle is one of the most dramatic, and because lemmings are a main food source for predators, there are related oscillations in their numbers.

---

[41]  Stonehouse, *Polar Ecology*, p. 15.
[42]  *Arctic* 44 no. 2 (June 1991) iii–iv, 95–164 is devoted to "Life in the Polar Winter – Strategies of Survival."
[43]  Stonehouse, *Polar Ecology*, p. 160.

Humankind, originating in the tropics, came into the polar regions of Siberia more than 20,000 years ago, crossed into North America by way of Beringia around 10,000 years ago, and spread south and east. They evolved ways of living on and off the land, sea, and ice through hunting and the use of animal skins. Their patterns of life were remarkably stable, and survived in many ways unchanged into the last century.[44]

In 1818, this world was almost entirely unknown to Europeans and their North American descendants. In the following chapters, we shall see how that situation changed over the ensuing century.

[44] Good introductory accounts are K. Birket-Smith, *The Eskimos* (London, 1959), and K. Rasmussen, *The People of the Polar North: A Record*, G. Herring, ed. (London, 1908).

# 1

Science and the Navy

The first Elizabethans and their fathers were a hard-headed, adventurous, entrepreneurial crew. They sought to expand trade, dominion, empire, knowledge, horizons, and profits. Guns and ships, clocks and navigational devices, and merchant banks and joint stock companies were the chief instruments of their expansion.[1] Spices, gold, silks, new lands, and n w knowledge were there for whoever was skilled, courageous, powerful, or lucky enough to take them. Geographers, mathematicians, astrologers, ships' captains and ship builders, instrument makers, poets, philosophers, diplomats, adventurers, and investment bankers all sought to share in that explosion of European confidence, power, and knowledge that was the triumphal outcome of the Renaissance. That expansion came later to England than to Italy, but it came with all the vigor of a late bloom. The Sidneys made Penshurst in Kent a primary focus of that flowering,[2] and thither came Richard Chancellor, navigator extraordinary, and Sir Hugh Willoughby, sea captain and the "brave Lord" of the ballads. They were responding to a problem posed in 1494 for all of northern Europe when the pope, in his treaty of Tordesillas, divided control of the sea lanes between the maritime powers of Spain and Portugal: those who wanted to trade with the Moluccan or Spice Islands of the Orient sought a northern route. In 1553 Chancellor and Willoughby set out "for the search and discovery of

---

[1] C. M. Cipolla, *Clocks and Culture 1300–1700* (London, 1967); *Guns and Sails in the Early Phase of European Expansion* (London, 1965).

[2] Philip Sidney is perhaps best known as a patron of literature: see John Buxton, *Sir Philip Sidney and the English Renaissance*, 2nd ed. (London, 1964). For the Sidneys's involvement with science and mathematics at Oxford, see M. Feingold, *The Mathematicians' Apprenticeship: Science, Universities and Society in England, 1560–1640* (Cambridge, 1984) pp. 124–5.

the northern part of the world," and particularly for a northeast passage to the Indies.[3] Here was the origin of the Muscovy Company. The search for a northwest passage to China and India came soon after.[4] Martin Frobisher, friend of Francis Drake and his equal in determination and swagger, was the leader of the first British Northwest Passage expedition that made significant inroads into the Arctic.[5] Frobisher's aims were glory and profit; a northwest trade route to China would make its discoverer wealthy beyond the dreams or avarice, and finding that passage was, he proclaimed, "*still* the only thing left undone, whereby a notable mind might be made famous and remarkable."[6] Frobisher did not find the passage, although he did get as far as Frobisher Bay at the southeast end of Baffin Island. He thought the bay offered a through passage, and named it Frobisher Straits, after himself. He encountered the Eskimos, whom he thought were inhabitants of Cathay. He brought one of them back with him, together with rocks that were rumored to contain gold. For a while it seemed that Frobisher might have made his own and his backers' fortunes. But in 1578 when Frobisher lost most of his ships during his third expedition to Baffin Island, he returned home only to meet another disaster. The golden glint in the rocks from the shores of Frobisher Bay had turned out to be the gleam of mica,[7] and that was the end of the Cathay Company. A decidedly unromantic attitude to the Arctic is apparent on the maps of his discoveries: the new and unknown

[3] The ships were separated, and Willoughby and his crew fetched up in Russian Lapland, where they died of scurvy in 1554. The voyage is described in *The Principal Navigations Voiages and Discoveries of the English Nation by Richard Hakluyt Imprinted at London, 1589*, reproduced and introduced by D. B. Quinn and R. A. Skelton, 2 vols., Hakluyt Society Extra Series No. 39 (Cambridge, 1965) vol. 1 pp. 263–70 (Willoughby), 270–92 (Chancellor).

[4] For a general history of Northeast and Northwest Passage attempts up to the early nineteenth-century English voyages, see John Barrow, *A Chronological History of Voyages into the Arctic Regions: Undertaken Chiefly for the Purpose of Discovering a North-East, North-West, or Polar Passage Between the Atlantic and Pacific: From the Earliest Periods of Scandinavian Navigation, to the Departure of the Recent Expeditions, Under the Orders of Captains Ross and Buchan* (London, 1818; reprinted David & Charles, Newton Abbot, 1971).

[5] John Cabot's earlier voyages (1497 and 1498) got as far as the eastern (Atlantic) coast of the continent. There were also French, Portuguese, and Spanish voyages of discovery, as well as the less publicized explorations of fishermen who headed for the Grand Banks after Cabot's news spread.

[6] Quoted in Pierre Berton, *The Arctic Grail: The Quest for the North West Passage and the North Pole 1818–1909* (Toronto, 1988) p. 16.

[7] W. A. Kenyon, *Tokens of Possession: The Northern Voyages of Martin Frobisher* (Toronto, 1975); J. Payne, ed., *Voyages of the Elizabethan Seamen to America*, 2nd ed. (Oxford, 1893) vol. 1 pp. 88–95 (Frobisher's 1st voyage 1576), 96–132 (2nd voyage 1577), 133–92 (3rd voyage 1578).

lands were marked "meta incognita," unknown value; and the meaning was entirely commercial.

The question of a northwest passage remained open, and other voyages were soon under weigh. John Davis, the most accomplished navigator of his day,[8] wrote after his first arctic expedition that "the north-west passage is a matter nothing doubtful . . ."[9] Next came Henry Hudson, with arctic experience under the Dutch, who in 1607 under the Muscovy Company had made the first attempt to reach the North Pole, attaining near Svalbard (formerly Spitsbergen) a more northerly latitude than anyone else until Constantine Phipps in 1773. Hudson's final voyage was sponsored by London merchants still eager for a northwest passage to Cathay; he discovered and explored part of Hudson Bay, where he and five of his men were marooned by a mutinous crew.

Navigation depended on commerce to sponsor such costly voyages, in hopes of large returns. But navigation, and therefore commerce, also depended on science. Following the coastline and taking bearings of fixed points along the coast was fine for much of European navigation, and could even take one gradually down the coast of Africa, in a line of exploration seized on by the Portuguese. But transoceanic navigation was altogether a different matter, depending on celestial navigation. Astrolabe, mariner's astrolabe, and cross staff, and the calculations that followed observation, required expertise[10] that was lacking in mid-sixteenth-century England but well developed in Spain; indeed, that expertise was the key to Spanish voyages to South America, and hence to the gold that maintained the Spanish Empire. If the English were going to generate similar wealth, they needed to acquire similar skills. So they persuaded Sebastian Cabot, Chief Pilot of Spain, to return to England and there share his learning. One of Cabot's pupils was navigator with Chancellor when he opened the northern sea route to Russia.

Within the half century, Cabot's lessons had been well used. The relevant parts of applied mathematics were taught at the universities.[11] England had its own instrument makers, aptly called mathematical practitioners.[12] At

8   L. P. Kirwan, *The White Road: A Survey of Polar Exploration* (London, 1959) p. 25.

9   Davis to Walsingham, 3 October 1585, quoted in Kirwan, *The White Road*, p. 26.

10  E. G. R. Taylor, *The Haven Finding Art: A History of Navigation from Odysseus to Captain Cook* (London, 1956); D. W. Waters, *The Art of Navigation in England in Elizabethan and Stuart Times*, 2nd ed. (Greenwich, 1978).

11  Feingold, *The Mathematicians' Apprenticeship*, especially chap. 4, pp. 122–65; L. Cormack, *Non Sufficit Orbem: Geography as an Interactive Science at Oxford and Cambridge 1580–1620*, Ph.D. thesis, University of Toronto (Toronto, 1988).

12  E. G. R. Taylor, *The Mathematical Practitioners of Tudor and Stuart England* (Cambridge, 1954). *Annals of Science* 48 no. 4 (July 1991) is a special issue on early surveying;

Gresham College in London, instruction was offered in science, or natural philosophy, and in mathematics. There is not much evidence that sea captains took advantage of these lectures, but the college was successful as a focus of research. William Gilbert, who made the first attempt to create a comprehensive understanding of geomagnetism and to show how it could serve navigation, was associated with the college through its first professor of geometry, Henry Briggs.[13] The mathematician and astrologer John Dee[14] had brought the first Mercator globes to England. Now, at his home in Mortlake near London, Dee was sought after by learned and curious travelers.

Science was important for navigation; so was experience. Thus it was that Robert Bylot, one of the mutineers against Henry Hudson, went back to the Arctic as pilot under Thomas Button. Before he went, he was given scientific instruction, and orders to make a variety of observations, including those of magnetic variation and of the predicted solar eclipse.[15] Important refinements in navigation based on science were made by William Baffin, who was the first to measure the refraction of the sun's rays, and the first to take such refraction into account in interpreting his astronomical observations. He also recorded the ship's course to the nearest degree, rather than to the nearest compass point. He measured longitude by deducing the time of the moon's transit of the meridian from the sun's position; he then used sophisticated mathematics, solving a spherical triangle. As a recent commentator stated, "if Baffin could not find longitude astronomically, no one at that time could."[16]

Baffin's skill is one of the best indications of the penetration of navigation by science; another is the list of instruments that Thomas James of Bristol took on his northwest passage search with Luke Foxe of Hull in 1631–2.[17] He took a four-foot quadrant divided into minutes of arc, a

---

the article therein by S. Johnston, "Mathematical Practitioners and Instruments in Elizabethan England," pp. 319–44 is especially useful here.

[13] F. R. Johnson, "Gresham College: Precursor of the Royal Society," *Journal of the History of Ideas* 1 (1940) 413–38 at 427; Feingold, *The Mathematicians' Apprenticeship*, pp. 166–89, "Gresham College and its Role in the Genesis of 'London Science' "; I. Adamson, "The Administration of Gresham College and its Fluctuating Fortunes as a Scientific Institution in the Seventeenth Century," *History of Education* 9 (1980) 13–25. See also Adamson's *The Foundation and Early History of Gresham College*, Ph.D. thesis (Cambridge University, 1975).

[14] P. J. French, *John Dee: The World of an Elizabethan Magus* (London, 1972).

[15] Peter Broughton, "Astronomy in Seventeenth-Century Canada," *Journal of the Royal Astronomical Society of Canada* 75 (1981) 175–208.

[16] Ibid., p. 183.

[17] *The Voyages of Captain Luke Foxe of Hull, and Captain Thomas James of Bristol, in Search of a North-West Passage, in 1631–32 . . .* , ed. Miller Christy, 2 vols. (London: the

five-foot sextant, cross staffs and back staffs, meridian compasses and various compass needles, a watch and a clock, and mathematical tables and books, as well as *"Study Instruments* of all sorts." Compass variation was observed "as often as conveniently we could," as well as astronomical observations for determining latitude and longitude.[18] Latitude was generally obtained by measuring the meridian altitude of the sun at noon, that is, its angular elevation above the horizontal. Latitude could thus be determined within a few minutes of arc; longitude, especially at sea before the invention of accurate chronometers, was much more difficult. Observation of the altitude of the sun or a star at least two hours from the meridian gave one a measure of local time. A good chronometer, once these became available in the last decades of the eighteenth century, would then give one the difference between local time and time at a reference point. The Greenwich Observatory and meridian would later serve as such a reference. Meanwhile, however, navigators looked up and used the heavenly clockwork as their measure of time. The moon's movements against the background of the stars provided the most obvious clock, but there were problems here, because astronomers did not develop an accurate account of lunar motions until the end of the eighteenth century.[19]

On land, observers do not need accurate tables, as long as they are able to arrange observations of a given event, for example, an eclipse of the moon, from two different places. Lunar eclipses are particularly convenient, since they are visible simultaneously for the entire nighttime hemisphere of the earth. The difference in observed times of the event at the two places gives the longitudinal difference between them.[20] Thus noting the local time of an eclipse, and comparing it, on one's return, with the time of the eclipse at the reference point (e.g., Greenwich) gives one the longitude;

Hakluyt Society, 1894) vol. 2 pp. 604–6. Broughton, "Astronomy in Canada," pp. 185–8 discusses James's observations.

[18]  *The Voyages of . . . Foxe . . . and . . . James* vol. 2 pp. 607–11.

[19]  Both good chronometers and the *Nautical Almanac* came late in the century. D. Howse, *Nevil Maskelyne: The Seaman's Astronomer* (Cambridge, 1989) pp. 9–17 discusses "The longitude problem"; pp. 84–96 give a good account of the *Nautical Almanac*; pp. 91–3 provide a concise account of determining longitude by lunar observations, and state, in summary: (1) determine local apparent time by the elevation of the sun or a star at least two hours from the meridian; (2) eliminate parallax and refraction from the measured lunar distance (the angle between the moon and the sun or star), using tables, to obtain the true lunar distance; (3) from the true lunar distance, find Greenwich apparent time, also using tables; (4) the difference between local and Greenwich apparent time gives the longitude. See also chap. 4.

[20]  Broughton, "Astronomy in Canada," p. 177.

the difference in times corresponds to the difference in longitude at the rate of fifteen degrees per hour.

The skills needed for successful navigation and exploration were both theoretical and practical, and in this they were representative of the best of the new science. Gilbert worked with dip needles, other needles serving as azimuth compasses, and models of the earth, and collected magnetic data from varied sources. In his treatise *De Magnete,* published in 1600, he proposed the systematic collection of data about the earth's magnetism, because it was of intrinsic theoretical importance and might also serve as an aid to navigation. Galileo was exploring the interaction of mathematics and experiment in the study of nature.[21] Observation and experiment were keys to the new way of doing science.

## FRANCIS BACON

Francis Bacon,[22] lawyer, politician, and philosopher of science at the turn of the century, rightly identified these keys, laying particular stress upon the former. He wrote extensively about the reform of natural knowledge, proposing an encyclopedic program of observation, and stressing the role of insightful questions. He argued that the sciences should occupy a position of great dignity in society, because they contributed equally to mental illumination and material amelioration. He wrote an unfinished fable about an island community, the "New Atlantis," dedicated to learning, and centering about a college instituted "for the interpretation of nature." One of the ordinances of the college, known as Solomon's House, was that every twelve years two ships should set out on several voyages in search of knowledge.[23] The college itself contained two "long and fair galleries," one displaying examples of "all manner of the more rare and excellent inventions," the other containing statues of the principal inventors, including the inventor of ships, and Columbus, for his discovery of the West Indies.[24]

---

[21] This is a major theme throughout Stillman Drake, *Galileo at Work: His Scientific Biography* (Chicago, 1978).

[22] P. Rossi, *Francis Bacon: From Magic to Science* (London, 1968).

[23] Note Bacon's title page (to the 1620 edition of his *Instauratio Magna*, reproduced in *The Works of Francis Bacon*, ed. J. Spedding, R. L. Ellis, and D. D. Heath, 14 vols. [London 1857–74; reprinted Stuttgart, 1961–3] vol. 1 p. 119) with ships going to and fro between the pillars of Hercules (the Straits of Gibraltar) into the ocean of undiscovered truth, a metaphor much used by his successors, including Isaac Newton.

[24] *The Works of Francis Bacon* (1859) vol. 3 pp. 165–6.

Elsewhere, in a short essay "In Praise of Knowledge," Bacon identified three inventions as having recently transformed the world: printing, source of a revolution in knowledge; artillery, which had led to a similar revolution in warfare; and the compass needle, which had transformed "the state of treasure, commodities, and navigation."[25]

In his enumeration of subjects appropriate for study, he included tides, currents, and waves.[26] He proposed that other aspects of the sea should also be noted, including "its Saltness, its various Colours, its Depth; also of Rocks, Mountains and Vallies under the Sea and the like."[27]

## THE ROYAL SOCIETY

Bacon's writings on the new philosophy were not greatly influential in his lifetime. But the new science, in its early development in England, often approximated closely to Bacon's prescription. When the Royal Society of London was founded in the 1660s,[28] it seized upon Bacon as precursor, prophet, and almost patron saint. Thomas Sprat wrote a voluminous history of the society, in which he apostrophized Bacon as one "who had the true Imagination of the whole extent of the Enterprize, as it is now set on foot." In his books, "there are every where scattered the best arguments, that can be produc'd for the defence of Experimental Philosophy; and the best directions, that are needful to promote it."[29] Sprat wrote his big book so soon after the society received its royal charter, and he co-opted Bacon's authority, because the new science and its advocates stood in need of a defense against the criticisms that they soon provoked. These criticisms can be summarized under three headings: that the new philosophy, or science, was inclined to atheism; that it was ridiculous; and that it was useless. To which Sprat, the virtuosi, and the ghost of Francis Bacon replied in the spirit of the New Atlantis, that science was a religious activity, dignified, and preeminently useful. Exploration across the oceans depended on the new science, and was valuable because it offered prospects of utility, or more crassly, of profit. Columbus had featured in the pantheon of the House of Solomon,

[25] Ibid., vol. 8 (1862) pp. 125–6.    [26] Ibid., vol. 4 (1858) p. 266.
[27] Ibid., vol. 5 (1858) pp. 443–58.
[28] M. C. W. Hunter, *Science and Society in Restoration England* (Cambridge, 1981); *Establishing the New Science: The Experience of the Early Royal Society* (Woodbridge, Suffolk, and Wolfeboro, New Hampshire, 1989).
[29] Thomas Sprat, *The History of the Royal Society of London, For the Improving of Natural Knowledge* (London, 1667; reprinted St. Louis, Missouri, 1966) p. 35.

and Columbus, even before Bacon urged it, had been the first to contribute to a knowledge of currents in the Atlantic. By the 1660s, one and a half centuries of geographical discovery had contributed information and added weight to the empirical discoveries of philosophical navigators and natural historians.[30] Marine science was an important part of the expansion of knowledge.

In Wadham College, Oxford, where debates took place and knowledge was shared during the Commonwealth, foreshadowing the meetings of the Royal Society of London, and in some of the earliest meetings of that optimistic and ambitious body, the sea was a significant topic. Christopher Wren proposed "Inventions for better making and fortifying Havens, for clearing Sands, and to sound at Sea."[31] At the new Royal Society, the first indication of interest in marine research was on 14 June 1661, when documents were registered containing *Propositions of some experiments to be made by the Earl of Sandwich . . .* during his expedition to the Mediterranean in that year. He was to investigate the depth of the sea, variations in salinity, the pressure of sea water, tides and currents at Gibraltar, and luminescence[32] – a wide-ranging list of observations for a military expedition.

In the following year, Laurence Rooke, Professor of Geometry at Gresham College, drew up at the request of the society a set of "Directions for Sea-men, bound for farre Voyages," proposing a comprehensive research program covering geomagnetism, meteorology, hydrography, and oceanography.[33] This was added to and then revised with the help of Robert Hooke, one of the most ingenious inventors of apparatus and experiments, and a key figure in the Scientific Revolution in Britain. The new Directions were in fact a directive

to be lodged with the Master *Trinity House,* to be recommended to such, as are bound for far Sea-Voyages, and shall be judged fit for the performance: who are also to be desired, to keep an exact *Diary* of such observations and Experiments, and

---

[30] Margaret Deacon, *Scientists and the Sea 1650–1900: A Study of Marine Science* (London and New York, 1971) pp. 47, 53.

[31] Christopher Wren, "A catalogue of New Theories, Inventions, Experiments, and Mechanick Improvements, exhibited by Mr. Wren, at the first Assemblies at Wadham College in Oxford, for Advancement of Natural and Experimental Knowledge, called then the New Philosophy: Some of which, on the Return of Public Tranquillity, were improved and perfected, and with other useful Discoveries, communicated to the Royal Society," in Christopher Wren jr., *Parentalia* (London, 1750) p. 198.

[32] Deacon, *Scientists and the Sea*, p. 74, based on Thomas Birch, *The History of the Royal Society of London for Improving of Natural Knowledge from its First Use* (London, 1756–7; reprinted New York and London, 1968) vol. 2 p. 111.

[33] Royal Society, *Register Book* vol. 1 pp. 149–52; see Deacon, *Scientists and the Sea*, p. 75.

deliver at their return a fair Copy thereof to the Lord High Admiral of *England*, his Royal Highness the *Duke of York*, and another to *Trinity-house*, to be perused by the said *R. Society*.[34]

Here was the beginning of an official alliance of science and the navy that still operates today, and that has played a major role in the history of scientific exploration, including nineteenth-century expeditions to the arctic archiplelago to the north of Canada.

The instructions themselves were pretty durable, and were specific about the instruments to be used, and the observational and mathematical techniques required. The first requirement was "to observe the Declinations and Variations of the Compass or Needle from the Meridian exactly, in as many Places as they can, and in the same Places, every several Voyage." At the same time, magnetic needles in dip circles were to be used. This was at once a practical and a theoretical matter. Declination or variation, the horizontal angle between magnetic north and geographical north, most easily measured by taking a bearing of the pole star, and dip, the vertical angle between a freely suspended compass needle and the horizontal plane, were known to vary with change of place on the earth's surface, as well as with the passage of time. Once their local magnitudes, and the way in which they changed with time were known, it might be possible to use them, as William Gilbert had already suggested, for the determination of longitude at sea, a most urgent problem of practical navigation; and through the same observations, "all will be reduced to Rules, and so from hence Philosophical or Natural Knowledge, will probably be enlarged by a happy discovery of the true cause of the *Verticity,* or *Directive* faculty of the Loadstone; one of the *Noblest* and most *abstruse Phaenomena,* that falls under the cognizance of humane Reason."[35]

Next came the observation of tides, and the mapping of coastlines and estuaries, with bearings taken of prominent points. What we comprehend under the headings of hydrography and oceanography were certainly part of seventeenth-century marine science. The Directions of 1667 continued with the requirement "to sound the deepest Seas without a Line." This was to be done by means of a specially constructed instrument, consisting of a varnished wooden sphere, to which a lead or stone weight was attached in such a manner as to release the sphere when it touched bottom. The sphere would then rise to the surface, and the time that it took would give a mea-

[34] "Directions," *Philosophical Transactions of the Royal Society of London* no. 24 pp. 433–48, 8 April 1667.
[35] Ibid., p. 437.

sure of the depth. This early sounding device was more ingenious and sanguine in conception than it was reliable in practice.

Comprehensive meteorological records were to be kept; so were observations on the density of sea water, at different places and different depths.
Once again, an instrument was described – a bucket whose lid was kept
open by the movement of descent, but was closed by the pressure of water
on its lid as soon as it started its ascent.

### EDMOND HALLEY AND THE VOYAGES OF THE *PARAMORE*

Of all these areas of scientific observation, the magnetic work, seeking to
understand complex and abstruse phenomena, excited the greatest interest
within the Royal Society. Not only was there the appeal of profound theoretical advances, but the intensely practical aims of navigation, including
the vexatious problem of determining longitude at sea, made the magnetic
inquiry one of the first importance. We now know that fluctuations in the
earth's magnetism are not as simple and regular as Rooke, Moray, and
Hooke had hoped, and the use of magnetic variation to determine longitude
has never proved reliable. But in the latter half of the seventeenth century it
seemed entirely reasonable to pursue that inquiry. Edmond Halley was the
leading figure in the enterprise.[36] Samuel Pepys, diarist, Fellow of the Royal
Society of London, and to a large extent creator of the modern Royal Navy,
wrote of "Mr Hawley – May he not be said to have the most, if not to be the
first Englishman (and possibly any other) that had so much, or (it may be)
any competent degree (meeting in them) of the science and practice (both)
of navigation?"[37] Certainly Halley's voyages in the *Paramore* have been
widely regarded as the first major naval expeditions mounted for scientific
purposes. They were certainly the first dedicated to magnetic researches.

Halley was one of the most vigorous fellows of the Royal Society. He undertook various scientific voyages, including one to St. Helena to begin a
catalogue of southern stars; he was a good mathematician and natural philosopher, and not only encouraged Newton to publish his ideas, but saw the
*Principia* through the press; and, since his school days in London, he had
carried out observations in magnetism, and theorized about their interpretation. Secular variation, the variation of the earth's magnetism with time,

---

[36] C. H. Cotter, "Captain Edmond Halley, RN, FRS," *Notes and Records of the Royal Society of London* 36 (1981) 61–78.

[37] J. R. Tanner, ed., *Samuel Pepys's Naval Minutes* (London, 1926) p. 420, quoted in Norman J. Thrower, ed., *The Three Voyages of Edmond Halley in the 'Paramore' 1698–1701*,
2 vols. (London: The Hakluyt Society, 1981) vol. 1 p. 15.

had been described in 1635 by Henry Gellibrand.[38] Halley published his own observations, and suggested that the existence of four magnetic poles would account for variation and its secular change.[39] In the following decade, he returned to the subject with a more elaborate hypothesis about the internal structure of the earth.[40]

At the same time, plans were afoot within the Royal Society and the Admiralty for Halley "to undertake a Voyage . . . to incompass the whole Globe from East to West," partly in order to construct maps that would be helpful to trade, and also "to endeavour to gett full information of the Nature of the Variation of the Compasse over the whole Earth, as Likewise to experiment what may be expected from the Severall Methods proposed for discovering the Longitude at Sea."[41] Halley's instructions, when the *Paramore* was commissioned for him in 1698, stressed magnetic observations and the longitude.

This first voyage took him to the south Atlantic, touching the West Indies and South America at Paraiba, returning via Madeira. His journal does indeed record compass variation, and latitude and longitude, as well as meteorological and natural history observations. Scientifically, it was a success; politically, as far as relations between science and the navy went, it was a disaster. It had been Halley's first naval command, and in order to control the crew, he had requested and been given a lieutenant as his second in command. Edward Harrison was the man appointed, and relations between him and Halley were poor. Halley's complaints about him after the voyage led to a court martial of Harrison and all the other officers on board.

The report of the court was that Halley's subordinates had not disobeyed him, "tho there may have been some grumbling among them." The result was an acquittal, accompanied by a reprimand of Harrison and the rest. Halley was now doubly disgruntled. He felt the verdict was wrong, and he had only just learned that Harrison was the author of a work on finding the longitude that the Royal Society had, some years previously, asked Halley to assess. Halley had reported adversely on Harrison's method, and was now

---

[38]  H. Gellibrand, *A Discourse Mathematical of the Variation of the Magneticall Needle together with its Admirable Diminution Lately Discovered* (London, 1635).

[39]  E. Halley, "A Theory of the Variation of the Magnetical Compass," *Philosophical Transactions of the Royal Society of London* 13 (1683) 210.

[40]  E. Halley, "An Account of the Cause of the Change of the Variation of the Magnetical Needle; with an Hypothesis of the Structure of the Internal Parts of the Earth," *Philosophical Transactions* 17 (1692) 563–78.

[41]  Admiralty orders to the Navy Board, National Maritime Museum, Greenwich, MS ADM/A/1797, published in Thrower, *The Three Voyages of Edmond Halley*, vol. 1 p. 252.

convinced that troubles on the *Paramore* had been caused by Harrison's lingering resentment.[42] What the Admiralty long remembered was that a Fellow of the Royal Society in his first naval command had court martialed all his officers. Once Halley's expeditions were over, the navy would be inclined to keep science in a properly subordinate place on shipboard.[43]

Halley's second voyage, which had different officers, and was geographically more ambitious but otherwise had similar instructions to his first voyage, was free of the grumbling that had afflicted the earlier voyage. Leaving in September 1699, Halley took the *Paramore* through the south Atlantic to the latitude of the Falkland Islands, and through the north Atlantic to Newfoundland. He had made around fifty magnetic observations on his first voyage, and now added another hundred. He incorporated these in a highly original and impressive isogonic chart, which was an eminently satisfactory achievement.

## JAMES COOK

In the first half of the eighteenth century, British achievements in arctic exploration were negligible compared with the Russians, most notably Bering and his great explorations. But the second half of the century saw renewed exploration, this time combining the theoretical goals of the Royal Society's instructions with practical scientific wisdom.

James Cook[44] was largely responsible for the new approach to exploration. Cook had made the charts of the St. Lawrence that were a necessary prelude to Wolfe's assault on Quebec. From 1762 to 1767 he had carried out marine surveys of Newfoundland and Labrador; he met Constantine Phipps and Joseph Banks at St. John's, when these gentlemen were engaged in surveying and botanical and zoological collecting.[45] Phipps was to go

---

[42]  Ibid., pp. 281–6.

[43]  G. S. Ritchie, *The Admiralty Chart: British Naval Hydrography in the Nineteenth Century* (London, 1967) pp. 7–9.

[44]  The definitive work on Cook is J. C. Beaglehole, *The Journals of Captain James Cook*, Hakluyt Society Extra Series no. 37, 4 vols. (London, 1974), vol. 4, *The Life of Captain James Cook*. These are supplemented by A. David, ed., assisted by R. Joppien and B. Smith, *The Charts and Coastal Views of Captain Cook's Voyages: The Voyage of the Endeavour, 1768–1771* . . . , Hakluyt Society Extra Series no. 43 (London, 1988). Cook's scientific role is described in Tom and Cordelia Stamp, *James Cook: Maritime Scientist* (Whitby, 1978).

[45]  A. M. Lysaght, *Joseph Banks in Newfoundland and Labrador, 1766. His Diary, Manuscripts and Collections* (Berkeley and Los Angeles, 1971). The definitive biography is H. B. Carter, *Sir Joseph Banks 1743–1820* (London, 1988).

on to lead the first Royal Naval expedition aiming at the North Pole,[46] and Banks, as a future president of the Royal Society of London, was to be Britain's autocrat of science, "HM Minister of Philosophic Affairs."[47] In 1766 the Royal Society, prompted by Alexander Dalrymple,[48] had urged the Admiralty to support a voyage to Tahiti in order to observe the transit of Venus across the sun in 1769. Cook was chosen as commander of the expedition, the first of three great voyages that would transform geography and marine science.

The transit of Venus was the occasion for the voyage, but the Admiralty and the Royal Society wanted more than that. Once the transit observations were complete, the Admiralty's sealed and secret instructions told Cook that he was to look for a southern continent, to make charts and observations of the seas and coasts, and besides:

You are also carefully to observe the Nature of the Soil, and the Products thereof; the Beasts and Fowls that inhabit or frequent it, the fishes that are to be found in the Rivers or upon the Coast and in what Plenty; and in case you find any Mines, Minerals or valuable stones you are to bring home Specimens of each, as also such Specimens of the Seeds of the Trees, Fruits and Grains as you may be able to collect, and Transmit them to our Secretary that We may cause proper Examination and Experiments to be made of them.

You are likely to observe the Genius, Temper, Disposition and Number of the Natives, if there be any, and endeavour by all proper means to cultivate a Friendship and Alliance with them. . . .

You are also with the Consent of the Natives to take possession of Convenient Situations in the Country in the Name of the King of Great Britain. . . .[49]

Here was an ambitious agenda, in which trade and navigation, empire and science were all to be mutually supporting. Cook had his own requirements for the voyage, chief among them the nature of his vessel, which had to be both strong and small enough to be brought ashore for repairs.

In science, Cook had demonstrated his own competence. He also took with him a scientific complement: Charles Green, an astronomer, in charge of the transit observations; Daniel Carl Solander, a pupil of Linnaeus; and Joseph Banks, a wealthy nobleman and amateur naturalist. The Admiralty had given Cook his instructions; the Royal Society, tactfully, offered merely *Hints* for the consideration of Cook and his scientific gentlemen. These

---

[46]  See the next section of this chapter.

[47]  This is the title of part III, p. 351, of Carter's biography, *Sir Joseph Banks*.

[48]  Howard T. Fry, *Alexander Dalrymple (1737–1808) and the Expansion of British Trade* (Toronto and Buffalo, 1970).

[49]  Reproduced in Beaglehole, *The Journals of Captain James Cook*, vol. 1, pp. cclxxxii–cclxxxiii.

hints were especially concerned with natural history and ethnology, as well as with the clock needed for astronomical and navigational work.[50] As for natural history, a contemporary observed: "No people ever went to sea better fitted out for the purposes of Natural History. They have got a fine Library . . . ; they have all sorts of machines for catching and preserving insects; all kinds of nets, trawls, drags and hooks for coral fishing. . . ."[51]

If Cook's first expedition was well equipped, his second expedition was even better outfitted, with apparatus including a Hadley's sextant by Ramsden, the finest instrument maker of his day; a chronometer that was a copy of John Harrison's number 4 instrument, an improved version of the one with which he had won the prize for determining longitude at sea;[52] and a device for determining the temperature of sea water at different depths.[53]

Cook's epic voyages are mainly remembered for their discoveries in the southern oceans, but his instructions for the voyage of the *Resolution* and *Discovery* in 1776–80 were to investigate the possibility of a northwest passage on the west coast of America, east to Baffin Bay. In August 1778 they passed through Bering Strait and headed northeast, but encountered impenetrable pack ice, and turned back at Icy Cape. They returned to the South Pacific, where Cook was killed in an affray with natives; a second attempt on the passage in 1779 again ran into the arctic pack ice.

Cook had discovered less of the Arctic than some of his predecessors, but he had established the pattern for scientific exploration that was to guide the Royal Navy for much of the following century. The ships provided for these expeditions had to be as far as possible appropriate for the rigors of waves and ice, and capable of being repaired away from home. The navigational apparatus used had to be the best, as did the other astronomical, geophysical, and oceanographic equipment. Newly invented apparatus, and similar apparatus of varying designs, would be submitted to field trials. There would be an overall program of research embracing the physical and life sciences, devised in close liaison with the Royal Society. The expedition would be under the command of a naval officer, not a scientist. Halley's unfortunate example ensured henceforth the general subordination of science to naval control; in Cook's second expedition, the result was that Joseph Banks, deprived of the space and retinue that he demanded,

---

[50] Beaglehole, *The Journals of Captain James Cook*, vol. 1, appendix II, pp. 514–19; D. Howse, "Captain Cook's Marine Timekeepers," *Antiquarian Horology* (1969) 190–9.

[51] Quoted in Kirwan, *The White Road*, p. 62.

[52] The importance of good chronometers has already been indicated above, and will be further discussed in chapters 2 and 4.

[53] Beaglehole, *The Journals of Captain James Cook*, vol. 3, part II, pp. 1498–9 and 1501–4.

was replaced by the Forsters, father and son, both naturalists skilled in sketching.[54] A disciplined scientific program underlined the need for systematic collections in natural history and for competent artists like the Forsters to draw newly obtained specimens before they deteriorated. The earliest arctic voyages had economic more than territorial ambitions; Halley's voyages had both scientific and military objectives, the latter served through hydrographic investigations. By the late eighteenth century, territorial claims were well to the fore; Cook's surveys had facilitated the British conquest of New France, and with geography recognized as a science by the Royal Society, the interdependence of territorial claims and scientific activity was ensured.

### CONSTANTINE PHIPPS

Most but not all of these factors were in place when Constantine Phipps set out for the North Pole in 1773. He sailed too soon to be heir to Cook's full legacy in naval science, but the lessons and model of Cook's first Pacific voyage were there, and served considerably to shape this new venture.

Of more immediate importance for the planning of Phipps's expedition was the notion that ice could be formed from fresh water, both inland and at sea where it ran off from coasts and river mouths. The Swiss geographer Samuel Engel had argued from this premise that away from coasts, the polar ocean would be free from ice, and therefore navigable.[55] The Royal Society took cognizance of the argument, and wrote to Lord Sandwich, First Lord of the Admiralty, stating that they had "lately had under consideration the probability of Navigation being practicable nearer the North Pole than has been generally imagined," and that such a voyage would be valuable to natural knowledge, and might also offer a route to the East Indies.[56]

In consequence, an expedition toward the North Pole was approved. Phipps volunteered to lead it, and was entrusted with two ships, *Racehorse* and the inauspiciously named *Carcass*. Israel Lyons was appointed by the Board of Longitude to make astronomical and nautical observations,

---

[54]  *The* Resolution *Journal of Johann Reinhold Forster*, Michael E. Hoare, ed., 4 vols. (Hakluyt Society, 1982).

[55]  Samuel Engel, *Mémoires et Observations Géographiques et Critiques sur la Situation des Pays Septentrionales d'Asie et d'Amérique d'après les Relations les Plus Récentes* (Lausanne, 1765; enlarged German edition, 1772).

[56]  Royal Society Council Minutes, 1769–82, vol. 6 p. 158, quoted in Ann Savours, " 'A Very Interesting Point in Geography': The 1773 Phipps Expedition Towards the North Pole," *Arctic* 37 (1984) 402–28 at 403.

and to test chronometers; his equipment and instructions largely echoed those of Cook's second and third voyages.[57] Phipps received instructions concerning botany and natural history from the comparative anatomist John Hunter and from Joseph Banks, to whose friendship he wrote, "I am indebted for very full instructions in the branch of Natural History."[58] Alas, the greater part of the natural history collections were lost in a storm on the homeward voyage. The surviving sketches made from his specimens are unusually fine.[59]

For scientific work in general, Phipps observed, "I took care to provide myself with all the best Instruments hitherto in use, as well as such as had been imperfectly, or never tried."[60] Among these instruments was a self-registering thermometer invented by Lord Charles Cavendish.[61] Other apparatus and observations included a seconds pendulum for measuring the ellipticity of the earth, sounding equipment, magnetic compasses and Nairne and Blunt's dip needle, and meteorological instruments, Ramsden's manometer among them.[62] All in all, there was an impressive range of scientific work and of apparatus for it, but ice prevented the expedition from advancing beyond Spitsbergen, and the polar route to the Indies remained unconfirmed. The dream of an open polar ocean was, however, to prove remarkably durable.

[57] Savours, "The 1773 Phipps Expedition," appendix A, pp. 423–4, reproduces Lyons's Instructions, and the schedule of instruments that he was to use.
[58] British Library MS Kings 224, "Constantine Phipps Voyage Towards the North Pole 1773," p. 9, MS of Phipps, *A Voyage Towards the North Pole* (London, 1774).
[59] British Library MS Kings 225.     [60] Ibid.
[61] Deacon, *Scientists and the Sea*, p. 191.
[62] Phipps, BL MS Kings 224 pp. 9, 75, 79, 81. Nairne and Blunt were makers of fine scientific instruments; Jesse Ramsden was arguably the finest instrument maker of the eighteenth century. For these craftsmen, and for others of the period, see E. G. R. Taylor, *The Mathematical Practitioners of Hanoverian England* (Cambridge, 1966).

# 2

The Navy and the Northwest Passage after the
Napoleonic Wars: 1817–1834

The Royal Navy had its work cut out once war broke out with France.
There were scientists who sought to maintain a free exchange of knowledge
during the wars. The Royal Society sent the latest *Nautical Almanack* to
France, thereby giving the enemy the best available aid to navigation; the
French responded in kind. Joseph Banks was instrumental in procuring the
liberty of imprisoned French and British natural philosophers.[1] Napoleon,
at the height of the war, granted a passport to Humphry Davy, Britain's
leading scientist, and even awarded him a medal for his discoveries. In those
days it was possible to agree with John Hunter that "the sciences were never
at war."[2] But it was not possible to spare vessels and crews from active duty
for the perilous but militarily irrelevant exploration of the far North. In the
war years, so far was the navy from polar service that Samuel Taylor Coleridge's "Rime of the Ancient Mariner" was the best-known public presentation in Britain of the polar regions, drawing as it did on arctic narratives[3]
to fashion in imagination the unknown Antarctic:

[1]  In 1804 Sir Joseph Banks sent Delambre the *Nautical Almanac* for 1807; among other scientists for whom he sought freedom was the French geologist Déodat de Dolomieu, who
was captured at Taranto: see H. B. Carter, *Sir Joseph Banks 1743–1820* (London, 1988)
pp. 375–6, 389.
[2]  G. R. de Beer, *The Sciences Were Never At War* (London, 1960). Matching the French
award of a prize to Davy was the Royal Society's prize to Malus in 1810 for his discovery
of polarized light. There were, as Prof. D. M. Knight has reminded me, limits to neutrality.
Flinders, for example, was held by the French on Mauritius, and Banks was engaged in a
six-year tussle to free him (Carter, *Sir Joseph Banks*, p. 420). We shall encounter Flinders
later in the chapter in connection with Robert Brown's Australian botanizing, and his own
work on correcting for a ship's magnetism.
[3]  John Livingston Lowes, *The Road to Xanadu: A Study in the Ways of the Imagination*
(Cambridge, Mass., 1927).

And now there came both mist and snow,
And it grew wondrous cold:
And ice, mast-high, came floating by,
As green as emerald.

And through the drifts the snowy cliffs
Did send a dismal sheen:
Nor shapes of men nor beasts we ken –
The ice was all between.

The ice was here, the ice was there,
The ice was all around:
It cracked and growled, and roared and howled
Like noises in a swound![4]

The other image that kept the polar regions alive in popular imagination came from Phipps's voyage. Horatio Nelson, at the age of fourteen, had managed to be appointed midshipman on the *Carcass*. Playing truant one night on the ice, he challenged a bear; his musket misfired – the powder flashed in the pan – and he was left with the butt as his only weapon until one of the ship's guns scared off the animal. As Britain's leading hero in the naval war against Napoleon, Nelson was fair game for dramatizing hagiographers. But the extensive scientific work of Phipps's expedition was forgotten in the heat of war. Naval science and arctic exploration were temporary casualties, and earlier geographical discoveries became suspect. Baffin Bay vanished from many maps.[5]

Then came the peace. Napoleon was exiled in 1815; the Royal Navy was now unchallenged. It was also underemployed. In 1812, at the height of the wars, the British Parliament voted funds for 113,000 seamen. By 1816, with France defeated, the vote fell to funds for 24,000 seamen,[6] and most officers found themselves ashore at half pay. Thus by 1817, ninety percent of Britain's naval officers were unemployed, but their numbers, unlike those of the seamen, had increased to 6,000.[7] They were anxious, even desperate, for occupation and opportunities for promotion. The army was little better off. One thing that the services, the navy and the ordnance department of the army especially, could do in those years was scientific surveying. The renewal and expansion of the Ordnance Survey, begun in 1784, was an obvious move.

[4]  *Coleridge Poetical Works* (Oxford, 1967) pp. 188–9.
[5]  Berton, *The Arctic Grail: The Quest for the North West Passage and the North Pole, 1818–1909* (Toronto, 1988) p. 18.
[6]  *Encyclopaedia Britannica* 11th ed. (1911) vol. 19 p. 305.
[7]  Berton, *The Arctic Grail*, pp. 18–19.

Figure 1. Nelson and the bear, oil by Richard Westall (1781–1850). National Maritime Museum, London.

Sir Francis Beaufort, naval surveyor and meteorologist, who later became the navy's greatest hydrographer of the century,[8] wrote in 1816 to John Wilson Croker, F.R.S., since 1809 Secretary to the Board of Admiralty, urging that coastal and interior surveys of the kingdom be combined. He sent a copy of the letter to his friend William Edgeworth, into whose family he married. Edgeworth replied that Beaufort's arguments were unanswerable:

And much as it thwarts my interest or rather my ambition, yet I must acknowledge that the military surveyors are the proper body to be employed.

However, I think if it was suggested to Mudge[9] that two active companies might be formed of the present body, now employed in Scotland, that he or rather government, would catch eagerly at the idea, *for all they want are situations, in which they can place the unemployed officers, without the appearance of jobbing.*[10]

Edgeworth and Beaufort were not, at this juncture, thinking of the Arctic. But resuming arctic exploration, using technically trained officers from the navy and the ordnance to carry out hydrographic surveys, was attractive

[8]  A. Friendly, *Beaufort of the Admiralty: The Life of Sir Francis Beaufort 1744–1857* (London, 1977).

[9]  William Mudge (1762–1820), major general and author of geodetic works.

[10]  W. Edgeworth to Sir Francis Beaufort, 11 June 1816, Huntington Library MS FB 1243. My italics.

precisely for the reason that Edgeworth had advanced – employment "without the appearance of jobbing." The trouble was that enough was known of the Arctic to underline the difficulties of searching for a northwest passage. James Rennell, a geographer and naval officer, saw the objections clearly:

> The *NW and Northern* Passages are much talked of. How can any one suppose, that a ship can make her way from Baffin's Bay to Behring's Strait, in the short Summer of the Arctic Region, in less than 3 Months; when the Whalers are a Month or 6 weeks, in boring thro' the *loose* Ice, 3 or 4 degrees, to get to the Whaling Station: and it is 30 such degrees, that a ship is to go. If she be caught by winter, adieu – Nor is it probable that the ice is loose.[11]

In normal years, Rennell's objections would have been entirely valid. But when Rennell wrote, and he wrote with an unusual degree of knowledge and respect for the whalers, the reason for the popularity of discussions about arctic exploration was based upon information from the whalers themselves, and in particular upon information from William Scoresby, a most remarkable whaling captain from Whitby.[12]

## WILLIAM SCORESBY

Scoresby, born in the year of the French Revolution, had first visited Greenland under his father's command at the age of eleven; from the age of thirteen until his early thirties, he spent every summer on a whaling voyage. In 1806, he and his father reached the record latitude of 81°30′ N. In 1810 he took over the whaler *Resolution* from his father, and each winter sent Joseph Banks information about the high Arctic. In winter, from the age of seventeen, he studied at Edinburgh University, where his teachers included Thomas Charles Hope, Professor of Chemistry, and John Playfair, Professor of Natural Philosophy. Scoresby met Banks when he was eighteen. After the siege of Copenhagen, he had completed a brief spell in the navy by bringing captured Danish warships to Portsmouth. On his way back to Whitby via London, he took the opportunity of following his father's advice, and presented himself at Banks's house in London. There, he was soon put at his ease, "and I was enabled to converse very freely on the phenomena of the

[11] J. Rennell to Sir Francis Beaufort, London, 5 January 1818, Huntington Library MS FB 1534.

[12] Tom and Cordelia Stamp, *William Scoresby, Arctic Scientist* (Whitby [1975]). Additional information about Scoresby's dealings with Joseph Banks is from Carter, *Sir Joseph Banks*, pp. 505–7.

Arctic regions which have no parallel in any other country."[13] He met leading scientists at Banks's breakfasts. Back in Edinburgh, he was friendly with Robert Jameson, the leading Wernerian geologist in Britain.[14] He took Jameson's class in natural history, and was encouraged by him "to persevere with scientific pursuits."[15] Scoresby did so, to admirable effect. His scientific papers ranged over meteorology, the natural history of the Greenland whale, *Balaena mysticetus,* ice formations, snow crystals, the "anomaly in the variation of the magnetic needle as observed on shipboard,"[16] atmospheric electricity, refraction in cold climates, and many other topics central to the scientific exploration of the Arctic.[17] Had he been a naval officer and not a whaler, his name might well have been the foremost in the field.

Scoresby assumed his first command when he was twenty-one; the whale fishery of Spitsbergen was his usual summer destination. He presented his arctic observations to the Wernerian Society of Edinburgh, and maintained a scientific correspondence with Banks. Banks invented and had built for him an instrument for measuring the temperature of arctic seas at different depths.[18] The instrument broke under pressure in deep water, and Scoresby devised an improved version, which worked well until the line broke and it sank to the bottom. He carried out extensive observations on polar ice, which were read to the Wernerian Society. In that paper, he dismissed the notion of an open polar ocean, for his experience made him consider it "too improbable to render it necessary to hazard any opinion concerning it."[19]

Nonetheless, he found in his whaling voyage of 1817 that "a remarkable diminution of the polar ice had taken place, in consequence of which I was able to penetrate in sight of the east coast of Greenland, in the parallel of

[13] Quoted in Stamp, *William Scoresby,* p. 33.

[14] A. G. Werner (1749–1817) regarded geological formations in terms of successive precipitations and depositions from the waters covering the earth. Thus he proposed a historical framework for the interpretation of stratigraphic successions, and for the classification of minerals. See Rachel Laudan, *From Mineralogy to Geology: The Foundations of a Science, 1650–1830* (Chicago and London, 1987) chapters 5, 7, and 8. A. M. Ospovat, "The place of the *Kurze Klassifikation* in the work of A. G. Werner," *Isis* 58 (1967) 90–5.

[15] Jessie M. Sweet, "Robert Jameson and the Explorers: the Search for the North-West Passage Part I," *Annals of Science* 31 (1974) 21–47.

[16] *Philosophical Transactions of the Royal Society of London* 109 (1819) 96–106.

[17] For a list of Scoresby's scientific publications, and a brief account of his scientific work, see A. McConnell, "The Scientific Life of William Scoresby Jnr, with a Catalogue of his Instruments and Apparatus in the Whitby Museum," *Annals of Science* 43 (1986) 257–86.

[18] Banks to Scoresby 8 Sept. 1810, quoted in Stamp, *William Scoresby,* p. 49.

[19] Ibid., p. 52; Scoresby, "On the Greenland or Polar Ice," *Memoirs of the Wernerian Society for Natural History* 2 (1811–16) 261–338.

74°. A situation which for many years had been totally inaccessible."[20] In response to an inquiry from Banks, Scoresby sent him his printed *Treatise on the Northern Ice,* and added, in a covering letter, the remark that

Had I been so fortunate to have had the command of an expedition for discovery, instead of fishing, I have little doubt but that the mystery attached to the existence of a north west passage might have been resolved. There could have been no great difficulty in exploring the eastern coast of Greenland.

I do conceive that there is sufficient interest attached to these remote regions to induce Government to fit out an expedition.[21]

Banks was impressed by Scoresby's treatise, and by his optimism about the passage. In November, writing again to Banks, Scoresby cautiously qualified that optimism. He thought, from "attentive observation of the nature, drift and general outline of the polar ice," it might not be open again for ten or twenty years.[22]

## JOHN BARROW

Banks told Lord Melville, First Lord of the Admiralty, about the ice-free seas that summer. He urged that efforts be made

to endeavour to correct and amend the very defective geography of the Arctic Regions more especially on the side of America. To attempt the Circumnavigation of old Greenland, if an island, as there is reason to suppose. To prove the existence or non-existence of Baffin's Bay, and to endeavour to ascertain the practicability of a Passage from the Atlantic to the Pacific Ocean, along the Northern Coast of America.[23]

Here was an ambitious program, a blueprint for arctic geographers and seekers for the Northwest Passage. John Barrow,[24] second secretary of the Admiralty, advanced similar arguments, seizing on Scoresby's information,

[20]  Quoted in Stamp, *William Scoresby,* p. 64.      [21]  Ibid., p. 66.

[22]  Ibid., p. 67. If anything, Scoresby was too sanguine even in his pessimistic forecast: ice conditions deteriorated over the following decades, until they were at their worst when John Franklin sailed on his ill-fated final voyage. Scoresby was an unusually gifted observer of ice conditions, but he was not unique; Dutch and British whalers have been credited as the first dedicated observers of ice conditions in Davis Strait and Baffin Bay [Lamson in D. L. VanderZwaag and C. Lamson, eds., *The Challenge of Arctic Shipping: Science, Environmental Assessment, and Human Values* (Montreal and Kingston, 1990) p. 4].

[23]  Banks to Robert Dundas, 2nd Viscount Melville, First Lord of the Admiralty, 29 Nov. 1817, reproduced in Carter, *Sir Joseph Banks,* p. 508.

[24]  C. Lloyd, *Mr. Barrow of the Admiralty* (London, 1970).

while ignoring his warning that ice conditions would probably be worse in the following season.[25] Barrow's most ambitious travels had been to China, as comptroller of Lord Macartney's household during that nobleman's abortive British embassy to the emperor; the sciences had featured prominently in the embassy's arsenal of cultural artifacts.[26] But he had also once been to Greenland on a whaling ship, and now made the Arctic his personal crusade. He was convinced, unlike Scoresby, that there was an open polar ocean – did not the Russians winter comfortably in Spitsbergen, while they found Novaya Zemlya to the south less hospitable?

There was an added argument – Russian activity in the North. The great northern voyages of the previous century had been Russian, and very recently Lieutenant Kotzebue of the Imperial Russian Navy had sailed through Bering Strait and along the coast to the north and east, into the North American Arctic. Russia already held the main part of Alaska, and now it looked as if Russians would tackle the Northwest Passage. This would have territorial implications, as well as those affecting trade and glory. Barrow warned that "it would be somewhat mortifying, if a naval power but of yesterday should complete a discovery in the nineteenth century, which was so happily commenced by Englishmen in the sixteenth."[27]

Barrow gave no credit to Scoresby. The renewed search for a northwest passage was to be a naval affair. Barrow acknowledged with some embarrassment that

the last four expeditions, fitted out for discovery in this quarter, brought no accession to that knowledge of the geography of those seas and islands which had been acquired two hundred years before. We have heard it hinted, with sufficient illiberality, that the chief cause of failure was owing to their being under the command of naval officers. . . . [But the failure of a few does not] militate, in the slightest degree, against the employment of officers of the royal navy on this service: for in the instance alluded to, it so happened that one of them was suspected of having acted under the influence of his old masters, the Hudson's Bay Company, who were averse from all interference with what they are disposed to consider their exclusive privilege; another was addicted to drinking; a third took fright at the ice; and a fourth was totally incapacitated by a violent attack of fever. The circumstance most to be apprehended by the appointment of naval officers is that of attempting too much

25   For differences between Scoresby and Barrow, see Constance Martin, "William Scoresby, Jr. (1789–1857) and the Open Polar Sea – Myth and Reality," *Arctic* 41 (1988) 39–47.

26   J. L. Cranmer-Byng, *An Embassy to China: Being the Journal Kept by Lord Macartney during his Embassy to the Emperor Ch'ien-lung 1793–1794, with an Introduction, Notes and Appendices* (London, 1962). J. L. Cranmer-Byng and T. H. Levere, "A Case Study in Cultural Collision: Scientific Apparatus in the Macartney Embassy to China, 1793," *Annals of Science* 38 (1981) 503–25.

27   [J. Barrow], *Quarterly Review 18* (1817) 219–20.

rather than too little; but as the navigation among ice is itself a science, to be learned only from practice, prudence will necessarily dictate that every ship employed on this service shall be supplied with an experienced Greenland fisherman, to act as pilot in those seas.[28]

The navy scarcely needed that kind of defense. But Barrow had indicated several factors that were to characterize polar voyages in the coming decades. First was the professional exclusivity of the navy, at least in relation to civilians; this exclusivity underlay naval ambivalence toward the Hudson's Bay Company.[29] Even when the navy depended for logistical support upon the company, as it did in John Franklin's overland expedition, they undervalued that support. The company, in its turn, expected a degree of mobility and self-sufficiency in its employees that the navy could by no means match. Barrow did recognize that ice navigation was a science, but that did not mean that whalers could ever have command; they would be limited to the role of pilot. Similarly, naval commands, after Halley's fiasco with his officers on the *Paramore*, would not go to scientists, and the scientific effectiveness of successive expeditions was accordingly weakened. When Scoresby, experienced in arctic sailing, correspondent of Joseph Banks, friend of Robert Jameson, future fellow of the Royal Society of London, and corresponding member of the Institut de France, expressed hopes that he might lead the arctic expedition that he had first proposed to Banks, Banks gave him no encouragement, having already learned from previous brushes the limits of his influence at the Admiralty. It was soon clear, after initial evasiveness, that the commands (there were to be two expeditions) would go to naval officers.

Barrow had already made the argument for national glory. But science was to be co-opted in the enterprise. Banks, corresponding with Scoresby in the winter of 1817–18, had indicated that one reason for being interested in the decrease of the polar ice was the light that it might shed on the English climate; a succession of unusually cold springs and summers had much reduced the cider crop of apples for at least sixteen years. Glaciology, oceanography, hydrography, meteorology, and agriculture might be connected in ways that, once understood, could be of material as well as intellectual benefit. The ethos of scientific improvement was very much alive in the early nineteenth century:[30] the notion that science could be profitably

[28] Ibid., p. 213.

[29] See Hugh N. Wallace, *The Navy, the Company, and Richard King: British Exploration in the Canadian Arctic, 1829–1860* (Montreal, 1980).

[30] M. Berman, *Social Change and Scientific Organization: The Royal Institution, 1799–1844* (Ithaca, New York, 1978).

applied to economically significant activities reinforced the demand for arctic exploration. Since geography itself was very much one of the sciences in the early nineteenth century, the link between science and exploration was in any case direct. The Admiralty and the Royal Society of London were to be yoked in an unequal but symbiotic partnership as the blanks on the arctic map were gradually filled in.[31]

THE 1818 EXPEDITION OF JOHN ROSS: GEOMAGNETISM

Encouraged by the Admiralty and the Royal Society, Parliament in 1818 passed an act offering substantial rewards for the discovery of a northwest passage, or for the nearest approach to the pole, in the twin interests of commerce and science.[32] A full traverse of the passage, from the Atlantic to the Pacific by a northern sea route, would net the successful navigator £20,000 – this at a time when a school teacher's annual salary was around £50. Men and ships were made available for two expeditions, one (in which Lieutenant John Franklin, R.N., who had been with Flinders on H.M.S. *Investigator,* was second in command) heading for the North Pole by way of Spitsbergen; the other, the first of the major nineteenth-century expeditions in search of the Northwest Passage, was commanded by John Ross in H.M.S. *Isabella,* accompanied by William Edward Parry in H.M.S. *Alexander.* Ross and Parry were to have very different but highly prominent arctic careers. Both of their ships were transports, reinforced against the ice but scarcely designed for arctic seas. Ross was unhappy with them at the time: "when I arrived in London I was concerned to discover that the ships (by that time half finished) were totally unfit for [arctic] service; but my remonstrances were too late, and I was told that if I did not choose to accept the command some one else would . . ." In retrospect, he also regretted that all the junior officers had been appointed without his being consulted, with the sole exceptions of his nephew James Clark Ross, and the purser. Ross would have preferred officers experienced in northern ice and in wintering, as well as skilled in navigation and seamanship. "I would certainly have employed Mr. Scoresby."[33]

[31]   M. B. Hall, *All Scientists Now: The Royal Society in the Nineteenth Century* (Cambridge, 1984) pp. 199–215 looks at the Royal Society's encouragement of exploration, including (pp. 200–5) arctic exploration.

[32]   *A Bill for more effectually discovering the Longitude at Sea, and encouraging attempts to find a Northern Passage between the Atlantic and Pacific Oceans, and to approach the Northern Pole,* 9 March 1818.

[33]   John Ross, *Narrative of a Second Voyage in Search of a North-West Passage: and of a Residence in the Arctic Regions during the Years 1829, 1830, 1831, 1832, 1833 . . . In-*

Ross had had two seasons in the Baltic, which was as close as any living Royal Naval officer had come to the Arctic. He had entered the Royal Navy as a volunteer at the age of nine, transferred to the Merchant Service after two and a half years, next served in the East India Company, and then returned to the Royal Navy; now, at age forty, he had had thirty-one years of sea-going experience. He had distinguished himself in action, having been thirteen times wounded. He had studied astronomy under William Wales, the astronomer of Cook's second voyage. He was also skilled in surveying. He was a good choice for the command.[34]

When Ross left England in 1818, his primary mission was geographical discovery. He was also directed to make scientific observations; these would form the secondary but by no means trivial part of his mission.[35] Not only the Admiralty, but also the Royal Society provided instructions and guidance.[36] An attempt was to be made "to discover a Northern Passage, by sea, from the Atlantic to the Pacific Ocean." In order to supply necessary expertise in ice navigation, Ross's ships were accompanied by a master and mate of whale-fishing vessels, "well experienced in those seas," to offer advice but not to command.

The expedition offered a great opportunity to contribute to "the advancement of scientific and natural knowledge." Accordingly, in consultation with the Royal Society, the Lords of the Admiralty had caused to be placed on board "a great variety of valuable instruments."[37]

cluding the *Reports of Commander, now Captain, James Clark Ross, R.N., F.R.S., F.L.S., &c.* and *The Discovery of the Northern Magnetic Pole* (London, 1835) p. xi.

[34] Ernest S. Dodge, *The Polar Rosses: John and James Clark Ross and their Explorations* (London, 1973); Berton, *The Arctic Grail*, p. 22.

[35] J. Ross, *A Voyage of Discovery, Made under the Orders of the Admiralty, in His Majesty's Ships Isabella and Alexander, for the Purpose of Exploring Baffin's Bay, and Inquiring into the Probability of a North-West Passage* (London, 1819) pp. 9 et seq. Ross stated on 14 July 1818 that "It is my Orders and Directions that the Officers of the respective Watches do pay particular attention, to the Log, Courses, Winds, Signals, also all the objects enumerated in the Meteorological part [under the headings] Courses by Compass/ Variation on each Course/ Winds/ Leeway/ Soundings/ Temperature [of] Air [and] Sea/ Height of the Barometer [and] Sympiesometer/ Thermometer/ Hygrometer/ Cyanometer/ Bearing/ [and] Distance." [John Ross, "Occurrences and Meteorological Remarks HMS Isabella 1818," Scott Polar Research Institute, Cambridge, MS 546, entry of 14 July 1818.] The cyanometer was presumably on the pattern of Arago's, with an arbitrary scale of blues on a strip of porcelain, to compare with the blue of the sky; the question of atmospheric refraction in cold climates was of long-standing interest. The sympiesometer was an improved version of Hooke's air barometer, useful at sea. [Gerard Turner, *Nineteenth-Century Scientific Instruments* (Berkeley and Los Angeles, 1984) p. 232.]

[36] *Instructions for the Adjustments and Use of the Instruments Intended for the Northern Expeditions: Printed by Order of the Royal Society* (London, 1818).

[37] J. Ross, *Voyage*, pp. 9–10.

Figure 2. John Ross by Mary F. Hamilton ca. 1834. National Archives of Canada C-123839.

They had also, on the society's recommendation, ordered Edward Sabine of the Royal Artillery, a division of the army's Ordnance, to accompany the expedition. Sabine, who had been represented as "a gentleman well skilled in astronomy, natural history, and various branches of knowledge," was to be the expedition's scientific factotum.[38] Civilian natural philosophers were not welcome; members of the other armed service could at least be expected to understand naval discipline, and Sabine was accordingly accepted. Born in Dublin in 1788, he had entered the Royal Military Academy at Woolwich in 1803. His active service had been relatively brief, and based in Upper and Lower Canada, where in 1813 and 1814 he was engaged in resisting American attacks. Underemployed like so many officers after the Napoleonic Wars, he pursued scientific studies, notably in ornithology, astronomy, and magnetism. He became a fellow of the Linnean Society, and in April 1818, just before his appointment to Ross's expedition, he had been elected to the Royal Society. John Barrow was among the signatories to his election certificate; Barrow's patronage was to be one of the keys to an arctic posting.[39]

Sabine's scientific responsibilities on the expedition were comprehensive. So were Ross's instructions. I shall discuss these relatively thoroughly, because they were the model for subsequent naval arctic expeditions until the last quarter of the century, and the instruments issued to Ross, and essential to his scientific program, were correspondingly significant.

Prominent among Ross's instructions were those concerning geomagnetic observations. There were obviously going to be navigational problems in the far North: just where was the magnetic pole, and how would the compass behave near it? How many magnetic poles were there, and were they stationary or moving? Robert Hooke, active in theory and experiment in the early decades of the Royal Society, had proposed that the magnetic pole was in motion. Edmond Halley had concluded that there were four magnetic poles. There were so many unresolved questions that the subject was as enticing to natural philosophers as it was problematic for navigators. At the end of the Napoleonic Wars, the German scientific explorer Alexander von Humboldt set about obtaining government support for the study of terrestrial magnetism. The shared interests of natural philosophers and the armed forces, with their disciplined and organized body of potential observers, was later to lead to a chain of colonial magnetic observatories as part of an international scientific effort, well described as the magnetic crusade.[40] Ross was told that

---

[38] Ibid., p. 10.

[39] T. H. Levere, "Edward Sabine (1788–1883)," *Arctic* 38 (1985) 146–7. Royal Society of London, Election Certificates vol. 6 no. 375, read 15 Jan. 1818, balloted for 16 April 1818.

[40] John Cawood, "Terrestrial Magnetism and the Development of International Collaboration in the Early Nineteenth Century," *Annals of Science* 34 (1977) 551–87, and "The Magnetic Crusade: Science and Politics in Early Victorian Britain," *Isis* 70 (1979) 493–518.

Amongst other objects of scientific inquiry, you will particularly direct your attention to the variation and inclination of the magnetic needle, and the intensity of the magnetic force; you will endeavour to ascertain how far the needle may be affected by the atmospherical electricity, and what effort may be produced on the electrometer and magnetic needle on the appearance of the Aurora.[41]

Around 1600 William Gilbert had suggested that plotting magnetic vectors across the earth's surface, and noting how these varied with time, might provide a reliable means of navigation. Ross's instructions, like those of all his nineteenth-century successors in the North, required him to observe three variables: dip or inclination, declination or variation, and intensity.

Dip or inclination is the vertical angle between the horizontal plane at a given point and a freely suspended magnetic needle. It can be measured using a dip circle, an instrument invented around 1575 by Robert Norman, an Elizabethan navigator and instrument maker. By the time of Ross's voyage, dip circles were a staple of the instrument maker's catalogue and workshop. Dip circles provided by the navy in the early nineteenth century were generally of inferior construction, as Sabine complained.[42] The expedition therefore carried several dip circles made by well-known instrument makers. Sabine singled out for praise one made by Nairne and Blunt.[43] Nairne's dip needles had proved their worth over the years; one had accompanied Captain Cook on the *Endeavour,* and another had been used by Israel Lyons on Captain Constantine Phipps's voyage toward the North Pole in 1773.[44] On Ross's voyage, Nairne's needle was "the only one which could be depended on."

Declination is the angle between the needle and geographic north, located by observing the pole star. Declination had been known in Europe

---

[41]   J. Ross, *Voyage,* p. 10.

[42]   A. McConnell, *Geomagnetic Instruments before 1900* (London, 1980) p. 20.

[43]   This instrument was "similar in construction to one made by the same artists, and described by the Hon. Henry Cavendish in the 66th volume of the Philosophical Transactions": Sabine, "Observations on the Dip and Variation of the Magnetic Needle, and on the Intensity of the Magnetic Force, Made During the Late Voyage in Search of a North West Passage," *Philosophical Transactions* 109 (1819) 132–44 at 132. The Science Museum, London, has a dip circle by Nairne and Blunt circa 1775, Inv. no. 1900–129; a dip circle used by Henry Cavendish, Inv. no. 1930–903; and a dip circle by Nairne and Blunt for observations at sea, Inv. no. 1876–806. For general background, see Robert P. Multhauf and Gregory Good, *A Brief History of Geomagnetism and A Catalog of the Collections of the National Museum of American History* (Washington, D.C., 1987); A. McConnell, *Geophysics & Geomagnetism: Catalogue of the Science Museum Collection* (London, 1986), and *Geomagnetic Instruments.*

[44]   E. G. R. Taylor, *The Mathematical Practitioners of Hanoverian England 1714–1840* (Cambridge, 1966) pp. 50, 54–5.

almost as long as the magnetic needle itself, dating from around 1300. In 1818 apparatus for its measurement varied from pocket compasses to precision instruments in which large magnetic needles were suspended from a single thread. Once again Cavendish's instrument was among the finest and most sensitive of its day.[45] Ross's ships carried a variety of compasses. Two azimuth compasses,[46] designed by the military surveyor and engineer Henry Kater, F.R.S.,[47] were supplied; because they needed to be carefully leveled, they were of limited use when the ship was moving.[48] Under these circumstances, the compass designed by Ralph Walker (1749–1824) proved superior, "but its card being heavy, it ceased to traverse when the variation was 110°, and the dip 86°."[49] Walker had submitted his compass to the Board of Longitude in the 1790s, with support from George Adams, who both made and sold it. Walker advocated the use of his compass in determining longitude at sea, but the astronomer Nevil Maskelyne recommended against Walker's method.[50]

Steering compasses were obviously essential for navigation. Henry Constantine Jennings (1731–1820) supplied his new insulated steering compass, for which he had just taken out a patent. It did away with the local attraction of the ship's iron, but its card was heavy, and its needle short and not very powerfully magnetized, so that it ceased to act when the variation was large. Much more satisfactory, and indeed the best altogether, was Alexander of Leith's steering compass, four of which were carried; the card and needle were well proportioned, the friction was counteracted by "ingenious suspension," and the instrument was well adapted for use on boat or ship, especially when in motion. It traversed when all the other steering compasses carried by the expedition failed.[51] The number of instruments intended to perform the same function makes it clear that Ross's voyage was seen as an ideal laboratory in which to test instruments, many of which had previously enjoyed only theoretical advantages.

[45] Science Museum Inv. no. 1930–902.
[46] McConnell, *Geomagnetic Instruments*, plate 4. An azimuth compass is a mariner's compass fitted with vertical sights, for taking the magnetic azimuth of a star, i. e., the angle between the magnetic meridian and the great circle passing through the star.
[47] Taylor, *Mathematical Practitioners*, pp. 342–3. Captain Kater surveyed in India, and invented a variety of geodetic instruments.
[48] J. Ross, *Voyage*, pp. xix and Appendix p. cxxiv.       [49] Ibid., p. cxxiv.
[50] R. Walker, *A Treatise on Magnetism with a Description of the Meridional and Azimuth Compass* (1794); Taylor, *Mathematical Practitioners*, p. 300.
[51] J. Ross, *Voyage*, pp. xix and Appendix p. cxxv. Other compasses carried were: one Crow's boat compass, and two Burt's patent binnacle and compass. Taylor, *Mathematical Practitioners*, pp. 236, 385, 387.

The third magnetic vector, force or intensity, varies with space and time, and may be represented either by separate horizontal components, or by a single measure of total force. It was measured using a circle, in which the dip needle was moved from its position by applying another standardized needle above or below it. The disturbance caused the dip needle to oscillate, and the period and amplitude of the oscillations were determined by observation, both in the plane of the meridian, and at right angles to it. Comparison of these data with corresponding ones obtained with the same instrument in London gave one a relative measure of the magnetic force.[52]

The reference to aurora in relation to atmospheric electricity and to geomagnetism was an attempt to elucidate the nature of these phenomena, which were variously and speculatively considered to be electrical in nature, related to magnetism, and at once part of meteorology and of geophysics.

## HYDROGRAPHY, METEOROLOGY, AND OTHER GEOPHYSICAL SCIENCES

Magnetic surveys were important for navigation. But the principal justification for the voyage, in the eyes of the Admiralty, was the search for a northwest passage. Crucial evidence for the existence of such a passage was the summer current flowing down Davis Strait from the North: "hence Baffin's Bay cannot be bounded by land, as our charts generally represent it, but must communicate with the Arctic Ocean."[53] This was also John Barrow's conviction,[54] and one source of his enthusiasm for the enterprise. Tracing ocean currents would therefore be essential, and Ross was given detailed instructions for oceanographic and hydrographic work. "The strength and direction of the current should be tried once in twenty-four hours."[55] Temperature fluctuations, at the surface and at different depths, were also possible keys to the movement of the seas, and Ross was instructed to record these, as well as the atmospheric temperature. Temperature was measured with James Six's self-registering thermometer,[56] an

[52] *Instructions . . . Royal Society* (1818) pp. 14–16.     [53] J. Ross, *Voyage,* pp. 1–2.
[54] Ibid., p. 2; marginal note in John Barrow Jr.'s copy in the Baldwin Room, Metropolitan Toronto Library, BR 919.8 R59.21.
[55] J. Ross, *Voyage,* pp. 2, 10.
[56] J. Ross, *Voyage,* p. 2; A. McConnell, "Six's Thermometer: A Century of Use in Oceanography," in *Oceanography: The Past,* pp. 252–65; "Historical Methods of Temperature Measurement in Arctic and Antarctic Waters," *Polar Record, 19* (1978), 217–31; J. Six,

instrument still popular today. Six also made a deep-sea thermometer.[57]

Differing salinities could also be clues to the movements of fresh and salt water, and so temperature measurements were to be matched by measurements of salt content at considerable depths, because "snow or ice water may float at the surface, and salt water beneath." Humphry Davy accordingly designed a water bottle, a copper vessel with a stop cock "opened by a piston moving in consequence of the compression of air when the instrument is sunk in the sea." One could preset the pressure at which the valve would open, admitting water; as the bottle sank, the piston moved further, and closed the bottle. The temperature of the water was also to be taken: "If the current be ice cold and comparatively fresh, there can be little hope of reaching a deep sea in that direction."[58] Unfortunately, the bottle did not close properly, so that water escaped, or was mixed with that nearer the surface as it came up.[59] The importance of depth measurements, as well as of tidal observations, was stressed:

You are to attend particularly to the height, direction, and strength, of the tides, and to the set and velocity of the currents; the depth and soundings of the sea, and to the nature of the bottom; for which purposes you are supplied with an instrument better calculated to bringing up substances, than the lead usually employed for this purpose.[60]

Ross was issued with a sounder-sampler designed by John McCulloch,[61] and also with Burt's buoy and nipper, an inflatable canvas bag bearing an arm with a spring-loaded sheave and nipper. Line ran out until the lead reached bottom, when the strain came off the line and the nipper held it against the sheave, so that depth could be determined.[62] The sounder-sampler, which the Admiralty had lauded, proved inadequate; Ross complained of the many fruitless attempts he made to obtain bottom samples in deep water, and went on to invent his own instrument that the ship's armorer made for him. His "deep sea clamm" was a mechanical grab with jaws that sank into the ocean floor. When the clamm was hauled up, the jaws with their load of bottom sediment were lifted into a cast iron vessel,

"An Account of an Improved Thermometer," *Philosophical Transactions of the Royal Society of London*, 72, 72–81.
[57] A. McConnell, *No Sea Too Deep: The History of Oceanographic Instruments* (Bristol, 1982) pp. 22–4.
[58] Davy in *Instructions*, p. 35.  [59] J. Ross, *Voyage*, p. cxxix.  [60] Ibid., pp. 10–11.
[61] McConnell, *No Sea Too Deep*, p. 40; A. Rice, "The Oceanography of John Ross's Arctic Expedition of 1818: A Reappraisal," *Journal of the Society for the Bibliography of Natural History*, 7 (1975) 291–319.
[62] McConnell, *No Sea Too Deep*, pp. 30–1.

sealing it. The results were eminently satisfactory to Ross. For example, in Baffin Bay on 1 September 1818, "soundings were obtained correctly in one thousand fathoms, consisting of soft muds, in which there were worms, and, entangled on the sounding line, at the depth of eight hundred fathoms, a beautiful caput medusae . . ." Alas, the sounding was not as satisfactory as Ross believed; brittle stars are bottom-dwelling creatures, so at least 200 fathoms of line were lying on the bottom.[63] But Ross's soundings were generally accepted uncritically over the ensuing decades, to the confusion of oceanography.[64]

There were further instructions about oceanography and hydrography, in which Sabine was to provide material assistance. The Admiralty's confidence in his abilities was predicated in part upon his reputation as "an officer well versed in . . . the practical use of instruments."[65]

Navigation was going to depend not only on magnetic and hydrographic work, but also on conventional measurements of solar and terrestrial angles. These, however, might be vitiated by changes in the atmosphere's refraction at low temperatures. Refraction had been a problem for astronomers, and had been the subject of the Bakerian Lecture of the Royal Society of London in 1799.[66] William Hyde Wollaston, chemist, entrepreneur, and inventor, invented and described for the expedition a new sector and micrometer for measuring the dip of the visible horizon at sea. Ross was instructed to

cause the dip of the horizon to be frequently observed by the dip sector invented by Dr. Wollaston; and ascertain what effect may be produced by measuring that dip across fields of ice, as compared with its measurement across the surface of the open sea.

You will also cause frequent observations to be made for ascertaining the refraction, and what effect may be produced by observing an object, either celestial or

---

[63]  J. Ross, *Voyage*, pp. 10, cxxxiii–cxxxvi, 178; *A Description of the Deep-Sea Clamms, Hydraphorus and Marine Artificial Horizon* (London, 1819). McConnell, *No Sea Too Deep*, p. 42.

[64]  A. J. Rice, "Oceanography of Ross's Expedition," pp. 291–319.

[65]  John Barrow, *A Chronological History of Voyages Into the Arctic Regions; Undertaken Chiefly for the Purpose of Discovering a North-East, North-West, or Polar Passage Between the Atlantic and Pacific: From the Earliest Periods of Scandinavian Navigation, to the Departure of the Recent Expeditions, Under the Orders of Captains Ross and Buchan* (London, 1818; reprinted Newton Abbot, 1971) p. 367.

[66]  S. Vince, "Observations on an unusual horizontal refraction of the air," *Philosophical Transactions of the Royal Society of London* 89 (1799) 13–23. It was to become one of Scoresby's subjects of observation; see, e.g., his "Description of Some Remarkable Atmospheric Reflections and Refractions Observed in the Greenland Sea," *Transactions of the Royal Society of Edinburgh* 9 (1823) 353–64.

Figure 3. Deep-sea clamm, in John Ross, *A Voyage of Discovery, Made Under the Orders of the Admiralty, in His Majesty's Ships Isabella and Alexander, for the Purpose of Exploring Baffin's Bay, and Inquiring into the Probability of a North-west Passage* (1819). Royal Ontario Museum.

terrestrial, over a field of ice, as compared with objects over a surface of water; together with such other meteorological remarks, as you may feel you have opportunities of making.[67]

The dip sector was used on the outward and homeward passages, but in Baffin Bay, the "great inequality of refraction on the horizon" ruled out its use. The dip micrometer was simply not used.

Accurate navigation also depended upon the accurate measurement of time, which was possible thanks to Harrison's invention of the marine

[67] J. Ross, *Voyage*, p. 10.

chronometer; longitude at sea could now be determined accurately. Local time, generally noon, was determined astronomically by observing when the sun was at its meridian, and this time could be compared by the chronometer with the time at a standard location, generally Greenwich.[68] The difference between the two times gave a measure of longitude, one hour's difference corresponding to a difference in longitude of fifteen degrees. There were seven chronometers on Ross's expedition.

There was also a clock, which had demonstrated its reliability under Captain Cook.[69] Its "pendulum . . . , cast in one solid mass, vibrates on a blunt knife-edge, resting in longitudinal sections of hollow sections of agate."[70] Then there was a transit instrument; Henry Kater, whose azimuth compasses we have already encountered, provided instructions for the use of this apparatus. Neither this transit instrument, nor the variation transit made by Dollond, were used after the expedition left Waygatt (Waigat) Island[71] off the west coast of Greenland. The clock with its special pendulum was the main piece of apparatus used in determining the length of the seconds pendulum in high latitudes. Time by the clock was compared with time by the chronometers, whereas the error of the chronometers was determined using the variation transit.[72] Such observations would refine knowledge of the shape of the earth. They were incorporated by Sabine with subsequent data in a volume published in 1825.[73]

The measurement of speed through the water was important, and three patent logs were carried: Bains's, Massey's, and Jennings's, the last with a mercurial log glass patented the year after the voyage, which was regarded by Ross as "very superior," and recognized by a silver medal from the Royal Society of Arts.[74] Edward Massey's device, a towed rotator with a recording device, patented in 1802, was "the first effective model to be commercially produced."[75] Ross noted that it

---

[68]  Greenwich tradition was in fact lunars, using the *Nautical Almanack*, rather than chronometers; the navy used both. See chapter 4.

[69]  D. Howse, "Captain Cook's Marine Timekeepers," *Antiquarian Horology* (1969) 190–9.

[70]  J. Ross, *Voyage*, p. xvii.          [71]  J. Ross, *Voyage*, Appendix p. cxxvii.

[72]  *Instructions*, pp. 4–14.

[73]  E. Sabine, *An Account of Experiments to Determine the Figure of the Earth, by Means of the Pendulum Vibrating Seconds in Different Latitudes; as well as on Various Other Subjects of Philosophical Inquiry* (London, 1825). Henry Kater made a reversible pendulum that was particularly accurate for determining the length of a seconds pendulum; see Turner, *Nineteenth-Century Scientific Instruments*, pp. 49–50, and McConnell, *Geophysics & Geomagnetism*, pp. 16–17.

[74]  J. Ross, *Voyage* p. cxxxi; Taylor, *Mathematical Practitioners*, p. 236.

[75]  Turner, *Nineteenth-Century Scientific Instruments*, pp. 267–8.

performed extremely well, but from a defect in the materials with which it was made, and which we were not able to replace, we could not use it, after it was damaged, but I am of opinion that this instrument would be of great use, particularly to surveying vessels, as it is capable of measuring a distance with great accuracy.[76]

Robert Bains had been granted a six-months' patent, "pending specification," for a perpetual log. Ross found that this instrument also performed very well.[77]

There were a few other specialized instruments. Kater designed an altitude instrument, intended to facilitate observations of the sun's altitude when the horizon was obscured by mists. Ross was diplomatically negative in his comments: "This is likely to become a valuable instrument; – it requires practice, and Mr. Bisson and Mr. Ross made great progress in it; but it is not sufficiently near the truth to be depended on . . . ; the general opinion was, that it was on too small a scale."[78] Nor were they able to use the electrical apparatus, essentially an electrometer, which Davy had designed to explore the electrical character of the atmosphere in the polar regions. The instrument was built not only because of the analogy of the aurora to electrical light, but also because of Davy's speculation that the earth might have electrical as well as magnetic poles. However, as the ship was not frozen in, there was no opportunity to perform the electrical experiments.[79]

The Royal Society proposed more experiments than could be carried out, and invented more instruments than were used. Some items, notably Davy's water bottle, were too complicated, and failed to perform adequately. But the navy had collaborated in an enterprise that combined scientific with geographical exploration, and the advancement of natural knowledge with national prestige. The extreme conditions of arctic navigation demanded rugged, simple, and reliable apparatus to answer a host of questions, some of theoretical and others of eminently practical import, for example the possible extension of the whale fishery. Ross's was the first of a series of nineteenth-century voyages in which oceanographic, geophysical, and meteorological knowledge of the arctic regions was enhanced.

It was also an expedition in which significant natural history observations were made. Ross had been instructed, with Sabine's help, to make collections of specimens from the animal, vegetable, and mineral kingdoms. The specimens were of scientific interest, and the cause of public

[76] J. Ross, *Voyage*, p. cxxix.
[77] Taylor, *Mathematical Practitioners*, pp. 382–3. The oceanographic apparatus carried on the voyage has been well described by Anita McConnell in *No Sea Too Deep*.
[78] J. Ross, *Voyage* p. cxxix.    [79] Ibid., pp. cxxix–cxxx; *Instructions*, pp. 26–34.

merriment. George Cruikshank produced a cartoon, "Landing the Treasures, or Results of the Polar Expedition," published on 18 January 1819,[80] showing a vial of red snow (colored by algae along the Greenland cliffs),[81] and other delights for the British Museum. Sabine is portrayed with a seabird skewered on his bayonet; it was a newly discovered species, Sabine's gull, *Larus Sabini,* first seen on 25 July 1818.[82] The expedition also brought back an ivory gull, among many other specimens. The second and third appendixes of Ross's narrative of the voyage present zoological and geological data, and include some naive but remarkably convincing sketches by John Ross of a musk ox, arctic foxes and hares, and the Inuit, around Thule, who had never seen white men before, and whom Ross called Arctic Highlanders. Ross obtained a spear made of bone, and a knife made from iron, which the chemist William Wollaston showed contained nickel, and which he rightly identified as of meteoric origin.[83]

Sabine had disclaimed to Ross any special competence in natural history, except in ornithology.[84] None of the members of the expedition was competent in geology, and the geological collection was accordingly indifferent.[85] Ross was aided in natural history by the surgeon and assistant surgeon on *Isabella,* who provided anatomical data, for example observations on the unusual trachea in long-tailed ducks.[86] Ross's manuscript journal of the voyage includes a list of his own observations of marine invertebrates, which were carefully recorded and subsequently corrected by Dr. W. E. Leach of the British Museum, who was able from Ross's notes to identify new species, including a gastropod of the genus *marganita,* and a crustacean, both named after Leach. But the notes and specimens between them were not always enough. Ross wrote in frustration:

[80]  Rice, "Oceanography of Ross's Expedition," pp. 292–3; further information about the cartoon is in M. D. George, *Catalogue of Political and Personal Satires Preserved in the Department of Prints and Drawings in the British Museum* (London, 1949) vol. 9 1811–19.

[81]  Cryoalgae grow in the relatively warm microenvironment of tiny air cells below the surface of the snow, and appear as red (and sometimes green) patches when the snow melts: see R. W. Hoham, "Unicellular Chlorophytes – Snow Algae," in E. R. Cox, ed., *Phytoflagellates* (New York, 1980) pp. 61–84. The contemporary account was Franz Bauer, "On Some Experiments on the Fungi which Constitute the Colouring Matter of the Red Snow Discovered in Baffin's Bay," *Phil. Trans. Royal Society of London 110* (1820).

[82]  The species was discovered by Edward Sabine. See W. Swainson and J. Richardson, *Fauna Boreali-Americana* vol. 2 (London, 1831; reprinted New York, 1974) pp. 428–9.

[83]  Sweet, "Robert Jameson," 41.          [84]  J. Ross, *Voyage,* p. xl.

[85]  Sweet, "Robert Jameson," 42. The geological specimens were described by John Macculoch.

[86]  Ibid., p. xlix.

Figure 4. George Cruikshank, "Landing the Treasures, or Results of the Polar Expedition!!! at Whitehall 17 Dec. 1818." Courtesy of the Trustees of the British Museum.

Figure 5. Sabine's gull (Xeme) by Thomas Lewin, in Ross, *Voyage* (1819). Public Archives of Canada C-119867.

An endless variety of the class *Acalephae* were brought home and lent to the Museum, so much contracted by the Spirit, as to render it impossible for Dr. Leach to make out their genera. Observations on these animals whilst living accompanied by accurate drawings . . . are quite necessary to render the preserved specimens of any degree of use, and it is to be regretted that no naturalist capable of performing this indispensable part of his duties accompanied the expedition.[87]

[87] John Ross, *Journal of the Voyage of the Isabella to the Arctic Regions . . . From 13 April 1818–14 October 1818*, Huntington Library MS HM 708.

Figure 6. "The Royal Navy meets the Esquimaux," drawn by John Backhouse, and published in Ross, *Voyage* (1819). Metropolitan Toronto Library.

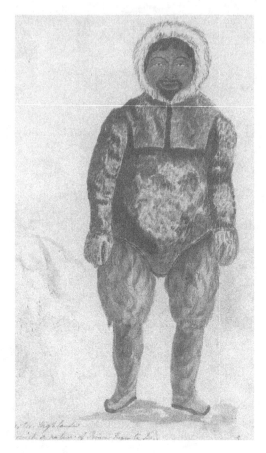

Figure 7. "Arctic Highlander. Ervick a native of Prince Regents Bay," sketch by John Ross. Public Archives of Canada C-100068.

In this respect Ross was less well served than either Cook or Phipps had been, and the Admiralty was to learn, at least in principle, from this omission.[88]

Important as all this scientific work was, there remained the one science that was of the greatest importance to the Admiralty, and that was geography. If Ross had discovered the Northwest Passage, he would have made his fortune, and guaranteed a stellar career for himself. He began well

---

[88]  See *A Manual of Scientific Enquiry: Prepared for the Use of Officers in Her Majesty's Navy; and Travellers in General,* J. W. F. Herschel, ed., 2nd. ed. (London, 1851; reprinted Folkestone, 1974, with introduction by D. M. Knight) pp. 379, 413. Aspects of the manual related to arctic science are discussed in chapter 4.

enough, following the Greenland coast of Davis Strait, initially in the company of a whaling fleet. He passed through and named Melville Bay after the First Lord of the Admiralty, and then had to explore Smith, Jones, and Lancaster sounds, discovered by Bylot and Baffin in 1616. Ross merely explored Smith Sound, but only briefly, deciding that it offered no passage. He was similarly brisk about Jones Sound, going no further than the entrance, and deciding that it was an inlet blocked by mountains. Then came Ross's great misfortune. He entered Lancaster Sound, and thirty or eighty miles in,[89] waiting for fog to clear and for Parry in the poor-sailing *Alexander* to catch up, he saw mountains where there were none, blocking the sound – he may have seen a *fata morgana* or other mirage, a not infrequent phenomenon of refraction over cold seas. He named them the Croker Mountains, after the First Secretary of the Admiralty. But none of his officers saw them, and Parry, who held his peace at the time, was later to express incredulity.[90]

When the ships returned to England, they carried with them scientific data and specimens, and geographical information confirming Baffin's and Bylot's discoveries, which had ceased to be believed. But the Croker Mountains were his undoing. When Ross published his narrative, Barrow reviewed it scathingly in the *Quarterly Review*.[91] Edward Sabine entered the lists against Ross.[92] Failing to find the passage was regrettable; seeing mountains where no other officer had seen them was either unfortunate, or bad judgment; but arguing with one's subordinates was downright unseemly. The Admiralty Board held a court of inquiry, Ross was retired on half pay, and he never again received a naval command.[93] Nevertheless, with the backing of the distillery magnate and philanthropist Sheriff Felix Booth, he would later manage one of the most successful nineteenth-century expeditions,[94] spending four years at sea or in the ice. During that expedition, his nephew, James Clark Ross, who in 1818 had been midshipman on *Isabella,* discovered the north magnetic pole. But John Ross's Royal Naval career was over. James Clark Ross, Sabine, and Parry, who had all been

[89] Sabine's figure was thirty miles, Ross's was eighty.

[90] Alan Cooke and Clive Holland, *The Exploration of Northern Canada 500 to 1920: A Chronology* (Toronto, 1978) p. 139; they offer admirable summaries of all the expeditions to the Canadian Arctic up to 1920, and I have used their work gratefully throughout this book.

[91] *Quarterly Review* 21 (1819) 213–62.

[92] Sabine, *Remarks on the Account of the Late Voyage of Discovery to Baffin's Bay, Published by Captain J. Ross, R.N.* (London, 1819).

[93] Dodge, *The Polar Rosses*, chap. 4; a summary is in Berton, *The Arctic Grail*, pp. 30–4.

[94] See the penultimate section of this chapter.

Figure 8. "Passage through the ice. 16 June 1818," in Ross, *Voyage* (1819). Metropolitan Toronto library.

with him in 1818, were among the officers who later contributed most to arctic science; and Parry was the first to do so.

## WILLIAM EDWARD PARRY

Parry was the son of Dr. Caleb Hillier Parry, F.R.S., of Bath, one of whose patients. was the niece of Admiral Cornwallis. Parry entered the navy in 1803 at the age of twelve, on Admiral Cornwallis's flagship. He subsequently served in the North Sea and the Baltic, and was promoted to lieutenant in 1810. He was promptly assigned to a frigate protecting the Spitsbergen whale fishery, and on shipboard developed an interest in mapping the stars. "Astronomy," he wrote in 1811, "is a delightful science."[95] In 1813 he went to the North America station, and that winter wrote *Nautical Astronomy by Night,* published in 1816.

*On Ross's expedition of 1818.* Parry returned to England in 1817, where he soon interested himself in the plans afoot for a northwest passage expedition. He wrote to a friend of his father's that he was "ready for hot or cold," and that friend showed the letter, together with *Nautical Astronomy,* to John Barrow, who was much taken with it, and recommended Parry to Lord Melville for the forthcoming arctic expedition. Parry was called to London, met and was approved by Barrow; he also secured an introduction to Joseph Banks, who "shook hands with me very cordially, said he was glad to become acquainted with a Son of Dr. Parry's, for whom he entertained the highest respect, and was glad to find I was nominated to serve on the Expedition to the North West."[96] Banks discoursed to him about the Greenland ice, its connection with the weather, properties of whale oil, and more besides.

Parry made good use of his time in London. Barrow had introduced him to Sir George Hope, who "is *the* man at the Admiralty; he is the man upon whom the really-naval part of every out-fit falls." Now Parry was off to the Hydrographical Office, "to copy some late information respecting Greenland, transmitted by a clever man of the name of Scoresby, Capt^n of a Greenland whaler." On Boxing Day, he planned to visit Sir William

[95] Parry, letter dated Leith, Sept. 1811, in E. Parry, *Memoirs of Rear-Admiral Sir W. Edward Parry, Kt., F.R.S. etc. Late Lieut.-Governor of Greenwich Hospital* (London, 1857) p. 35.

[96] Ann Parry, *Parry of the Arctic: The Life Story of Admiral Sir Edward Parry 1790–1855* (London, 1963); W. E. Parry to his parents, London, 23 Dec. 1817, Scott Polar Research Institute (SPRI) MS 438/26/18.

Figure 9. Captain William Edward Parry. National Archives of Canada C-11567.

Herschel, Britain's great observational astronomer of the day, and the builder and owner of the world's largest telescope. William Herschel was kindness itself, and his son John, soon to be a leading scientist and spokesman of science, and later editor of the Admiralty's *Manual of Scientific Enquiry,* favored him with instructive conversation. John Herschel gave Parry a letter of introduction to Captain Kater. Parry followed up swiftly; Kater was out at the time but a couple of days later invited him to breakfast on New Year's Eve. Parry explored Banks's library, where he found "a good deal of information as to the voyages of the old Navigators to the

North-west"; and while he was in the library, Banks came in with Scoresby, to whom he introduced him. Scoresby was "indeed a scientific man."[97]

On 30 December, Parry visited Major Rennell, the geographer, who "does not think the scheme of finding the North-West passage a practicable one, although he approves the attempt." He also finally met Captain John Ross, "a good-tempered, affable man in his manner, and we were acquainted of course immediately." Ross had already chosen the *Isabella,* and he and Parry agreed to go down to Deptford on the following day to select Parry's smaller vessel. Ross, he noted approvingly, was "clever in the surveying way, and is a good seaman. . . . [He] means to attend the first meeting of the Royal Society: I ought to do so too."[98] Clearly scientific preparations were as important as other naval ones in Parry's mind. It was equally clear that he moved easily and with the advantage of connections in the naval and especially in the scientific circles of the capital.

The expedition sailed in April, and on 25 July Parry wrote to his parents, before H.M.S. *Isabella* and *Alexander* left the Greenland whalers behind:

[I]t may be said that this season has been just like any other season; for the *whimsicalities* (as I cannot help calling it) of the ice are such, that it is impossible to say, from the appearance of the fields of it at one moment how it will be in ten minutes afterwards. . . . The story of the disappearance, or diminution of the ice, however, is perfectly without foundation.[99]

Still, he expected the expedition to perform wonders, and he had already written a paper on magnetism, which he sent to his parents and to Barrow, with a covering letter. He informed his parents that

Since I wrote that paper, the variation of the compass has increased to 89°!! – so that the North Pole of the needle now points nearly due *West!* The *Dip* of the needle is about 84°40'. As the needle is supposed to direct itself constantly to the Magnetic Pole, it follows that this pole must now be West from us, and as the dip is not far from 90°, it follows also that it must be placed somewhere not very far from us in that direction. The greatest variation observed by Baffin here 200 years ago (and the greatest, as he says, in the world) was 56°, so that an amazing increase has taken place during that interval.[100]

Parry, excited by the observation of such unanticipated magnetic phenomena, rejoiced in sailing closer to the magnetic pole than any previous

[97] Parry to his parents, London, 29 Dec. 1819, SPRI MS 438/26/19.
[98] Parry to his parents, London, 30 Dec. 1819, SPRI MS 438/26/20.
[99] Parry to his parents, H.M.S. *Alexander,* Davis Strait, Lat. 75°30″ N, 25 July 1819, SPRI MS 438/26/22.
[100] Ibid.

explorer; he constantly speculated about magnetism, which he found the most interesting subject of the whole voyage. Perhaps the recent decrease in magnetic variation in London, after several years of relative constancy, might be related to changes in the weather, and to the state of the Greenland ice? Magnetic observations were most assiduously taken, with help from several officers. For example, at 75°45″ N, and around 79° W, Sabine took dip observations; Parry and his officers noted the variation, which Sabine and the other officers of *Isabella* then confirmed. They all noted the increasing sluggishness of the needle as they sailed north (heavy cards were a hindrance), and they went through most elaborate tests to identify the causes and measure the deviation of compasses on board.

*Alexander*'s complement of instruments was not generous; much of the key scientific apparatus was on *Isabella*. Parry accordingly had provided additional apparatus himself, including two repeating circles, a self-registering thermometer, three sextants, a theodolite, and an artificial horizon. He was determined to prosecute as best he could the scientific instructions provided by the Royal Society. Even where he did not duplicate apparatus, he made sure that he understood it, and his private journal contains a discussion of the use of the seconds pendulum and other instruments. He was cautious in using Massey's patent log to detect currents, since he had only the one to lose. He was greatly frustrated by the impossibility of taking soundings while under weigh on *Alexander*, because of its poor sailing capabilities, "which will not allow me to heave to at any time, or even to luff the ship up in the wind, for that purpose." But when on occasion they were able to anchor or tie up to an obliging iceberg, Sabine, sometimes joined by James Clark Ross, was quickly out on the ice to set up the portable observatory and determine not only magnetic vectors, but also other geophysical ones, using the transit instrument, seconds pendulum, and the rest of their scientific arsenal. Meteorological observations were taken throughout the exploratory portions of the voyage.

Parry encouraged and sometimes assisted in collecting, that is, shooting birds, which his assistant surgeon then skinned; identification was confirmed where possible by reference to Pennant's *Arctic Zoology*.[101] Parry had been busy collecting; at the end of June he prepared a box of mineral specimens for his father, as well as one of birds together with a few minerals for Sir Joseph Banks. The red snow that was to be a cause for merriment to

---

[101]  Thomas Pennant, *Arctic Zoology,* 2 vols. (London, 1784–5) and supplement (1787) (reprinted New York, 1974).

George Cruikshank was diligently collected just short of 60° N, and Parry noted that it was produced "by a red vegetable matter."[102]

Parry was omnivorous in his scientific curiosity, and the lack of trained scientific personnel on board (if one partially excepts Sabine and the medical officers) made it appropriate for him to engage in a host of observations. He was sanguine about the expedition's prospects, and about its health – scurvy had not raised its head after six months at sea. He was all the more disappointed at the decision to turn back after Ross's sighting of the suppositious Croker Mountains. However, as the optimistic commander of Ross's other ship, he had enough arctic experience to make him an obvious choice as leader of the next naval expedition to the North.

*Parry's first expedition 1819–1820.* Parry was appointed to the command of *Hecla* and *Griper* in January 1819. Just as he had done before his voyage under Ross, he began his preparations on scientific, naval, and political fronts. In February, he, Sabine, and the surgeons of his expedition attended a course of lectures on mineralogy, given by one Mrs. Lowry, who came recommended by Captain and Mrs. Kater.[103] By mid-February, the ships were completely manned, to Parry's entire satisfaction. He and his officers had been able to choose from among the men who sailed with Ross, and in the simultaneous arctic voyage east of Greenland:

It is a service which is now very popular among the seamen. Good wages, good feeding, and good treatment are not always to be had, poor fellows! and as we bear a tolerably good character in all these respects, we have been quite overwhelmed with volunteers.[104]

As for preparations on the political front, Joseph Banks and John Barrow were high on the list of those he consulted, and both of them urged him not to respond to Ross's strictures against his officers. At the same time, "the Royal Society have made a most handsome report to the Admiralty, of Sabine's zeal and ability in the performance of his duties on the last voyage – not as Naturalist, which he was not, but which the blundering Ross has called him, but as Astronomer." Sabine, no stranger to the usefulness of connections, was hoping for promotion, and planning to persuade the Duke of Wellington to recommend him for Parry's voyage.[105]

---

[102] Parry, Private Journal, *Alexander*, 1818, 3 vols., Naval History Library MS 31, vol. 1 pp. 12–24, 44, 130, 190; vol. 2 pp. 16, 70, 173–4; vol. 3 pp. 2–14.
[103] Parry to his parents, London, 15 Feb. 1819, SPPRI MS 438/26/30.
[104] Parry to his parents, London, 18 Feb. 1819, SPRI MS 438/26/31.
[105] Parry to his parents, Woolwich, 2 April 1819, SPRI MS 438/26/36.

He got the appointment. He also, with encouragement from Barrow, found it politic to provide the refutation of Ross that Parry had declined, and published a pamphlet with the unambiguous title of *A Refutation of Certain Mis-Statements in a Book entitled a Narrative of a Voyage of Discovery by Captain Ross* (London, 1819).

Parry's instructions from the Admiralty were to head straight for Lancaster Sound, to confirm or refute Ross's findings. If there was a passage, he was to take it, and proceed to the Pacific Ocean. Otherwise, he was to try in succession Jones, Smith, and Cumberland sounds. "When the season shall be so far advanced as to make it impracticable to go farther, I am, at my own discretion to return to England or to find a place for securing the ships for the winter – which latter I shall of course do, and I am sure we can make ourselves very comfortable." Sabine was once again scientific factotum, and the scientific program was closely similar to that of Ross's expedition, as were the scientific instruments provided – including Ross's deep sea clamm.[106] The importance of accurate drawings of coastlines and of natural history specimens was underlined, and two young lieutenants, Frederick William Beechey and Henry Parkins Hoppner, both of whom were to have distinguished arctic careers, were recommended to Parry as being skilled draftsmen.[107] The Admiralty gave science a higher priority in this expedition than they had in Ross's, perhaps because they were less sanguine about the prospects of sailing the Northwest Passage. Parry was to understand

that the ascertaining the correct position of the different points of the land on the western shores of Baffin's Bay, and the different observations you may be enabled to make with regard to the magnetic influence in that neighbourhood, supposed to be so near the position of one of the great magnetic poles of the earth, as well as such other observations as you may have opportunities of making in Natural History, Geography, &c., in parts of the globe, &c., little known, must prove most valuable and interesting to the science of our country; and we, therefore, desire you to give your unremitting attention, and to call that of all the officers under your command, to these points; as being objects likely to prove of almost equal importance to the

---

[106]    Parry to his parents, H.M.S. *Hecla*, Deptford, 22 April 1819, SPRI MS 438/26/40; "Official Instructions," in W. E. Parry, *Journal of a Voyage for the Discovery of a North-West Passage from the Atlantic to the Pacific: Performed in the Years 1819–20, in His Majesty's Ships Hecla and Griper . . . with an Appendix, Containing the Scientific and Other Observations* (London, 1821; reprinted New York, 1968) pp. xix–xxix.

[107]    For Beechey's work, see S. S. Bershad, "The Drawings and Watercolours by Rear-Admiral Frederick William Beechey, F.R.S., P.R.G.S. (1796–1856) in the Collection of the Arctic Institute of North America, University of Calgary," *Arctic* 33 (1980) 117–67.

Figure 10. "Burnet Inlet, Barrow's Strait," engraved by W. Westall from a sketch by Lieut. Hoppner, in W. E. Parry, *Journal of a Voyage for the Discovery of a North-West Passage from the Atlantic to the Pacific: Performed in the Years 1819–20, in His Majesty's Ships Hecla and Griper* (1821). Fisher Library, University of Toronto.

principal one . . . , of ascertaining whether there exist any passage to the northward, from one ocean to the other.[108]

These instructions fully justified his preparation through the Royal Society. But the elements were less biddable than scientists: the voyage began with a frustrating wait for the wind to turn around, forcing Parry to have a steamboat tow them out to Northfleet. Once under weigh, they set about exploring the irregularities of the compass produced by the ships' iron, and began to take a range of observations similar to Ross's. In the process, they found that Six's self-registering thermometer, used for determining the temperature of water at different depths, could fail when subjected to sudden changes in temperature. Parry suggested improving the instrument "by making the lower end of each index a little larger, so as to prevent the passage of the mercury between it and the tube."[109] They were in Davis Strait in June, making soundings, detecting tides and currents, noting atmospheric refraction and consequent mirages, and determining magnetic dip, variation, and force. They dredged when they could, hauling up *clio borealis* and other invertebrates. And they ran into the ice that, breaking off from the polar pack, runs down the middle of Davis Strait.

Parry took the punishingly hard but shorter route westward through the ice, and broke out at the end of July; they entered Lancaster Sound on 1 August, and as they proceeded, found clear passage. They were able to remove the Croker Mountains from the map. As they went along, Beechey made a fine series of sketches of the coastlines that were of both topographical and geological value. The surgeons made observations of belugas and narwhals, and Sabine kept up his magnetic observations.

On 7 August, they observed

the curious phenomenon of the directive power of the needle becoming so weak as to be completely overcome by the attraction of the ship; so that the needle might now be properly said to point to the north pole of the ship. It was only, however, in those compasses in which the lightness of the cards, and great delicacy in the suspension, had been particularly attended to, that even this degree of uniformity prevailed. . . . For the purposes of navigation, therefore, the compasses were from this time no longer consulted.[110]

They realized that the sluggishness of the compass meant that they were near the magnetic pole, and accordingly they landed a party to make careful observations of dip (89°41′22″) and latitude and longitude, as well as collecting natural history specimens and noting tide and currents. As they

---

[108]  Parry, *Journal of a Voyage*, pp. xxv–xxvi.    [109]  Ibid., p. 4.    [110]  Ibid., pp. 37–8.

sailed west, they continued to collect rocks and fossils whenever they could go ashore, and to make magnetic, meteorological, and other observations. At the end of September, as the ice began to thicken around them, they reached what promised to be a good Winter Harbour, after cutting a passage through the ice and man-hauling the ships for two-and-a-third miles.

Once they arrived there, they set about preparing the ships, themselves, and their stores for winter. Masts were dismantled, except for the lower ones and "the Hecla's main-top-mast, the latter being kept fidded for the purpose of occasionally hoisting up the electrometer-chain, to try the effect of atmospherical electricity."[111] Heating and ventilation were looked after, and Parry showed his genius in his detailed planning to preserve the health and morale of the crew.[112] As soon as they arrived, Sabine had chosen a site for the observatory, almost half a mile from the ships and so well away from their iron. They had also built a house near the beach, "for the reception of the clocks and instruments."[113] The house caught fire late in February; everyone rushed out, and managed to extinguish the flames before they had reached the clocks, transit, and other instruments. Sabine's servant rescued the dip circle at the cost of frost bite, and had to have part of four fingers on one hand and three on the other amputated.[114]

As the winter progressed, the officers and crew maintained a program of magnetic and meteorological observations, including detailed descriptions of paraselenae (moon dogs). When the sun returned, they were able to add corresponding descriptions of parhelia, and Sabine made pendulum observations. They also kept the first winter record of natural history observations. The caribou migrated south before the end of October, "leaving only the wolves and foxes to bear us company during the winter": by March, even the wolves and foxes had vanished from the vicinity. Scurvy made its unwelcome appearance in January, and Parry and the surgeons drew upon their full arsenal of antiscorbutics, including mustard and cress that Parry grew in his cabin. Ptarmigan arrived in May, and they found the first tracks of musk-oxen and caribou returning with the spring.

---

[111]   Ibid., p. 101.

[112]   Parry has deservedly had a generally good press. His skill in sailing was impressive, but probably no more so than that of other competent naval commanders. But his success in bringing his crew through the first arctic winter deliberately encountered by Europeans was truly remarkable; even one of his rare detractors [A. G. E. Jones, "Rear Admiral Sir William Edward Parry: A Different View," *Musk-Ox 21* (1978) 3–10] respects Parry for that wintering.

[113]   Parry, *Journal of a Voyage,* p. 107.

[114]   Ibid, p. 149. Alexander Fisher, *A Journal of a Voyage of Discovery to the Arctic Regions . . . 1819–1820* (London 1824) p. 178.

Figure 11. David Caspar Friedrich, "The Shattered Hope," © Elke Walford, Kunsthalle, Hamburg.

That August, after spending ten months at Winter Harbour and in exploring Melville Island, they cut their way out to the sound, and pushed westward to the longitude of 113°48′, "the westernmost meridian hitherto reached in the Polar Sea." They discovered Banks Island, and as far as distance went, had gone most of the way through the Northwest Passage. But they inevitably encountered impenetrable ice, blown in, although they could not know it, from the Beaufort Sea: as Parry remarked, "there is something peculiar about the south-west end of Melville Island, extremely unfavourable to navigation." Still, he was very sanguine about the *existence* of the passage, although the shortness of the summer and severe climate had defeated them. Perhaps, he speculated, it would be better next time to try nearer the North American coast.[115] Now, however, they returned eastward, met whalers in Baffin Bay, and upon encountering Inuit on the coast of Baffin Island, traded with them for a variety of artifacts.

[115]   Ibid, pp. 296 et seq.

Both of Parry's ships entered the Thames in mid-November, to a hero's welcome, and to reward as well as acclaim. Meanwhile, Parry efficiently published his account of the voyage, with appendixes summarizing magnetic, tidal, pendulum, lunar, and other observations.

The natural history observations, however, were not so soon completed. Sabine described the birds, mammals, fishes, and marine invertebrates, using the second edition of Temminck's manual of European ornithology and Lamarck's invertebrate classification. But he depended on other naturalists to classify new invertebrate species, for example *Talitrus Edvardsii*, a shrimplike creature brought up while dredging, and land invertebrates, one of which William Kirby, a leading entomologist, included under *lepidoptera* and called *Bombyx Sabini*. Other specialists provided a Wernerian account of the mineral specimens – an arrangement pleasing to Jameson in Edinburgh – and wrote up the shells, and handled assorted technical realms of description and classification. Parry relied on a variety of nonmilitary experts, and one of them in particular did not share his sense of urgency. This was the botanist Robert Brown, a former surgeon in the army, who had been invited by Joseph Banks to go as the naturalist on Matthew Flinders's voyage to Australia in 1801–5. Brown was now librarian and curator to Joseph Banks, and later became keeper of the botanical collection at the British Museum; his work on the botany of Flinders's expedition was a major contribution to botany and plant geography (phytogeography), and of great importance in the English-speaking world.[116] Parry badgered

---

[116] There is a fine biography of Brown: D. J. Mabberley, *Jupiter Botanicus: Robert Brown of the British Museum* (Braunschweig and London, 1985). Captain Matthew Flinders, R.N., was an explorer, hydrographer, and cartographer. He described the expedition, including his imprisonment by the French, in *A Voyage to Terra Australis; Undertaken for the Purpose of Completing the Discovery of that Vast Country, and Prosecuted in the Years 1801, 1802, and 1803, in His Majesty's Ship the Investigator, and Subsequently in the Armed Vessel Porpoise and Cumberland Schooner. With an Account of the Shipwreck of the Porpoise, Arrival of the Cumberland at Mauritius, and Imprisonment of the Commander during Six Years and a Half in that Island* (London, 1814); Brown contributed an appendix, "General Remarks, Geographical and Systematical, on the Botany of Terra Australis." Brown's chief publication on Australian botany was his *Prodromus Florae Novae Hollandiae et Insulae Van-Diemen Exhibens Characteres Plantarum quas Annis 1802–1805 per Oras Utriusque Insulae Collegit et Descripsit Robertus Brown; Insertis Passim Aliis Speciebus Auctori Hucusque Cognitis, seu Evulgatism seu Ineditis Praesertim BANKSIANIS, in Primo Itinere Navarchi Cook Detectis*, Vol. I (London, 1810); see also *Supplementum Primum Prodromi Florae Novae Hollandiae: Exhibens Proteaceas Novas quas in Australasia Legerunt DD. Baxter, Caley, Cunningham, Fraser et Sieber; et quarum e Siccis Exemplaribus Characteres Elaboravit Robertus Brown* (London, 1830; reprinted New York, 1960, with introduction by W. T. Stearn). Mabberley identifies Brown's greatness as a botanist as due to his skill as an observer, the foundation of his discoveries on taxonomic research, his observations on affinities on a worldwide basis,

him, but it was not until December 1823, after Parry's second Northwest
Passage attempt, that Brown came through and the sheets were printed.[117]
Brown had had a fair challenge, with a number of new species to identify
and characterize, including *Parrya Arctica* and *Pleuropogon Sabinii*. He
was a meticulous observer, unequaled in minute dissection and microscopic
examination.[118] The volume appeared in 1824.[119] But Parry did not have
to wait that long for his election to the Royal Society, which came on 15
February 1821. The certificate accurately describes him as "a gentleman
who has distinguished himself by the discovery of a passage from the
Northern Atlantic into the Polar Sea"; the signatories included Edward
Sabine, John Herschel, John Barrow, and Humphry Davy, Banks's successor
as President of the Royal Society.[120] Parry had powerful support.

*Parry's second expedition 1821–1823.* Parry had made such a large hole
in the Northwest Passage that it seemed only sensible to send him back to
the Arctic for another try. That was certainly his own view. On 6 December
1820 he had written to his parents that

> I yesterday had a long conversation with Mr. Barrow, respecting another expedition.
> I have always been of opinion that the equipment of such a thing ought to be matter
> of very serious consideration before it is adopted, by which I mean that I *know* the
> difficulties of the *whole* accomplishment of the North-West Passage too well to
> make light of them. [We had been lucky in pitching on] Lancaster's Sound, where
> one outlet to the Polar Sea exists. Should another Expedition be determined on, it
> must be in a lower latitude [so as to find a route from the Polar Sea to the coast of
> North America].

and his championing of the natural system against Linnaeus's sexual system: see *Jupiter
Botanicus*, pp. 142–8, 400, 404–5.

[117] British Museum (Natural History), Robert Brown correspondence vol. 2, nos. 157–9.

[118] For Brown's use of the microscope, see Mabberley, *Jupiter Botanicus*, pp. 54, 404. Most
of Brown's proposed new species are not recognized as such today. Brown wrote: "I have
experienced much greater difficulty than I had anticipated in determining many of the
species; arising either from their extremely variable nature, from the incomplete state of
the specimens contained in the collection, or from the want of authentic specimens of
other countries, with which it was necessary to compare them." (Quoted in Mabberley,
*Jupiter Botanicus*, p. 232).

[119] *A Supplement to the Appendix of Captain Parry's Voyage for the Discovery of a North-
West Passage in the Years 1819–20 Containing an Account of the Subjects of Natural
History* (London, 1824). The collections made on Parry's voyages were arranged by
Brown and distributed in boxed herbaria, some of which are still to be seen at the British
Museum (Natural History): see Mabberley, *Jupiter Botanicus*, p. 234.

[120] Royal Society of London, Membership Certificates 1820–30, vol. VII.

To find such a route meant beginning again. Parry therefore expected the next expedition to fall short of accomplishing the passage, and so he was sure that the results would be disappointing to the public, for whom success alone counted.

Having stated all this, and a great deal more, I then told Mr. Barrow that I was still as ready as ever I was, to do my best towards the accomplishment of the object in view, and that I was the more inclined to it, from a fear of the Russians being beforehand with us.[121]

If any encouragement were needed, Parry's appeal to Barrow's fear of being beaten by the Russians provided it. Parry was given the command, and he began to think about the scientific program. Sabine had been the only officer seconded for specifically scientific duties in 1819, and he was not available this time, having more southerly magnetic prospects before him. The navy's sphere of exploration through the nineteenth century extended over all the oceans of the globe, and arctic officers might also be sent to the South Pacific, the Indian Ocean, or the China seas. Parry was worried that the Admiralty, following advice from the Royal Society, was in danger of neglecting arctic science on his next voyage, wrongly believing that determination of the earth's ellipticity had been settled, and that the other observations to be made on the voyage would not require a trained observer:

I fear the Royal Society intend recommending that *nobody* in . . . Sabine's capacity . . . shall attend the next Expedition, as they consider all *is* done that *can* be done *there*, as regards the pendulum. But there are so many branches of science, in which Sabine has been valuable to me, and to the Country, that to send nobody now will be to add a terrible weight to my shoulders – or rather, it will be the certain means of leaving undone a thousand useful things, to which my more immediate duties will not allow me to attend. However, I must do the best I can, and shall have some good assistance in Edwards . . . and Lyon.[122]

John Edwards had been surgeon on H.M.S. *Isabella* under Ross in 1818, and on H.M.S. *Hecla* in Parry's first expedition. He was a good amateur naturalist. There was striking continuity in arctic complements, with most arctic officers sailing on more than one expedition. This was to be true of George Francis Lyon, who was a newcomer to arctic service in 1821, having more nearly tropical experience in discovery; but he was chosen to command *Hecla*, Parry's second ship in the forthcoming expedition, while Parry commanded H.M.S. *Fury*, another vessel of the same class as *Hecla*. The use

---

[121] Parry to his parents, London, 6 Dec. 1820, SPRI MS 438/26/54.
[122] Parry to his parents, London, 5 Feb. 1821, SPRI MS 438/26/58.

of surgeons and officers as scientific observers was in the amateur tradition of the Royal Navy, but it was scarcely encouraging to science. Parry so far prevailed that the Reverend George Fisher, who had been on the Spitsbergen expedition in 1818, was appointed astronomer, and for good measure doubled as chaplain. Although he was not limited to following Sabine's work, where he did so, he had to follow Sabine's example. Fisher was a happy choice; not only was he a skilled astronomical observer, but he also possessed a variety of fine instruments that he brought on board, including an astronomical clock, three chronometers, a circular transit, and a hydrostatic balance. The Admiralty too, in response to scientific urging, came up with a fuller range of apparatus than heretofore, although, as Parry had foreseen, they dispensed with the pendulum for measuring the earth's ellipticity. Each ship carried a full set of instruments, including a forty-inch achromatic telescope made by Dollond, a variety of excellent surveying equipment, and apparatus for chemical and physical table-top experiments. The Admiralty's standard statement that they had caused to be put on board "a great variety of valuable instruments" was more than usually accurate. There was also more urgency in their recognition of science, to which "we therefore desire you to give your unremitting attention, and to call that of all the Officers under your command, . . . as being objects of the highest importance."[123] The urgency was perhaps because the Admiralty had listened to Parry enough to realize that they could not count on a successful navigation of the passage.

Parry drafted scientific orders for his officers, ranking first those observations relating to astronomy and navigation; latitude would be determined by the stars, by sun, moon and planets, longitude by lunars, chronometers, and the sun; the different methods would serve as checks upon one another. Then came meteorology, hydrography, and oceanography; observations on the aurora, lightning, and shooting stars; and natural history observations, to be reinforced where possible by the capture, delineation, and preservation of specimens. In every case, field notes were to be made: "It is my particular wish that nothing is trusted to memory, even for five minutes."[124] There was no official expedition naturalist. Edwards took on a good deal of the collecting work in that department, James Clark Ross, midshipman on *Fury*, superintended the taxidermy, and Parry uneasily assumed overall responsibility.[125]

---

[123]  W. E. Parry, *Journal of a Second Voyage for the Discovery of a North-West Passage from the Atlantic to the Pacific: Performed in the Years 1821–22–23, in His Majesty's Ships Fury and Hecla* . . . (London, 1824; reprinted New York, 1969) p. xxvii.
[124]  SPRI MS 438/8.          [125]  Parry, *Journal of a Second Voyage*, p. xiv.

Parry had been sent, as he wanted, to seek a passage to the south of Lancaster Sound, via Foxe Basin, between Baffin Island and Hudson Bay. He confirmed the eighteenth-century discovery of Repulse Bay, and explored the coast and islands to the north of it. The officers and crew dredged and took soundings where they could, collected specimens as they went, and in October settled into winter quarters on Melville Island, just south of Melville Peninsula. They erected the portable observatory, and promptly commenced making magnetic observations. Another building was constructed for other scientific work, including astronomy and studies of atmospheric refraction.[126] The astronomical clock unfortunately refused to go in extreme cold. An electrometer chain was hoisted up Fury's masthead, and a tide pole was inserted through the ship's fire hole in the ice.[127]

In February, a party of Inuit visited the expedition, and remained with them for much of the winter. Parry and his officers took advantage of the opportunity to make extensive observations on their customs and language. They also learned a good deal from them about the geography of Melville Peninsula, as Parry acknowledged.[128]

Parry's account of the Inuit aimed at plainness and accuracy, and it far exceeded previous naval reports. Captain Lyon's sketches provided admirable illustrations of Inuit activities and artifacts.[129] Parry and his officers measured them, described their clothing and its admirable effectiveness, their ornaments and tattoos. Parry carefully described snow houses and their construction, as well as domestic utensils and their use. He gave careful accounts of the Inuit's hunting and preparation of food, and observed their social structure, religion (of which he could make little), amusements, and toys. He recorded their songs, to some of which "the words . . . seem to be as interminable as those of 'Chevy Chace'." He also made a stab at giving an account of the language, although, as he despairingly remarked, "the language of the Esquimaux is so full of words, and so varied and peculiar in the formation of its sentences, that it would require a much larger acquaintance with these people, as well as far greater ability than mine, to give a satisfactory account of its grammatical construction."[130] Instead, he took David Crantz's recent account,[131] and sought to approximate his

[126] See Fisher's data in National Maritime Museum MS FIS 6a, 6b, 8, 56.

[127] Parry, *Journal of a Second Voyage*, pp. 124–5.

[128] Ibid., p. 276. For another example of Inuit geographical knowledge (including cartography) see the section of this chapter on "The Polar Rosses," especially n. 169.

[129] See also George Francis Lyon, *The Private Journal of Captain G. F. Lyon of H.M.S. Hecla, During the Recent Voyage of Discovery under Captain Parry* (London, 1824).

[130] Parry, *op. cit.*, pp. 551–2.     [131] Crantz, *The History of Greenland* (London, 1820).

findings to Crantz's. One often finds what one seeks; Parry nonetheless was surprised "to observe how slight a change time and distance have been able to effect in the language, as well as in the habits, of this widely-scattered nation."[132] He compiled a list of some 650 words, as well as a shorter list of place-names, the latter important for subsequent navigators. Only very recently has Canada begun to restore Inuktitut place-names to the map of the Arctic.

In July 1822, helped on their way by a map drawn for them by the Inuit, Parry's expedition discovered the entrance to Fury and Hecla Strait. Ice kept them out of it, but overland travel showed them that it led to a sea, and thus offered the prospect of a shorter northwest passage than did Lancaster Sound. They spent the next winter at Igloolik, near the entrance to Fury and Hecla Strait. The following summer, still barred by ice from passing the strait, and threatened by scurvy, they sailed home to England, whereupon the natural history specimens were promptly sent to Jameson and to John Richardson in Edinburgh, and to William Hooker in Glasgow. Jameson we have already met as the doyen of British Wernerian geologists and scientific mentor to a generation of arctic travelers. Hooker, protégé of Joseph Banks and at that date Regius Professor of Botany at the University of Glasgow, a gentle and great botanist, and father of the far less gentle but no less great botanist Joseph Dalton Hooker,[133] was to devote many years to the flora of North America, including the Arctic. And John Richardson, naval surgeon and naturalist, had already survived his first arctic expedition with John Franklin, and would later direct and largely write a magnificent four-volume fauna of British North America; with his assistant Thomas Drummond, he would supply many of Hooker's specimens for his flora.[134] Richardson and Hooker will feature prominently later, but for now, it is noteworthy that Parry wrote to Richardson using Franklin's name as an

132  Parry, *Journal of a Second Voyage*, p. 552.
133  For a general account of both Hookers, see M. Allan, *The Hookers of Kew 1785–1911* (London, 1967). For J. D. H., see W. B. Turrill, *Joseph Dalton Hooker: Botanist, Explorer, and Administrator* (London, 1963).
134  William Jackson Hooker, *Flora Boreali-Americana, or, the Botany of the Northern Parts of British America: Compiled Principally from the Plants Collected by Dr. Richardson & Mr. Drummond on the Late Northern Expedition under Command of Captain Sir John Franklin . . .*, 2 vols. (London, [1829]–1840). John Richardson, *Fauna Boreali-Americana*, 4 vols. (1829–37). Richardson, in writing up the ornithological results of Parry's expeditions, recognized that a gull procured by James Ross in 1823 was a new species, which he intended to name *Larus rossii*, Ross's gull; but William McGillivray at Edinburgh University Museum was the first to name it in print, as *Larus roseus*, to Richardson's considerable annoyance: see Michael Densley, "James Clark Ross and Ross's Gull – A Review," *Naturalist* 113 (1988) 85–102.

introduction,[135] and secured his cooperation. Indeed, Jameson, Hooker, and Richardson all obliged, and wrote appropriate appendixes for the scientific supplement to Parry's account of the voyage. The foundations of arctic natural history grew steadily stronger.

*Parry's third expedition 1824–1825.* In November 1823, Parry had an interview with Lord Melville, the First Lord. Melville offered him the post of hydrographer of the Navy. Parry asked for time to think about it. Their conversation then turned to the Northwest Passage, and Parry said that "*should* another Expedition be determined on, I trusted he would once more accept my services – for that I should not a moment hesitate in declining the other situation, and giving the preference to the more active employment. He said the one need not interfere with the other, or to that effect."[136]

Melville's offer to keep the hydrographer's situation open for Parry, "*even during àn Expedition,*" forced Parry to accept it. But "How I shall get through the work, and another equipment, *and* my book, I know not."[137] With Melville's support and Barrow's enthusiasm, Parry was all set for his final attempt on the passage.

Parry was reminded that his expedition was for discovery and science – just as Thomas Jefferson had pretended that Lewis and Clark had no imperial mandate in their expedition. Still, British interests were more than scientific. Barrow considered that it would be unfortunate to let in Russia now that Britain had opened the door. Relations with Russia were delicate in the North.[138] Eighteenth-century explorations had made the Bering Strait and adjacent lands Russian, and Russian claims had been reinforced in 1821, while Parry was on his second expedition, by an imperial ukase. According to that decree, foreign vessels were forbidden to approach within one hundred Italian miles of the coast of Russian America, now the state of Alaska. Britain and the United States protested, but the issues were not fully resolved until more than a decade after the Alaska purchase was completed near the end of the century. In the 1820s, feelings ran high, and put

---

[135] Parry to Richardson, 21 Nov. 1823, SPRI MS 1503/5/6.
[136] Quoted in Ritchie, *The Admiralty Chart*, p. 156.
[137] Ibid.
[138] There is a substantial literature on the history of Russians in Alaska and Alaskan waters, including F. A. Golder, *Russian Expansion on the Pacific 1641–1850* (Cleveland, 1914); T. E. Armstrong, *Russian Settlement in the North* (Cambridge, 1965), H. H. Bancroft, *History of Alaska*, vol. xxxiii in *The Works of Hubert Howe Bancroft* (San Francisco, 1886); and the volumes of the *Alaska Boundary Tribunal.* I am grateful to Gordon W. Smith of Ottawa for help here.

in question the safety of a successful Northwest Passage expedition, which would have to enter the Pacific Ocean by way of Bering Strait, and would at least want to stop off for fresh water. There was a remote possibility that Britain might be at war with Russia before Parry returned from his third voyage. The Admiralty was determined to maintain the neutrality and immunity of their scientific expedition. Even if Britain was at war, they told Parry, his ships, H.M.S. *Hecla* and *Fury*, would not be engaged in hostilities, "it being the practice of all civilized nations to consider Vessels so employed [in discovery] as excluded from the operations of War."[139]

Parry had Barrow's agreement that the attempt should be by way of Prince Regent Inlet, southwest from Lancaster Sound and discovered by Parry on his first expedition. "To give up the attempt before this point be tried, would indeed be to have opened the door, at great expense and labour, for some other nation to reap the honour & glory, and to triumph over us who have for two centuries and a half endeavoured in vain to accomplish it."[140] Pendulum experiments were to be resumed, and the Admiralty wrote to the Royal Society to ask for the clock in Sabine's possession, ignoring earlier advice that there was no longer any need for such experiments; Lieutenant Henry Foster, F.R.S., who had carried out survey work near the mouth of the Columbia, assisted with pendulum experiments in South America, and had, with Sabine, been to the coasts of Greenland and Norway, was responsible for astronomical and mathematical work. The Admiralty, in drafting Parry's instructions, were so far aware of the conservative nature of the scientific program that they merely referred Parry to the instructions for his previous expedition, stipulating that Foster should diligently follow Sabine's and Fisher's example.[141] Captain Hoppner, Commander of H.M.S. *Fury*, would be responsible for making drawings of the land, its people, and its flora and fauna, with the assistance of a midshipman with special responsibilities in this department, and of *Hecla*'s surgeon.

New requests in matters scientific came from Francis Beaufort, who wrote to Parry about atmospheric tides: "You are well aware that the barometer rises & falls four times in the 24 hours."[142] There was, however, some dispute about such tides in high latitudes. Parry responded by sending

139   Draft of Instructions for Captain Parry, of H.M.S. *Hecla*, 1824, MS PRO CO 6/16 f. 107.
140   Barrow, 14 Nov. 1823, National Library of Scotland MS 3845.
141   Draft of Instructions, ff. 95–110.
142   Beaufort to Parry, Manchester St., 29 April 1824, Public Archives of Canada (PAC) MG 24 H49.

Beaufort his arctic meteorological journal for 1822–3, which appeared to show only two atmospheric tides in twenty-four hours between Winter Island and Igloolik. Beaufort hoped that further observations by Parry would reveal the laws (i. e., the underlying regularities) of the phenomena.[143] Parry sailed north, and it was not until the beginning of July 1824, when the transport that accompanied *Hecla* and *Fury* with extra supplies was due to return to England from Davis Strait, that Parry found time to reply to Beaufort.[144] Now, for the first time on his arctic voyages, Parry was furnished with "barometers deserving the name." He now promised to have regular observations made for Beaufort's scrutiny.

The expedition was not a success. Delayed by weather and ice, they reached Lancaster Sound in September, wintered on the shores of Prince Regent Inlet, and lost *Fury*, which was driven ashore by ice in August 1825. The crews of both ships returned to England in *Hecla*.

In spite of geographical failure, they carried out a full program of scientific observations. They set up the standard observatory on shore as soon as the ships were secured for the winter. The interest of the observations that they undertook, "especially of such as related to magnetism, increased so much as we proceeded, that the neighbourhood of the observatory assumed, ere long, almost the appearance of a scattered village, the number of detached houses having various needles set up in them, soon amounting to seven or eight."[145] Using suspended instead of supported needles, they found that magnetic variation changed according to a pronounced regular daily cycle. They also found that the intensity of the earth's magnetic field increased from morning to afternoon, and decreased from afternoon to morning. They began to suspect an influence of the sun on the earth's magnetism, "although the exact laws of this influence may still remain to be discovered."[146]

While, as Parry put it, "unassisted Nature was thus developing, on a large scale, some curious facts on the subject of Magnetism," Foster repeated the experiments of Peter Barlow and Samuel Hunter Christie, in which magnets were applied to a compass needle to reduce its directive power. Foster found "that the *true* bearing upon which a needle exhibits its

---

[143] Beaufort to Parry, 13 May 1824, PAC MG 24 H99.

[144] Parry to Beaufort, *Hecla*, Davis Strait Lat. 69° N, 1 July 1824, Huntington Library MS FB 1490.

[145] W. E. Parry, *Journal of a Third Voyage for the Discovery of a North-West Passage from the Atlantic to the Pacific: Performed in the Years 1824–25, in His Majesty's Ships Hecla and Fury . . .* (London, 1826) p. 51.

[146] Ibid., pp. 52–3.

*minimum* variation" was the same on the shores of Prince Regent Inlet as it was in the south of England, "which would almost lead to a conclusion that this is a constant line all over the world."[147]

Barlow had been a vigorous critic of Admiralty compasses, reporting in 1820 that "half the compasses in the British Navy were mere lumber and ought to be destroyed." He proposed a compass with four or five parallel strips of magnetized steel fixed to a card, a form that was standard until Lord Kelvin introduced his compass in 1876.[148]

Barlow also addressed himself to the effect of iron used in the construction of ships on a mariner's compass. This had first been systematically studied by Matthew Flinders,[149] who showed that the effect was negligible when the ship was facing north or south, and at its greatest when the ship's head was facing west or east. He corrected the deviation by means of a bar of vertical iron. Scoresby's investigations of the phenomena were published in 1819.[150] Thomas Young investigated the phenomena in 1820,[151] and in 1824 Barlow introduced a correcting plate of soft iron.[152] Parry tested the device, submitting it to a

severe trial on the ship's arrival in Barrow's Strait, and Prince Regent's Inlet, where, from the extraordinary increase of dip, and the consequently augmented effect of the ship's iron upon the magnetic needle, the compasses had before been rendered wholly useless on board ship. Never had an invention a more complete and satisfactory triumph. . . . [W]hen I consider the many anxious days and sleepless nights which the uselessness of the compass in these seas has formerly occasioned me,

[147]  Peter Barlow, "Observations and Experiments on the Daily Variation of the Horizontal and Dipping Needles under a Reduced Directive Power," *Philosophical Transactions of the Royal Society of London [Phil. Trans.]* (1823) 326–41; Samuel Hunter Christie, "On the Diurnal Deviations of the Horizontal Needle when under the Influence of Magnets," *Phil. Trans.* (1823) 342–92. Parry, *Journal of a Third Voyage*, p. 53.

[148]  *Encyclopaedia Britannica* 11th ed., art. COMPASS. Kelvin's compass had eight short light magnetic needles suspended by silk threads from an outer aluminium ring around the compass card; the ring was connected by silk threads to an aluminium cap with a sapphire center to receive the pivot needle.

[149]  M. Flinders, "The Differences in the Magnetic Needle on Board the *Investigator* Arising from an Alteration in the Direction of the Ship's Head," *Phil. Trans.* (1805) 186–97; see also Flinders, *A Voyage to Terra Australis Undertaken for the Purpose of Completing the Discovery of that Vast Country, and Prosecuted in the Years 1801–2–3 . . .*, 2 vols. (London, 1814) vol. 2 appendix 2.

[150]  Carter, *Sir Joseph Banks*, p. 509.

[151]  Young, "Computations for Clearing the Compass of the Regular Effect of a Ship's Permanent Attraction" [unsigned], *Quarterly Journal of Science*, 9 (1820) 372–80.

[152]  Barlow, "Magnetic Experiments and Discoveries, Particularly as Rendered Applicable to the Correction of the Local Attraction of Vessels," *Edinburgh Philosophical Journal*, 11 (1824) 65–87.

I really should esteem it a kind of personal ingratitude to Mr. Barlow, as well as great injustice to so memorable a discovery, not to have stated my opinion of its merits. . . .[153]

A good deal of other work was done: astronomical observations for navigation and cartography, studies of atmospheric refraction at low temperatures, tidal observations, and the use of Kater's pendulum to determine once again the ellipticity of the earth, with impressive accuracy.[154] James Clark Ross wrote up the zoology, including the description of a new species of fish named after Parry,[155] who, with the confidence of a nonscientist, announced that "little is now left to be said on the subject" of northern fauna. William Hooker wrote an elegant botanical appendix describing fifty-two species, valuable material for his developing flora.

Some of the most impressive work during the expedition was by Samuel Neill, surgeon on *Hecla,* who provided a fine stratigraphic and mineralogical account, with a little paleontology thrown in, for the eastern shore of Prince Regent Inlet.[156] This, together with other observations and the specimens brought back to Britain on this and previous Northwest Passage expeditions, enabled Jameson to give a fine Wernerian summary, in terms of successive depositions of strata, and the subsequent erosion of the resulting landforms.[157] Werner's principal groups of rocks, from the oldest to the most recent, were called primitive, transition, secondary, and alluvial. The principal rocks of the explored portion of the archipelago were primitive and transition rocks; newer secondary rocks were relatively limited, whereas modern volcanic rocks were altogether lacking, and only traces of tertiary strata, which in Werner's classification were intermediate between secondary and alluvial rocks, were found around Baffin Bay. Jameson considered that the islands were formerly part of the North American continental mass, from which they had been separated by erosion, in which much secondary and tertiary rock was worn away. There was fossil evidence of a rich flora, primarily of cryptogams such as tropical tree ferns, in the transition and primitive rocks; fossil corals in the secondary limestones also indicated an environment like that of present equatorial oceans. Later,

[153] Parry, *Journal of a Third Voyage*, p. 56.
[154] The Royal Society's first inclination, not to repeat these experiments, had been a sound one; Sabine's report had been entirely adequate. See his "An Account of Experiments to Determine the Acceleration of the Pendulum in Different Latitudes," *Phil. Trans. 111* (1821) 163–190; Sabine published a more comprehensive statement in his *An Account of Experiments to Determine the Figure of the Earth* (1825).
[155] *Ophidium Parryii:* Parry, *Journal of a Third Voyage*, p. 109.
[156] Ibid., pp. 91–4.       [157] Ibid., pp. 149–51.

when the tertiary rocks, then more extensive than now, were being deposited, there were forests of dicotyledonous trees, as shown by fossils in coal measures. There was very limited evidence of older volcanic action, but there were many useful ores, including those of iron, copper, and titanium. Thus Jameson was able to portray a dynamic history of the Arctic, as well as indicate its potential utility to industrial civilization. Even if the passage turned out, as seemed increasingly probable, to be useless as a trade route to the East, it might still prove a paying proposition – such at least was the implication of Jameson's summary.

Parry, back in London, continued to press upon Barrow and the Lords of the Admiralty the cause of polar exploration, for geography and science. As he had told Melville, he preferred arctic exploration to the hydrographer's office, which he nonetheless occupied. It was not long before he sailed again for the North, this time via Spitsbergen. In 1829, he resigned as hydrographer, to be replaced by the admirable Beaufort, who was charged with the duty of "selecting and compiling all such information as may appear to be requisite for the purpose of improving the Navigation, and for the guidance and direction of the Commanders of His Majesty's ships in all cases wherein any knowledge may be found necessary."[158] Parry continued his crusade; in 1846, he and Barrow were still encouraging one another and the world of science about the practicability of a sea route to the North Pole, with all the benefits that might arise from it.[159]


JAMES CLARK ROSS


James Clark Ross entered the navy in 1812 at the age of twelve. He served as a midshipman under his uncle John Ross on H.M.S. *Isabella* in 1818. In the light of the disagreement about who had seen the suppositious Croker Mountains, and following the public friction between John Ross and Sabine, an inquiry was held. Before John was called in, James was asked, among other things, about observations on magnetic dip and force. A major question between Sabine and his commander was the authorship of those observations. Following the inquiry, James wrote to his uncle that he had told the inquiry that the observations were taken by Sabine.

[158]  Barrow to Beaufort, 14 May 1829, Huntington Library MS FB 99.
[159]  John Barrow, *Voyages of Discovery and Research in the Arctic Regions from the Year 1818 to the Present Time* (London, 1846); W. E. Parry and J. Barrow, "1. On the Practicability of Reaching by Sea the North Pole. 2. What a Visitor to the Pole of the North Might Obtain in the Way of Science," *Edinburgh New Philosophical Journal* 40 (April 1846) 294–301.

I told them I had seen you take one set, in which I assisted as with Sabine. They asked me if you had taken any more than this one set and I said not *that I remembered*. They asked me if you had been accustomed to attend while Captain Sabine was observing. I answered that you had been present on 3 or 4 occasions. I was then asked if on these occasions you interfered in the conducting of the experiments. *I said I believe not*.[160]

He had thus made public his support for Sabine and was therefore attractive to John Ross's critics. John Ross was pushed aside by the navy, whereas his nephew enjoyed both arctic and antarctic naval commands in the ensuing decades. But first he sailed under Parry in 1819, once again as a midshipman. He was a lieutenant on H.M.S. *Fury* in Parry's third arctic expedition, where his miscellaneous responsibilities included making a botanical collection, as well as collecting and stuffing quadrupeds and, especially, birds. He also took a fair number of magnetic observations.[161] Then he was Parry's second-in-command in an attempt at the North Pole by sailing to the east of Greenland, and he had a similar range of scientific responsibilities as before. Nothing scientifically dramatic emerged from his labors until, with past quarrels buried, he joined John Ross as his second-in-command on a privately sponsored Northwest Passage expedition in 1829–33.

## THE POLAR ROSSES AND THE NORTH MAGNETIC POLE
### 1829–1833

John Ross had been virtually in forced retirement for the decade following his voyage of 1818, and in that decade he completed the building of his home, dabbled in phrenology, and pondered the utility of steam power for ships. The Admiralty, conservative to a fault, did not welcome the advent of steam power, which, Lord Melville considered, "was calculated to strike a fatal blow to the naval supremacy of the Empire." But John Ross was persuaded of its virtues. He considered that steam would have special virtues in the Arctic, where, unlike sail, it could propel the ship up every lead and opening in the ice. Because the winds that cleared the ice were often from the north, ice-free sailing was often into a head wind, so that here too steam

---

[160]  J. C. Ross to J. Ross, London, 30 April 1819, SPRI MS 486/4/2.
[161]  Dodge, *The Polar Rosses*, chap. 5; J. C. Ross to unknown correspondent, 25 April 1826, copy in SPRI MS 621/2.

had the advantage. Again, steam did not require such a deep draft as sail, and it could drive a ship through ice that would defeat sails.[162]

Ross put this to the Admiralty, proposing that they send him to the Arctic in steamships; unsurprisingly, the Admiralty's memory, and Barrow's in particular, was long enough to dismiss the proposal. Ross spent some time before he persuaded Felix Booth, gin distiller and philanthropist extraordinary, to sponsor the expedition. Booth provided £18,000, and Ross contributed a further £3,000.

Thus provided with financial backing, Ross was determined to vindicate himself after his first voyage and its humiliating aftermath. Steam was one requirement; small size, and especially shallow draft, was another. He found both in the steam packet *Victory*, and replaced the paddle wheels with a set that could be swiftly raised out of the water. Then he equipped the ship with three years' provisions of food and fuel, as well as scientific and navigational instruments, most of them his own. The Admiralty and Colonial Secretary also lent instruments.

After obtaining passports to guarantee neutrality in case of war, Ross set out to try for a passage by way of Lancaster Sound and Prince Regent Inlet.

They had a lot of trouble with the steam engine. Ross still defended "the principle, which was judicious," but was mightily provoked: "There seemed indeed no end to the vexations produced by this accursed machinery."[163] Scarcely less vexatious, when they put in to shore, was the constant devilish torment of Greenland's mosquitoes. But they made good speed up the Greenland coast, across Baffin Bay and into Lancaster Sound. On 11 August they entered Prince Regent Inlet. There, they overlooked Bellot Strait, and set up winter quarters at Felix Harbour in Lord Mayor Bay on the Boothia Peninsula: Felix Booth was getting due credit. At the end of September, Ross decided, with the enthusiastic agreement of every officer, that *Victory*, plagued by its unsatisfactory engine, would henceforth be a sailing ship. On 30 September arrangements were made to dismantle the boilers.

Ross had with him a good store of food, and this was supplemented by their discovery of *Fury*'s preserved and canned supplies. This, as Ross recognized, was just as well, since abundant food was needed to combat cold and the diseases that followed in its train, including scurvy. The natives knew this,

---

[162] J. Ross, *A Treatise on Navigation by Steam* (London, 1828); Dodge, *The Polar Rosses*, pp. 112–13.
[163] J. Ross, *Narrative*, pp. 15, 53.

their consumption of food . . . being enormous, and often incredible. . . . It would be very desirable indeed if the men could acquire the taste for Greenland food; since all experience has shown that the large use of oil and fat meats is the true secret of life in these frozen countries, and that the natives cannot subsist without it. . . . I have little doubt, indeed, that many of the unhappy men who have perished from wintering in these climates . . . might have been saved if they had been aware of these facts, and had conformed, as is so generally prudent, to the usages and the experience of the natives.[164]

George M'Diarmid, surgeon of the *Victory,* agreed with him; the natives' diet was in harmony with their environment, and, with its high fat content, was well suited to resisting cold and to generating bodily heat.[165] The importance of preserving vitamins by eating the meat raw was only realized much later.

Fortunately for Ross and his men, Inuit hunters would provide them with fresh meat, and with furs to renew their garments. As Ross observed, "The Esquimaux . . . could travel easier than we, could house themselves with a hundredth part of the labour, could outdo us in killing the seal, could regale on abundant food where we should starve because we could not endure it . . ."[166] Ross was a good observer, and wrote valuable accounts of the Inuit, including the "Boothians" whom he met in the Boothia Peninsula. He was not uncritical, but, given the Eurocentricity of his class, he was remarkable for the sympathy and respect that the Inuit aroused in him; he treated them as individuals, not specimens for the naturalist or anthropologist. Not the least of his achievements, borrowing in large part from earlier work in Danish Greenland, was the construction of a vocabulary of some 2,500 Esquimaux words, together with a substantial trilingual English–Danish–Esquimaux phrase book.[167]

Ross, like Parry, was concerned with mental as well as physical health. Exercise, school (difficult for two illiterates among the seamen), meteorological and other observations all contributed to health and the alleviation of monotony. By December, with the ship well prepared and winter routines established, they were ready to erect the magnetic and astronomical observatories on shore.

The astronomical observatory was larger than Parry's had been, so that "the breath of the observers was not so ready to condense on the instruments. Our transit instrument was also on a much larger scale, being of thirty-six inches."[168]

[164] Ibid., pp. 200–2.
[165] Report of Dr. George M'Diarmid, in J. Ross, *Appendix,* p. cxxvi.
[166] J. Ross, *Appendix,* p. 10.    [167] Ibid., Part I.
[168] J. Ross, *Narrative,* p. 234.

On 9 January 1830, they were visited by a party of thirty-one Inuit, established friendly relations, and on the following day visited the Inuit in their "village" of twelve snow huts. The next day, the two groups went over their chart of the region together, and the Inuit leaders added detail and extended the coastline on it. "They, however, told us that one of their party was a much better geographer than themselves, and promised that we should see him."[169] Subsequent discussions further enriched the chart, and enabled Ross to conclude that "if there was any opening to the westward, it must be a very narrow one."[170] Any route to a western ocean would lie to the northwest.

Natural history collecting continued, and the flora collected "contained, as was believed, many new ones."[171] They made good progress in surveying the shore. Both Rosses made several trips across Boothia Isthmus, on one of which James discovered King William Island: believing it to be part of the mainland, he was mightily frustrated that a shortage of provisions prevented him from continuing westward to complete the survey of the North American coast.[172] Greater frustration was to come when they tried to leave Felix Harbour, warping the ship along, cutting through the ice, and making almost no headway. In October, "the whole month had been employed in making a worse than tortoise progress, the entire amount of which, after all our toils, was but eight hundred and fifty feet."[173] They were iced in for a second winter, hard by Felix Harbour.

The magnetic observatory went up on 7 December. Tidal changes were observed, and displayed a remarkable irregularity. "Every thing was out of rule: whatever the moon might effect, the counteracting causes, in winds, currents, ice, and perhaps more, set all calculations at defiance. It was a high or low tide whenever it chose to be; and that was nearly all we knew of the matter."[174]

This was a harder winter than the previous one. To their frustration at having scarcely moved the ships in a year was added their disappointment at not seeing the Inuit, who had moved on, and who could therefore not provide them with seals, caribou, and furs. The people who had helped them previously had waited for them further north, provided with salmon

[169]   Ibid., p. 255.                    [170]   Ibid., p. 263.
[171]   Ibid., p. 465; for the flora of other arctic expeditions, see W. J. Hooker, *Flora Boreali-Americana*, 2 vols. (London [1829]–1840).
[172]   For J. C. Ross's survey of Lord Mayor Bay on the east of Boothia Peninsula, see Rear Admiral James Ross and J. M. Savelle, "Round Lord Mayor Bay with James Clark Ross: The Original Diary of 1830," *Arctic* 43 (1990) 66–79.
[173]   J. Ross, *Narrative*, p. 480.          [174]   Ibid., p. 495.

Figure 12. "Ikmallik and Apelagliu" drawing a map in Ross's cabin, after sketch by Sir John Ross, in Ross, *Narrative of a Second Voyage in Search of a North-west Passage* (1835). Fisher Library, University of Toronto.

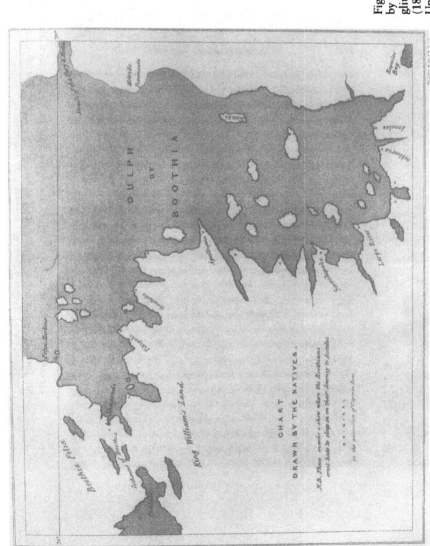

Figure 13. Chart drawn by Ikmallik and Apelagliu, in Ross, *Narrative* (1835). Fisher Library, University of Toronto.

and caribou, and were surprised that the ship had been locked in. In April a party of Inuit returned with fish and seals for sale: as John Ross recognized, he "could not well manage without the aid of the natives and the assistance of their dogs."[175]

Magnetic observations made during the previous winter, as well as those made on previous expeditions, had led James Clark Ross to suspect that the magnetic pole was not far to the north and west of the ship, and in the spring he went north along the west coast of Boothia Peninsula to Cape Adelaide. He was looking for the most direct evidence,

that, if possible, the observer might even assure himself that he had reached it, had placed his needle where no deviation from the perpendicular was assignable. . . .

These hopes were at length held out to us; we had long been drawing near to this point of so many desires and so many anxieties, we had conjectured and calculated, once more, its place, from many observations and from nearer approaches than had ever yet been made, and with our now acquired knowledge of the land on which we stood, together with the power of travelling held out to us, it at last seemed certain that this problem was reserved for us, that we should triumph over all difficulties, and plant the standard of England on the North magnetic pole, on the keystone of all these labours and observations.[176]

On 1 June 1831, they found a spot, at latitude 70°5′17″ and longitude 96°46′45″, where the angle of magnetic dip was 89°59′, almost vertical, and where the most delicately suspended horizontal needles showed no tendency to move, no matter how placed. They were very nearly, but not quite at the magnetic pole, and Ross knew it. He was wonderfully elated at coming so close to what we now know to be a migratory point on the earth's surface. He and his men erected a cairn to mark the site, and only regretted that it was not as large as the pyramid of Cheops, more fitting in its grandeur. "[I]t almost seemed as if we had accomplished everything that we set out to do; as if our voyage and all its labours were at an end, and that nothing now remained for us but to return home and be happy for the rest of our days."[177]

The location James Clark Ross had found coincided very nicely with the one projected on a magnetic chart prepared by Professor Barlow and

---

[175]  Ibid., p. 510; J. C. Ross was similarly aware of the vital help of the Inuit, "who had formerly been of such essential service to us, by means of their sledges and dogs, and by the great ease and expedition with which they raised our temporary encampments. We had nevertheless gained some experience." (Ibid., p. 521.)

[176]  J. C. Ross in J. Ross, *Narrative*, p. 550. He also published a formal account, "On the Position of the North Magnetic Pole," *Phil. Trans. Royal Soc. London* 124 (1834) 47–52.

[177]  J. C. Ross in J. Ross, *Narrative*, p. 555.

presented before the Royal Society in the spring of 1833, six months before the return of the expedition.[178]

That was the scientific highlight of the expedition,[179] and also approximately its midpoint in time. During the summer of 1831, they moved the ship just fifteen miles, and were then once more frozen in. That winter scurvy made its first appearance, but not severely. As before, they carried out regular meteorological and geomagnetic observations. John Ross remarked that the latter were likely to be of little value, because they and their needle were encamped on rocks.

What Saussure had originally shown, the observations of Dr. M'Culloch had extended far more widely, by demonstrating the influence of granite as well as many other rocks, not less than the basaltic ones, on the magnetic needle, so as to produce "deviations". . . . On instruments, and with experiments, so delicate as ours, this influence was likely to be destructive of all results.[180]

They kept up their scientific program, but Ross's pessimism about the value of their magnetic data was exceeded by the depressive effects of yet another year entombed in the ice. As he wrote in retrospect:

to see, to have seen, ice and snow, to have felt snow and ice for ever, and nothing for ever but snow and ice, during all the months of a year, to have seen and felt but uninterrupted and unceasing ice and snow during all the months of four years, this it is that has made the sight of those most chilling and wearisome objects an evil which is still one in recollection, as if the remembrance would never cease.[181]

Ross managed to keep his despair hidden from his men. But he decided that it was necessary to abandon *Victory;* they would drag the boats to Fury Point in order to use Parry's supplies there, and make their way east to Baffin Bay, in hopes of encountering a whaler to convey them home. They abandoned the bulk of their mineral collection at Northeast Cape, to lighten their load. The geological account that John Ross wrote on their return[182] was of very little use, lacking the comprehensive collection of specimens that would have enabled a skilled geologist to reconstruct the stratigraphy and paleontology of the region, and being couched in the broadest generalities about granite, limestone, and trapp rocks. But even

---

[178] Peter Barlow, "On the Present Situation of the Magnetic Lines of Equal Variation, and Their Changes on the Terrestrial Surface," *Phil. Trans.* (1833) 667–74.

[179] The data for magnetic and other observations, including an extensive account of the Inuit, are published in John Ross, *Appendix.*

[180] J. Ross, *Narrative*, p. 630.      [181] Ibid., p. 603.

[182] J. Ross, *Appendix*, p. ci.

without their mineral collection, they could not move; ice again frustrated their designs, and they spent yet another winter in Prince Regent Inlet. Finally, in August 1833, they made their escape, and were rescued in Lancaster Sound by the whaler *Isabella,* the same *Isabella* that Ross had first commanded in the Arctic. His achievement in keeping almost all his crew alive and in relatively good health after four arctic winters depended in part on the Inuit, in part on Parry's old supplies, and in part on leadership and seamanship of no common order.

On board *Isabella,* he wrote to the Admiralty, summarizing the expedition's achievements.[183] Of his nephew's contributions, he wrote that, besides acting as second-in-command, James had had the responsibility for astronomy, natural history, and surveying. The first and last were of greatest interest to the Admiralty, but the natural history results were exciting. The recent publication of the first parts of John Richardson's *Fauna Boreali-Americana* rendered unnecessary the detailed descriptions of birds and mammals, which Ross presented succinctly, ordered according to Georges Cuvier's *Règne Animal.*[184] Richardson himself wrote an appendix on four *salmonidae,* "*Salmo Rossii* (Ross's Arctic Salmon), *Salmo Alipes* (Long-finned Char), *Salmo Nitidus* (The Angmalook), and *Salmo Hoodii* (The Masamacush)." The insects were of some interest, and John Curtis, Fellow of the Linnaean Society, named new species of lepidoptera after Booth and Ross, the latter name conveniently shared by the leader and second in command of the expedition. The marine invertebrates were of greatest interest: even though the bulk of the collections had been abandoned with the *Victory,* the most remarkable specimens had been kept, especially those that appeared to be the types of new genera. The comparative anatomist Richard Owen, at that date conservator of the Museum of the Royal College of Surgeons, was able to demonstrate the existence of new genera not only by careful description of the exterior appearance of the specimens brought home, but also by careful and elegant dissection, "particularly valuable at a time when the internal organization of the inferior orders of animated nature has become so extensively used in their classification."[185] The technique parallels the recent dissection of fossils in unraveling the

[183] J. Ross, *Narrative,* pp. 732–5.
[184] G. Cuvier, *Le Règne Animal Distribué d'après son Organisation,* 4 vols. (Paris, 1817; 2nd ed. 5 vols., 1829–30). *Fauna Boreali-Americana,* 4 vols. (1829–37).
[185] J. C. Ross in J. Ross, *Appendix,* p. vi. For a discussion of marine invertebrate classification, see Mary P. Winsor, *Starfish, Jellyfish, and the Order of Life: Issues in Nineteenth-Century Science* (New Haven, 1976). For a more general discussion of nineteenth-century classification, see Louis Agassiz, *Essay on Classification,* E. Lurie ed. (Cambridge, Mass., 1962).

fauna of the Burgess shale.[186] Owen looked at a decapod crustacean that had been treated merely as an interesting species in the supplement to Parry's first voyage (*A Supplement to the Appendix of Captain Parry's Voyage* . . . 1819–20, p. ccxxxvi and plate 2), but argued that the "peculiar formation" of the second pair of legs of this singular animal required "a new genus, of which it is the only known specimen"; he called it *Sabinea Septemcarinata*, in honor of Sabine, who had discovered it on Parry's voyage.[187] Then, given more prominent billing, there was a cephalopod mollusc, of a new genus that Owen named *Rossia*, working with the single specimen brought back in spirits from Prince Regent's Inlet, which he carefully dissected; the original specimen, and the stages in dissection, were engraved by Curtis. For the specific name, based on the single specimen collected, Owen proposed *palpebrossa*, "taken from the remarkable development of the skin surrounding the eyeball, by means of which this animal evidently possesses the power of defending the eyes, as the pulmonated Vertebrata do by means of their more regularly formed eyelids."[188] Owen was showing himself a splitter, not a lumper.[189]

## CONCLUSION

John Ross reached London on 19 October 1833, and "on the next morning, caused myself to be presented to His Majesty at Windsor: receiving permission . . . to add the name of William the Fourth to the Magnetic Pole."[190]

In spite of this honor, Ross felt that his geographical findings were sufficiently negative for him to assert of the Northwest Passage that "there are now fewer temptations than ever to make any fresh attempt for solving this problem." As for the possible commercial gains from any passage, "[m]erchants risk much on commerce, it is true, but they are not given to hazard every thing, in opposition to the dictates of common sense, or in equal defiance of experience and probability."[191]

Ross's depression about the expedition's discoveries was not shared, but it was true that the Select Committee struck to inquire into and report on those discoveries was most impressed by James Clark Ross's discovery of

---

[186] S. J. Gould, *Wonderful Life: The Burgess Shale and the Nature of History* (New York and London, 1989).

[187] J. Ross, *Appendix*, p. lxxxii.      [188] Ibid., pp. xcii–xciii.

[189] Splitters increase the number of species in a particular scheme of classification by emphasizing the importance of differences; lumpers do just the opposite.

[190] J. Ross, *Narrative*, p. 726.      [191] Ibid., p. xvii.

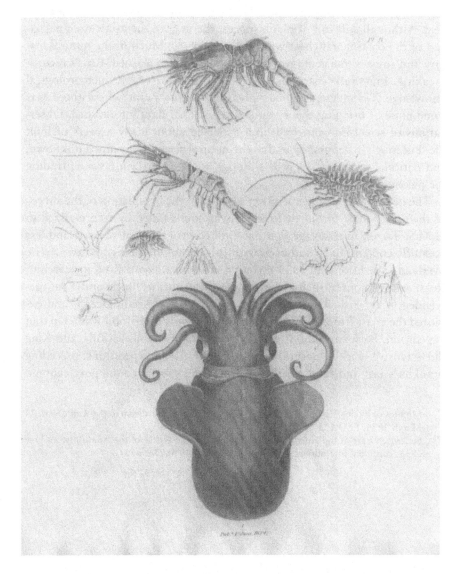

Figure 14. "*Rossia*, a new genus of cephalopod," in Ross, *Appendix to the Narrative of a Second Voyage* (1835). Metropolitan Toronto Library.

the location of the magnetic pole.[192] The expedition had achieved a good deal. Although still full of blank spaces, the map of the arctic archipelago and of the continental coast was taking shape.[193] Much of the natural history and some of the geology of the Arctic had been learned. Ross's record-breaking four full years in the North had extended meteorological knowledge. Geomagnetism had advanced to the point where there was some hope of bringing theory and observational data into accord. Observations of scientific value had been collected about many aspects of Inuit life. But much in geography and in the natural sciences remained unknown, and national pride still argued, although more weakly, in favor of finding the passage.

The next major naval expedition to attempt the completion of the survey of the arctic coast of North America was commanded by George Back in H.M.S. *Terror,* in a voyage that had limited geographical success, and less scientific content. Back had been with John Franklin on his first two arctic overland expeditions. In 1833 he had gone with Richard King in search of Ross, whose nonreturn had begun to cause unease. Back and King descended Great Fish River (now Back River) to the Arctic Ocean, and explored the coast from its mouth, returning the year after John Ross. On that expedition, Back carried out magnetic and auroral observations, and King did extensive work in natural history. The overland expeditions, preceding Franklin's last and fatal expedition, form the subject of the next chapter.

---

[192]  Minutes of Evidence before the Select Committee on the Expedition to the Arctic Seas, 24 March 1834, PRO CO 6/18 ff. 41–47.

[193]  See, e.g., Chart of the North Polar Sea, Hydrographic Office of the Admiralty, 29 June 1835, consulted in National Library of Scotland, MS 5877 No. 11.

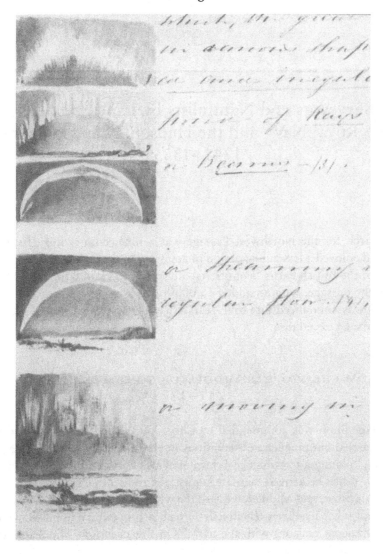

Figure 15. George Back, aurora borealis, 21 Dec. 1834. McCord Museum Montreal M2634.

# 3

────────────────────────────────

## Surveyors and Naturalists by Land and Sea: The Royal Navy and the Hudson's Bay Company 1741–1835

The search for the Northwest Passage was a nineteenth-century fixation that had enjoyed a less public period of gestation in the eighteenth century.[1] In 1745, the British Parliament offered a reward of £20,000 for the discovery of the passage, with an added £5,000 for reaching the North Pole: the reward was announced just four years after the first Royal Naval Northwest Passage expedition.

### FROM CHRISTOPHER MIDDLETON TO SAMUEL HEARNE: 1741–1772

That expedition was commanded by Christopher Middleton, who had already carried out magnetic observations in Hudson Bay.[2] Middleton sailed in 1741,[3] planning to winter at Prince of Wales Fort at Churchill on Hudson Bay. When he arrived there in August, the resident factor fired a volley across his bows, and Middleton raised the white flag of surrender. The company had what has been felicitously termed a proprietary impulse, which meant frequent tensions with the navy.[4] On this occasion, Middleton and

─────

[1] Glyndwr Williams, *The British Search for the North-west Passage in the Eighteenth Century* (London, 1962). Another bill to similar effect was passed in March 1818: see chapter 2.

[2] C. Middleton, "New and Exact TABLE Collected from Several Observations, Taken in Four Voyages to *Hudson's Bay* ... Shewing the Variations of the *Magnetic Needle* ... from the Years 1721, to 1725 ... ," *Phil. Trans.* 34 (1726) 73–6.

[3] Middleton, *A Vindication of the Conduct of Captain Christopher Middleton, in a Late Voyage on Board His Majesty's Ship the Furnace, for Discovering a North-west Passage to the Western American Ocean. In Answer to Certain Objections and Aspersions of Arthur Dobbs Esq.* (London, 1743; reprinted Johnson Reprint Corp., 1967).

[4] Newman, *Company of Adventurers* (Markham, Ontario, 1985) pp. 236–7.

his men spent a hard winter at Churchill; ten of them died of scurvy. They left Churchill in July, and headed north into unknown waters. They discovered Wager Bay, Repulse Bay, and the entrance to Frozen Strait, all three in the northwest of Hudson Bay, and they rightly concluded in each case that these routes offered no hope of a passage.[5] But however inaccessible, arctic lands and seas remained politically sensitive.

Both the Admiralty and the company were aware of Russian interests in the North and, less threateningly, of Spanish northward exploration along the Pacific coast. The passage would help to confer on its discoverers a legitimate title to northern sovereignty. The company had no intention of letting that go to outsiders. Since 1740, it had come under attack from Parliament because of its monopoly on trade. One M. P., Arthur Dobbs, who vigorously attacked Middleton in print for failing either to find the passage, or to prove that there wasn't one,[6] also claimed that the company was concealing the existence of a sea route from Hudson Bay to the interior.

The resulting publicity was an embarrassment to the company. Almost all the Arctic remained terra incognita, fueling suspicions about the company's secrecy. More than twenty years later, and perhaps partly to rebut Dobbs's claim, Moses Norton, governor of Prince of Wales Fort, proposed an overland expedition to the arctic shore. An additional motive was the search for a rich copper mine far in the North, hinted at by copper ore brought in by Indians; Norton dropped a rich piece of that ore on the desk of the company's governors in London in 1768, and received authority for the overland expedition. The man he chose for it was Samuel Hearne, who had entered the Royal Navy at the age of twelve, left it after the Seven Years' War, and joined the company in 1766. He and Norton cordially disliked one another, and this dislike may well have prompted Norton to send Hearne far away for a few years.[7]

Commerce, geography, and mining combined to encourage arctic exploration in these years. Science lent a hand too. The Royal Society was making plans for observations around the world of the transit of Venus of 3 June 1769. In 1768, they sent William Wales, astronomer and mathematician, to Hudson Bay for that purpose.[8] Cooke and Holland note that "it was

[5] Cooke and Holland, *The Exploration of Northern Canada*, pp. 61–2, and Middleton, *A Vindication.*

[6] A. Dobbs, *An Account of the Countries Adjoining to Hudson's Bay, in the North-West Part of America . . . With an Abstract of Captain Middleton's Journal, and Observations upon his Behaviour during his Voyage and since his Return . . .* (London, 1744; reprinted Johnson Reprint Corp., 1967).

[7] Newman, *Company of Adventurers*, pp. 249–50.

[8] Wales, "Astronomical Observations Made by Order of the Royal Society, at Prince of

probably at Wales's suggestion that the Royal Society requested the Hudson's Bay Company to supply specimens of the natural history of Hudson Bay, which resulted in a great advance in scientific knowledge of the region."[9] Wales's presence enabled Hearne to improve his surveying skills.

Hearne's first expedition, in 1769, took him just over 300 kilometers northwest of the fort, when his Indian guide deserted him, and he retraced his steps. On his second expedition, in 1770, he penetrated 600 kilometers in the same direction, and then turned back after the breakage of his quadrant. On the return journey they were forced to give most of their possessions to a party of Indians. Returning in severe distress, they were joined by Matonabbee, first among the Chipewyan chiefs, who provided succor, hospitality, and advice. He also offered to guide Hearne on another expedition in the following year. Hearne summarized his advice:

He attributed all our misfortunes to the misconduct of our guides, and the very plan we pursued, by the desire of the Governor, in not taking any women with us on this journey, was, he said, the principal thing that occasioned all our wants: "for, said he, when all the men are heavy laden, they can neither hunt nor travel to any considerable distance; and in case they meet with success in hunting, who is to carry the produce of their labour? Women, added he, were made for labour; one of them can carry, or haul, as much as two men can do . . . ; and, in fact, there is no such thing as travelling any considerable distance, or for any length of time, in this country, without their assistance."[10]

The navy, as we have seen, depended for success either upon the company or, later, upon the Inuit. The company, in turn, depended upon Indian guides, who in their turn depended upon their womenfolk.

In December 1770, furnished with a replacement quadrant, Hearne again set out, this time with Matonabbee as his guide. They reached the Coppermine River near Sandstone Rapids in July 1771, soon after their first encounter with musk-oxen since leaving the factory. Hearne recorded careful and extensive observations of these animals.

Wales's Fort, on the North-West Coast of Hudson's Bay," *Phil. Trans.* 59 (1769; published 1770) 100–36. Wales and his fellow astronomer on the expedition, Joseph Dymond, also kept a meteorological journal: Dymond and Wales, "Observations on the State of the Air, Winds, Weather, &c Made at Prince of Wales's Fort . . . ," *Phil. Trans.* 60 (1770) 137–77.

[9]  Cooke and Holland, *The Exploration of Northern Canada*, p. 84; G. Williams, ed., *Andrew Graham's Observations on Hudson's Bay 1767–91: With an Introduction by Richard Glover* (London, Hudson's Bay Record Society, 1969) pp. xx–xxi.

[10] Samuel Hearne, *A Journey from Prince of Wales's Fort in Hudson's Bay to the Northern Ocean Undertaken by Order of the Hudson's Bay Company for the Discovery of Copper Mines, a North West Passage, &c. In the Years 1769, 1770, 1771, & 1772* (London, 1795; reprinted Rutland, Vermont, and Tokyo, 1971) p. 55.

Though they are a beast of great magnitude, and apparently of a very unwieldy structure, yet they climb the rocks with great ease and agility, and are nearly as sure-footed as a goat; like it too, they will feed on any thing; though they seem fondest of grass, yet in Winter, when that article cannot be had in sufficient quantity, they will eat moss, or any other herbage they can find, as also the tops of willows and the tender branches of the pine tree.[11]

His descriptions were scarcely those of the formal zoologist, but they were accurate and thorough.

His journal also records much about the customs of the Indians who accompanied him, and others whom they encountered en route. On 17 July, at what he aptly named Bloody Falls, his Indian guides massacred a party of Inuit. On the next day, they reached the shores of the Arctic Ocean: Hearne was the first European to stand there. He surveyed the river's mouth, and determined its latitude. Unfortunately, by this time there was a thick fog and drizzling rain,

and finding that neither the river nor sea were likely to be of any use, I did not think it worth while to wait for fair weather to determine the latitude exactly by an observation; but by the extraordinary care I took in observing the courses and distances when I walked from Congecathawhachaga, where I had two good observations, the latitude may be depended upon within twenty miles at the utmost. For the sake of form, however, . . . I erected a mark, and took possession of the coast, on behalf of the Hudson's Bay Company.[12]

The company's sovereignty, not Britain's, was his concern. There is no doubt that he reached the ocean; his description of the river mouth is exact, and has been confirmed by subsequent observers, including John Franklin. But his dead reckoning placed the river's mouth over 300 kilometers north of its true position. Doubts about his dead reckoning were to provide much of the rationale for Franklin's first overland expedition.

Hearne briefly described the Copper Eskimos, their dress, weapons, hunting gear, "canoes," tents, and utensils, pointing out similarities between them and the Inuit of Hudson Bay, as well as differences: "there is one custom . . . – namely, that of the men having all the hair of their heads pulled out by the roots – which pronounces them to be of a different tribe from any hitherto seen either on the coast of Labrador, Hudson's Bay, or Davis's Straits."[13]

He decided that musk-oxen, arctic hares, and several other species remained in the far North year round. He found, for example, heaps of the

---

[11]  Ibid., p. 136.          [12]  Ibid., pp. 163–4.          [13]  Ibid., p. 170.

musk-oxen's dung on hills denuded of snow, but where there was no trace of their tracks in the moss, "this was a certain proof that these long ridges of dung must have been dropped in the snow as the beasts were passing and repassing over it in the Winter."[14] As for the mine that he had been sent to find, he found it merely a jumble of rocks and gravel, rather than a rich source of ore.

Hearne returned to Prince of Wales Fort by way of Great Slave Lake, completing his expedition in June 1772. He had found little to bring material benefits to the company or to Britain. But he had, as he wrote, "put a final end to all disputes concerning a North West Passage through Hudson's Bay," besides showing that the company was not averse from discoveries, nor from enlarging their trade. The latter scarcely needed demonstration.

The final third of Hearne's published journal describes the northern Indians, and the animals and birds of the North. His chapter on the Indians is of real anthropological value, ranging through social organization, marital customs, techniques and implements for hunting and fishing, clothing, amusements, beliefs and ceremonies, treatment of the aged, and more besides. He also prepared a vocabulary of their language, covering sixteen folio pages, but this he had lent to the corresponding secretary of the company, who died soon after, and it was lost among his effects.[15]

His natural history observations owed something to Pennant's *Arctic Zoology*,[16] which enabled him to name the birds properly; "for those by which they are known in Hudson's Bay are purely Indian, and of course quite unknown to every European who has not resided in that country."[17] He was able, however, to correct Pennant on one point. Pennant (p. 21) classes the moose with the "We-was-kish, though it certainly has not any affinity to it . . . The person who informed Mr. Pennant that the we-was-kish and the moose are the same animal, never saw one of them . . ."[18] Hearne then proceeded to describe the wapiti or elk, which is certainly not a moose, although both are deer. Pennant had compiled a valuable fauna; Hearne's account was the first coherent account of northern animals based upon personal observations, deriving primarily from his work in Hudson Bay, and to a lesser extent from observations made during his expeditions.

---

[14]   Ibid., p. 171.                              [15]   Ibid., pp. 304–57, xxxii.

[16]   Thomas Pennant, *Arctic Zoology*, 2 vols. (London, 1784–5) and supplement (1787) (reprinted New York, 1974).

[17]   Hearne, *Journey*, pp. xxxi–xxxii.        [18]   Ibid., pp. 360–1.

## JOHN FRANKLIN AND HIS NATURALISTS

*Franklin's first overland expedition 1819–1822.* Hearne may have satisfied himself, the Admiralty, and the company that there was no viable northwest passage near the coast. His survey was based upon the use of the compass as well as the quadrant and dead reckoning.[19] Here was a weakness: the compass, as Sabine and Parry had discovered on John Ross's expedition, was an unreliable guide in high northern latitudes. That was all that John Barrow needed to construct an argument for an overland expedition by the Royal Navy, to complement the oceanic expeditions that he was also considering. On 26 February 1819 he wrote to Henry Goulburn, under secretary for war and the colonies: one of the interesting scientific observations from Ross's expedition had been

an extraordinary and unlooked for degree in the variation of the magnetic needle, from which it is more than probable that the direction of the Coppermine River, of Hearne, and more particularly the point where it discharges itself into the Northern Ocean, are very erroneously marked down on the charts. In order to ascertain this fact, and also to correct and amend the very defective geography of the Northern parts of North America and more particularly to ascertain the direction and position of that line of the coast which lies between the mouth of the Coppermine river and the eastern extremity of the said continent, it has been suggested to my Lords Commissioners of the Admiralty that it would be highly desirable to send a party by land, under the command of an officer well skilled in astronomical and geographical science, and in the use of instruments, to ascertain the points above mentioned, and from whose discoveries and cooperation it is possible that the Naval Expedition shortly about to sail for that quarter might derive material benefit.

With this view a communication has been had with the Governor and Directors of the Hudson's Bay Company, who not only concur in the measure, but have liberally made an offer to carry out such officer and party in one of their ships of the present season, and to land them either at Fort York or Churchill.[20]

---

[19] Dead reckoning involves estimating position by using the compass to determine the direction of the route taken from a point of known latitude and longitude, and the linear distance from the latter point by noting the time traveled and the rate of travel. Thus a ten-hour's journey along a straight line, at an average speed of four miles an hour, would take one forty miles away from the starting point. Such estimates are unquestionably rough. The quadrant is an instrument for measuring angles of elevation, generally of a celestial body above the horizon, and when such a body is at its meridian (e.g., the sun at noon) it gives one an indication of latitude. Accurate determination of the time of passage across the meridian requires a transit instrument; otherwise the result is decidedly approximate. The determination of longitude requires either lunar tables, or accurate chronometers to establish the difference between local time and time at longitude 0° (conventionally the Greenwich meridian): see chapter 4.

[20] PRO CO 6/15 ff. 115–16.

The company had also undertaken to provide or arrange logistical support, for which it of course charged. It was not a philanthropic foundation. At the same time, agents of the rival North West Company were sent letters from London urging them to assist Franklin's expedition if they encountered it: "Lieutenant Franklin's object is one of a purely Public and Scientific nature, and has no connection whatever with any disputes or territorial claims in discussion betwixt us [the NWC] and the Hudson's Bay Co."[21]

Barrow had in mind a two-pronged attack on the problem of the passage, one by sea from the east by way of Lancaster Sound, the other from the south down the Coppermine and thence along the arctic coast of the continent. The two expeditions would both be aimed at the passage, and at charting the continental coast. These were properly hydrographic and naval concerns, in which the overland part of the expedition was merely a preliminary to maritime work; and so the overland expeditions were rightly the preserve of the navy.[22] John Franklin, who had seen vigorous active service during the wars, and latterly had been part of the attempt to reach the pole by a sea route to the east of Greenland, was chosen to lead the overland expedition. He was certainly an experienced navigator, familiar with instruments and so, by implication, far more likely than the unschooled Hearne to get things right. He was acceptable to the Royal Society, having furnished Joseph Banks with reports during the Spitsbergen expedition in 1818.[23] Besides whatever guides, interpreters, and porters the company might arrange for him, Franklin would take with him "a Surgeon two Mids . . . and a servant."[24] George Back and Robert Hood, midshipmen, were there for their skills in making sketches for natural history and cartography, and to assist in making and recording scientific observations. The naval surgeon was Dr. John Richardson,[25] who was to become a major figure in natural history, arctic exploration, and the navy's medical service. He now doubled as the expedition's naturalist; he had turned to botany after

21  Simon McGillivray to NWC agents, London, 21 May 1819, PRO CO 6/15 ff. 99–101.
22  When northern military expeditions were not aimed at coastal and hydrographic work, they could be properly carried out by the army, as happened with Lefroy's magnetic surveys: see chapter 4.
23  Franklin to Banks, 7 May 1818, Lerwick, Shetland, in McCord Museum, McGill University, MS M1974. For other observations of scientific interest from this expedition, see Franklin, 24 Nov. 1818, SPRI MS 248/325, and n.d. SPRI MS 248/294.
24  Franklin to Isabella Cracroft, 33 Fleet Street, 12 April 1819, SPRI MS 248/298/9.
25  R. E. Johnson, *Sir John Richardson: Arctic Explorer, Natural Historian, Naval Surgeon* (London, 1976).

the Napoleonic Wars, but claimed that although that was the branch of natural history in which he was most interested, he had made but little progress there.[26]

Franklin's instructions were to proceed from Hudson Bay, either northward to the arctic coast and thence westward to the mouth of the Coppermine, or westward to the river and thence northward to its mouth; they took the latter course. The main aim of the expedition was "to amend the very defective geography of the Northern part of North America." The explorers were also to keep a meteorological and a magnetic log, incorporating dip and variation:

and you will take particular notice whether any, and what kind or degree of influence the Aurora Borealis may appear to exert on the magnetic needle. And as it is yet a point in dispute whether the Aurora Borealis may actually be heard to make a rustling noise, like that of a fan or a flag as reported by various authors, but denied as being impossible by others, you will particularly attend to this circumstance, and make any other observations on this meteorological phaenomenon that may occur to you as being likely to lend to the future development of its cause, and the laws by which it is governed.

. . . . You are to give every facility to [Dr. Richardson] in collecting, preserving, and transporting, the various subjects of Natural History, which can be allowed consistently with the primary object of the Expedition. . . .[27]

Richardson took with him geological hammers, a portable microscope, a goniometer for measuring the angles between crystal faces (an aid to mineralogical identification), a blowpipe for the chemical analysis of minerals, and other impedimenta for a naturalist-explorer. Other equipment for the expedition included, besides surveying gear and meteorological and magnetic instruments, an electrometer for use during auroral displays.[28]

The expedition succeeded in some of its geographical aims. They reached and descended the Coppermine to the arctic coast, arriving there in July

[26]  Richardson to W. J. Hooker, 8 May 1819, MS Royal Botanic Gardens, Kew.

[27]  Lord Bathurst to Franklin, 29 April 1819, PRO CO 6/15 ff. 3–10; see also Franklin, "General Order 22 June 1819," SPRI MS 395/70/1.

[28]  Franklin, "The General List of Articles Supplied," SPRI MS 248/276. The four officers took notebooks in which to write their journals. Back's journal is in the McCord Museum, Montreal, MS M2716. The manuscript of Franklin's is partly in the SPRI MSS 248/278 and 248/277; for the published version, see Franklin, *Narrative of a Journey to the Shores of the Polar Sea, in the Years 1819, 20, 21 and 22 . . . With an Appendix on Various Subjects Relating to Science and Natural History* (London, 1823); Hood's journal has been published, *To the Arctic by Canoe 1819–1821: The Journal and Paintings of Robert Hood, Midshipman with Franklin*, C. Stuart Houston, ed. (Montreal and London, 1974). For Richardson, see *Arctic Ordeal: The Journal of John Richardson, Surgeon-Naturalist with Franklin, 1820–1822*, C. Stuart Houston, ed. (Kingston and Montreal, 1984).

Figure 16. Sir John Franklin by Negelen. National Archives of Canada C-5150.

1821. They went on to explore the coast in two canoes as far eastward as Turnagain Point on Kent Peninsula, a dangerous piece of seamanship in icy seas, yet far from the most demanding part of their journey. Until they were overwhelmed by the rigors of the return journey, they carried out most of their scientific program.

Part of their first winter was spent at Cumberland House. Back noted in his journal that

it was not our intention to remain inactive, but to obtain as many observations both astronomical and meteorological as we possibly could, as well as to make ourselves acquainted with the country through which we must pass – the language, manners etc of the natives – In the first and last of these we find ourselves miserably disappointed – for in attempting to make observations in winter, except such as do not occupy many minutes – you lose every sensation in the fingers – and if suffered to remain in contact with any metallic substances, becomes almost frozen to it, the certainty is that the skin remains behind.[29]

On the outward journey, they had collected specimens of birds and plants, and sent some back directly from Hudson Bay. That autumn, the British Museum acknowledged receipt of birds and plants, while the Royal College of Surgeons was indebted for the head of a walrus.[30] Richardson continued to collect assiduously, while Hood and Back finished accurate drawings of birds, plants, and other objects and views of interest.

On Christmas Eve 1819 Franklin wrote to Goulburn from Cumberland House, mentioning the discovery of an island containing strongly magnetic ore. Auroral observations had been few, and showed no effect on the electrometer;[31] nor did they hear any "quivering noise – but I then heard and have twice or thrice since from old residents in this Country that such noises have frequently been heard; none indeed with whom I have spoken doubt the Fact."[32]

Hood, who was the primary observer, sent the magnetic and auroral observations that he had made from February to May 1820. He noted that the magnetic variation followed a daily cycle, being at its maximum between 8 and 9 a.m., and at its minimum at 1 p.m., after which it gradually increased until the next morning's maximum. The most striking circumstance for him was that

though the needle varies and is stationary at the same hour here, in London, and at Sumatra, the laws which govern it are exactly reversed; the variation being greatest at the coldest period, instead of least, and vice versa. . . . The variation of the dipping needle were also observed, but they differed very unaccountably. The amount varied from 10' to 25', and the dip was generally least in the morning and greatest at 3 p.m.[33]

[29]   Back, MS Journal 23 Oct. 1819.          [30]PRO CO 6/15 ff. 176, 178.
[31]   In spite of this, Franklin reported that Hood had shown that the aurora were electrical phenomena: Franklin, *Narrative*, pp. 539, 542.
[32]   24 Dec. 1819, PRO CO 6/15 ff. 28–31. Old residents tell tall tales.
[33]   Hood to Goulburn, 11 June 1820, PRO MS CO 6/15 ff. 46–9.

Hood also observed that a magnetic needle was displaced up to 45' by the aurora, and only returned to the magnetic meridian after several hours.[34] Franklin subsequently confirmed this effect.[35] He reviewed the data, and found the variations of the dip needle to be so uncertain that he could not with confidence ascribe them to the aurora.[36] It is worth remarking that the position of a compass needle could be read with remarkable accuracy; using a reflecting microscope, one could read a Kater's compass to within 1'.[37] With such accurate and confusing data, it is little wonder that magnetism aroused increasing curiosity and debate among European natural philosophers.[38]

Natural history specimens, maps, drawings, and reports continued to make their way to the Colonial Office courtesy of the Hudson's Bay Company and its rival, the North West Company. It was just as well that the reports were sent piecemeal. In September 1821, they lost a canoe, and with it Franklin's portfolio with his summer's journal, "together with all the astronomical and meteorological observations made during the descent of the Copper-Mine River, and along the sea coast, (except for dip and variation)."[39]

That was the least of their troubles. The return journey was a brutal and painful one. Without food, reduced to only one canoe (which was so battered that they abandoned it the following week), and with winter coming on, they dropped from starvation and exhaustion. Richardson had already had to relinquish the burden of his summer's natural history collection. They ran out of food, and ate lichen from the rocks, and leather from their boots. In October, Hood was murdered by an Iroquois, whom Richardson subsequently executed. There was clear evidence of cannibalism, probably by Hood's murderer: Richardson, a trained surgeon, recognized a human bone in a stew. Some ten of the boatmen died of starvation. When the wretched crew returned to Fort Enterprise, they found it bare of provisions.

In December, near their end, they were rescued by Indians, who gave them dried meat and fat, and then they were met by a relief party from Fort Providence, with food, clothing, and letters from home.[40] The Hudson's Bay Company was not overly impressed with the navy's skill in wilderness travel.[41]

[34]   Hood in Houston, *To the Arctic by Canoe*, p. 100.
[35]   Franklin to Goulburn, Fort Enterprise, 16 April 1821, PRO CO 6/15 f. 59.
[36]   SPRI MS 248/277, Fort Enterprise, 16 April 1821.
[37]   Franklin, *Narrative*, p. 595.        [38]   See chap. 4.
[39]   Franklin, *Narrative*, p. 411.        [40]   Houston, *To the Arctic by Canoe*, Epilogue.
[41]   M. A. Macleod and R. Glover, "Franklin's First Expedition as Seen by the Fur Traders," *Polar Record* 15 (1971) 669.

Still, the expedition was not all loss. Hood's magnetic and meteorological observations were valuable. But the major scientific results of the expedition were due to John Richardson. He was the first medical officer in the Northwest to record the susceptibility of Indians to European diseases.[42] He gave clinical descriptions of the effects of starvation in the cold, as well as of hypothermia. Any well-trained doctor could have done as much. But few doctors could have matched his expertise in natural history. His botanical work was remarkable. On 1 October 1820 he recorded in his journal a list of forty-two lichens that he had observed around Fort Enterprise; his descriptions are admirable, ranking him with the leading lichenologists of his day, and include two new species.[43] In spite of his modest disclaimer to Hooker before the expedition, he was a first-rate botanist. The botanical appendix to Franklin's *Narrative* is a seminal report. There, and in the report of Franklin's second overland expedition, more than fifty new species are named by or for Richardson – enough, as one scholar has remarked, to "make a garden of respectable size."[44]

Richardson made regular meteorological and geological observations, which were important not only for natural history but also for assessing the economic potential of the land.[45] His special strength in geology, a science at that time still in the making, was in mineralogy, the identification and classification of minerals and rocks. He also paid attention to stratigraphy, and worked according to the modified Wernerian scheme promulgated in Britain by Robert Jameson. Following Jameson, Richardson erroneously placed the basalts that he identified in the Copper Mountains as precipitations from a universal ocean, rather than as volcanic in origin. He made a detailed and accurate inventory of the rocks and minerals encountered during the expedition, at least until its desperate final stage, but, as was typical within a science that sought to minimize the destructive impact of internecine theoretical debate, he offered no effective synthesis of his findings, and his few generalizations were so broad as to be uninformative to later geologists.[46]

Richardson's ornithological observations were considerably more extensive than is indicated in Joseph Sabine's appendix to Franklin's *Journey*, a

---

[42] Houston, *Arctic Ordeal*, p. 187.

[43] See John W. Thomson, Appendix E, in Houston, *Arctic Ordeal*.

[44] Ibid., Appendix D.

[45] Ibid., p. 194. Mineral resources and furs had since the Elizabethan age been the primary economic justification for northern exploration. Only later did the agricultural potential of land become important.

[46] W. O. Kupsch, "Appendix F. John Richardson's Geological Field Work," in Houston, *Arctic Ordeal*, pp. 273–316.

circumstance owing much to friction between Richardson and Sabine. Much of the ornithological content in Richardson and Swainson's *Fauna Boreali-Americana* derives from Richardson's collection and notes from this voyage, including accounts of previously undescribed birds: the yellow-billed loon, Wilson's phalarope, and the black-billed magpie.[47] His observations on mammals similarly include accounts of several species and subspecies not hitherto described: Richardson's ground squirrel, arctic ground squirrel (Parry's marmot), thirteen-lined ground squirrel, Franklin's ground squirrel, brown lemming, barren-ground grizzly, and stripped skunk.[48]

On their return, Franklin recognized Richardson's achievements. To him "were committed every branch connected with the Natural History of the countries through which the Expedition passed, whether it related to the descriptions, or the collecting of specimens of the Quadrupeds, Birds, Fishes, Plants or Minerals. These duties were in addition to his appointment as surgeon."[49] In the following year, Richardson, Franklin, and Parry were among those elected to the Wernerian Natural History Society of the University of Edinburgh, thanks in no small part to Jameson.[50]

*Prelude to Franklin's second overland expedition.* Franklin and Richardson were already contemplating another expedition together. Richardson's aim was quite precise, "namely to make such collections of plants and animals as may enable me to draw up a full account of the natural productions of these Northern latitudes," and he looked forward to more provision for the conveyance of his specimens. He was happy to submit himself to Franklin, but if Franklin were killed and another officer put in charge,

I wish to be left to my own judgments, as to the quarters I am to travel through in collecting and returning to Canada. . . . If my collections should be what I hope to make them, I wish to publish a separate work, not got up in the hurried way of the last Natural history appendix, but carefully got up as a Fauna & Flora of Hudson's Bay.[51]

He was already planning practical details, proposing that a good taxidermist be sent out with the Hudson's Bay ship in summer, urging upon the

[47]  Ibid., "Appendix A. Bird Observations by John Richardson during the First Franklin Expedition," pp. 223–44.
[48]  Ibid., "Appendix B. Mammal Observations," pp. 245–55.
[49]  Franklin to the Earl of Bathurst, Secretary of State, 8 November 1822, PRO MS Colonial Office 6 15 f. 83.
[50]  Jameson to Richardson, 7 Jan. 1823, SPRI MS Voss 1503/5/1.
[51]  Richardson to Franklin, 10 Jan. 1823, SPRI MS 1503/5/2, rough draft.

company the importance of obtaining good specimens, and sending Franklin instructions for drying plants that could be sent to England on a homeward-bound company ship.[52]

Sometimes Richardson felt "a kind of hankering desire" to return to the Arctic simply as a naturalist, rather than as a naturalist-explorer. With help from the Hudson's Bay Company, and with about £800 a year for four or five years, "the natural productions, mineralogy etc. of the country might be very much elucidated from New Caledonia and McKenzie's river down to the Red River Colony." This was a dream. No learned society would incur the expense, and governments were not that zealous for natural history. Still, he was keen to do it. As he contemplated the challenge, his ambitions and his budget grew; £1,000 a year would cover his salary and that of an assistant, as well as voyageurs, if the company provided provisions in winter. Under such circumstances, he would take his wife with him to Fort Enterprise.[53]

Richardson was right: a voyage purely for natural history was indeed a dream. But that summer, politics added a hand. The Russian imperial ukase of 4 September 1821, claiming the western Arctic as Russian, and forbidding it to other nationalities, had aroused the instant protests of the British government, as well as of the company. The ukase, if once acquiesced in, was "novel and monstrous," and its implied consequences were most alarming to Britain as a commercial and naval power.[54] The United States government had also protested; the *Times,* in a leader of 23 May 1822, insisted that when it came to the Northwest country between latitude 48° and 56° north, Britain would have to tell both Russia and the United States "that they have been disputing upon a point which did not concern them; and that she has not only the right and the possession, but the power to defend both, by whomever they be attacked." By the spring of 1823, when Russia had boarded a brig out of Boston, it had become clear that the ukase was to be taken seriously. The British government had already sent a note to the Russian ambassador, "declaring that *they could not accede either to the claims of Sovereignty, or the principle of maritime law laid down in it.*" Now, while maintaining that stance, they offered to enter into "an amicable negotiation" on the subject.[55] The public and the government were, in the summer of 1823, sensible of the political challenge of the North.

---

[52]   Richardson to Franklin, 4 Feb. 1823, SPRI MS Voss.
[53]   Richardson to Franklin, 28 July 1823, Kew MSS Richardson to Franklin 1823–42, no. 5.
[54]   Hudson's Bay Company Public Ledger, Dec. 1821, HBC A71/1, PAC microfilm HBC M9 ff. 61–3.
[55]   *The Courier,* 20 May 1823.

Barrow now saw himself vindicated for his long-term suspicions and jealousy of Russian intentions in the Arctic. Franklin wrote to him, brushing aside the failures of his previous expedition: these, he claimed, were partly the result of rivalry between the HBC and North West Company, now happily amalgamated, and partly caused by the unsuitability of canoes in arctic seas. Franklin proposed another expedition, this time to be undertaken in purpose-built boats down the Mackenzie and along the coast to the northwestern extremity of America. Like Parry before him, he flourished the Russian card: "The commercial and political advantage to be ensured, is the preservation of that portion of the country which is most rich in animals from the encroachment of Russia. . . ." This would prevent the establishment of "an other, and at some time perhaps a hostile power on any part of the Northern continent of America."[56]

Barrow needed no persuading, and he forwarded Franklin's letter to Lord Bathurst the following week. Bathurst was pleased to approve of Franklin's suggestions, and directed that preparations be made for another expedition.[57] In fact, three expeditions were in contemplation. Franklin would go as he had proposed, down the Mackenzie to its mouth and thence westward along the coast. Frederick William Beechey, a most efficient disciplinarian whose gentle parents were both respected artists, had begun his naval career at the age of ten, had grown up during the Napoleonic Wars, had been to the Arctic under Franklin in 1818, and on Parry's expedition in 1819. Now, in 1825, he was appointed to the *Blossom* to go via the Pacific to Bering Strait, and thence to Kotzebue Sound, there to meet Franklin and Parry.

We have already considered Parry's expedition.[58] Beechey's expedition was productive for science, with major collections made in the Pacific, and valuable but lesser arctic work.[59] From the standpoint of arctic science,

[56]   Franklin to Barrow, 26 Nov. 1823, PRO CO 6 16 ff. 2–5.
[57]   Franklin's note on a copy of his letter to Barrow of 26 Nov. 1823, SPRI MS 248/281/1.
[58]   As we have seen in chap. 2, Parry fell far short of the rendezvous.
[59]   For Beechey's expedition, see F. W. Beechey, *Narrative of a Voyage to the Pacific and Beering's Strait, to Co-operate with the Polar Expeditions: Performed in His Majesty's Ship Blossom, under the Command of Captain F. W. Beechey . . . in the Years 1825, 26, 27, 28,* 2 vols. (London, 1831); *The Zoology of Captain Beechey's Voyage: Compiled from the Collections and Notes Made by Captain Beechey, the Officers and Naturalist* [George Thomas Lay] *of the Expedition, during a Voyage to the Pacific and Behring's Straits . . .* (London, 1839) [containing accounts by Richardson on mammals; N. A. Vigors on ornithology; G. T. Lay and E. T. Bennett on fishes; R. Owen on crustaceans; J. E. Gray and G. B. Sowerby on molluscs; and W. Buckland on geology]. W. J. Hooker and G. A. W. Arnott, *The Botany of Captain Beechey's Voyage; Comprising an Account of the Plants Collected by Messrs Lay and Collie* [surgeon] *and Other Officers of the Expedition, during*

however, Franklin's expedition of 1825–7 was by far the most fruitful, in large part because of Richardson's work as a naturalist, and the indefatigable labors of his assistant, Thomas Drummond, who came to him on the recommendation of William Hooker.

*John Richardson, William Hooker, and Thomas Drummond.* Hooker served for several decades as the doyen of arctic botany. Richardson had corresponded with him before going off on Franklin's first expedition. In November 1823, the plants from Parry's second expedition went to Hooker for description: Richardson's notes on plants omitted in the published Appendix to Franklin's *Narrative* (1823) came in time to be useful to Hooker while he worked on those cryptogams and phaenogams. Parry's plants had come to Hooker because of Brown's dilatory behavior; Brown had been working on the plants from Parry's previous expedition when Parry returned from his latest one. Hooker was also in touch with Edward Sabine and his brother Joseph, asking for some of their arctic plants. Richardson had sent not only descriptions, but numerous specimens, not all of which were in good enough shape for proper examination. Relations between the two men were cordial enough for Hooker to invite Richardson to visit while he was working on Parry's plants.[60]

Hooker was soon supplying Richardson with ornithological books from his own library, and holding out such gems as the first edition of Temminck's *Manuel d'Ornithologie* and John Latham's *General Synopsis of Birds* as inducements to Richardson to visit him.[61] By the following summer, Hooker's encouragement had reinforced Richardson's botanical enthusiasms to the point where he was able to congratulate him on his decision to undertake

a Flora of Canada & the more northern parts of America. Your [own] researches in those countries, the toil & labour you will have gone through; all point you out as the only person under whose name such a Flora ought to appear: and much as I should be pleased at attempting such a thing under any other circumstances; as matters are, it *must* come from you. I need hardly tell you that all that ever I can do to assist you in it shall be done most cheerfully.

the *Voyage to the Pacific and Bering's Strait, Performed in His Majesty's Ship Blossom . . .* (London, 1841). A summary of the scientific accomplishments of the voyage is given in B. Gough, ed., *To the Pacific and Arctic with Beechey: The Journal of Lieutenant George Peard of H.M.S. 'Blossom' 1825–1828*, Hakluyt Society 2nd. Series vol. 143 (Cambridge, 1973) pp. 44–9.
[60] Hooker to Richardson, 20 Nov. 1823, Kew MSS Hooker to Richardson, 1819–43.
[61] Hooker to Richardson, 30 Nov. 1823, Kew MSS Hooker to Richardson, 1819–43.

...I had a letter from Parry two days ago, assuring me that all his best plants should be sent to me as soon as ever as he should return. Lady Dalhousie[62] offered to send me everything in her power from Canada & she is a very zealous Botanist. I had even urged her to collect for a Flora of Canada & she will be proud of doing all she can. . . . And I have now a zealous collector in Newfoundland a Dr. Morrison. I have a most excellent *poor* Botanist in view (Drummond of Forfar) who is most anxious to be sent out where he can be useful & I would subscribe something handsome to get him employed & shall probably write to Lady Dalhousie about him.[63]

The Canadian flora was to be the topic of numerous exchanges between them. Hooker had an excellent botanical network, maintained by tact, extensive correspondence, and the prospect for his correspondents of having a newly discovered species named after them. As in Drummond's case, he also sought to help needy botanists. We do not know if he did write to Lady Dalhousie about Drummond, but very soon Hooker was urging his virtues on Richardson. "I am very glad you have made up your mind to take some such a person with you as Drummond. I quite think as far as qualifications go that *he* would suit you. . . ." Drummond's main strength was in botany – he ran a botanical nursery, and first made his name by distributing sets of mosses, "Musci Scotici."[64] But he was also skilled as a general naturalist, who had stuffed birds and become "a good Scotch entomologist. . . . His great ambition was to be a means of introducing living plants that will bear this climate & I have told him that he should have the credit of introducing all that might be found by himself; but that as to the disposal of them & everything, they belonged to Government."[65] Nonetheless, Hooker hoped that Richardson would find a way for Drummond to keep some living plants and seeds for his own garden; he already had "by far, the best collection of living plants of any body in the kingdom & he cultivates the rarer kinds with a degree of success that would surprize you."[66] There was also a suggestion that Drummond, whose home life was not the happiest, would be better away on an expedition than in Forfar, where his weakness for liquor was damaging.[67]

---

[62] Lady Dalhousie was Christian Ramsay, née Broun, wife of George Ramsay, 9th Earl of Dalhousie, since 1820 governor-in-chief of British North America.

[63] Hooker to Richardson, 14 Aug. 1824, Kew MSS Hooker to Richardson, 1819–43, ff. 27–8.

[64] These are mentioned in C. D. Bird, "The Mosses Collected by Thomas Drummond in Western Canada, 1825-27," *The Bryologist* 70/2 (1967) 262–6.

[65] Hooker to Richardson, 23 Aug. 1824, Kew MSS Hooker to Richardson, 1819–43, ff. 29–30.

[66] Ibid., [2] Nov. 1824, ff. 37–8.

[67] See, e. g., Hooker to W. Swainson, Glasgow, 3 Jan. 1831, Linnean Society, Swainson correspondence.

Over the next few months, the Canadian *Flora*[68] and Drummond were the principal topics of correspondence. Hooker told Richardson that he should do the *Flora* himself, but expressed willingness to serve as a joint author: "I confess that I have a great partiality for the plants of North America."[69] Richardson insisted that the volume be in Hooker's name, and that he only be mentioned as a contributor.

> If you decline taking the principal charge, all that I can venture upon will be a Flora of Hudson's bay – I really know [so] little of botany, and am *known* to know so little, that my name as a principal would be injurious to the [progress] of the work and deter collectors from sending their discoveries. I will however labour during our journey as much as I can towards collecting materials . . . , and upon my return willingly undertake any portion that may be assigned to me. . . .[70]

Richardson prevailed. The *Flora* was to be Hooker's, with acknowledgment of the labors of Richardson, Drummond, and others, chief among whom was David Douglas, another Scottish botanist, Hooker's protégé, and major contributor to the botany of northwest North America. Douglas had already been sent by the Royal Horticultural Society, whose secretary was Joseph Sabine, to collect fruit trees and other plants in North America. In his second journey, coinciding with Franklin's, he was to botanize from Hudson Bay to the Columbia.[71]

By the fall of 1824 it was quite settled that Drummond would serve as assistant naturalist. Richardson wrote to Hooker that he found him

> every way qualified and if he is fond of a wandering life he will not find the task he has undertaken so unpleasant. The mosketoes are the only scourge of man in the district he will travel over, and the cold is so little to be dreaded that [he will] find the winter a very pleasant season. He is [very] likely to meet Douglas in his journey from the mouth of the Columbia to Hudson's bay, the route of the latter lying for a considerable distance down the Saskatchewan.[72]

Given Richardson's horrendous experiences on Franklin's first overland expedition, his breezy optimism about northern winters is remarkable, even for a Scot. So too is his confidence that Douglas and Drummond would

[68] For a general account of nineteenth-century Canadian botany see S. Zeller, *Inventing Canada: Early Victorian Science and the Idea of a Transcontinental Nation* (Toronto, 1987) Part III. For the second half of the century see also W. A. Waiser, *The Field Naturalist: John Macoun, the Geological Survey, and Natural Science* (Toronto, 1989).
[69] Hooker to Richardson, 23 Aug. 1824, Kew MSS Hooker to Richardson, 1819–43 ff. 29–30.
[70] Richardson to Hooker, 1 Oct. 1834, Kew MSS Hooker to Richardson, f. 137.
[71] W. Morwood, *Traveller in a Vanished Landscape: The Life and Times of David Douglas* (London, 1977).
[72] Richardson to Hooker, 1 Oct. 1824, Kew MS Miscellaneous Letters, 1818/30 XLIV 137.

meet along their vast exploring and botanizing routes.[73] Hooker wanted to ensure the most thorough coverage of the northern regions, and his letters to Richardson weave a giant web of botanical collecting.

You, if I understand right, will not attempt to reach the N.W. coast; but then will there be no collector with Captn. Franklin? What will be the probable extent of your journey when you leave Drummond? & will he not, by himself be ordered to make long and distant excursions? Is there any chance of Captn. Franklin's going as far south as the Columbia? or of Douglas' joining him at any point?[74]

Richardson tried to reassure Hooker about the problem of collectors with Franklin, who, like his officers, would be very busy, but would no doubt collect plants when possible; and he, Richardson, would instruct one or more of them in the best methods. Drummond, the expedition's botanist, would be stationed on the Saskatchewan near the northern extremity of the Missouri basin and would botanize along the slopes of the Rockies in latitude 52° N. Douglas would spend a season collecting for the Horticultural Society on the northwest coast through nearly ten degrees of latitude, would cross the Rockies in latitude 55°, and would, with luck, join Franklin's route at Ile-à-la-Crosse. Richardson would botanize between the Coppermine and Mackenzie rivers.[75] Botanical arrangements for the expedition were in good hands.

*Franklin's second overland expedition 1825–1827.* Overall arrangements were of course in Franklin's hands. Having learned from previous disaster, he was determined that his forthcoming expedition would be an unqualified success.[76] His first wife, Eleanor Anne Franklin, wrote to his sister Betsey:

I think I can venture to assure you, that as far as human calculation can extend, or human prudence can provide, there is no danger of his again encountering the sufferings of his last journey. [The boats are now building at Woolwich], the first division of stores is gone or going, and orders are sent out for the laying up of a stock of provision on all the route, added to which, the Companies, who from a desire of

[73]  They did meet. See notes 128 and 129 below.
[74]  Hooker to Richardson, 23 Nov. 1824, Kew MSS Hooker to Richardson 1819–43 ff. 39–40. There was indeed a chance of Douglas joining Franklin's expedition: he was to do so on the shores of Hudson Bay at the end of the impending expedition, and returned with the expedition to England in 1827.
[75]  Richardson to Hooker, 28 Nov. 1824, Kew MSS Miscellaneous Letters, 1818/30 XLIV 139.
[76]  For his arrangements with the HBC, see Captain J. Franklin's Second Arctic Expedition, Accounts and Miscellaneous 1824–8, HBC archives ref. E.15/1, reproduced in PAC microfilm HBC 4M21.

monopoly have never hitherto cooperated cordially in any plan for North Western Discovery, have their interest so much involved in the success of the Expedition that there can be no doubt of their warm and complete support.[77]

Franklin had also entered into correspondence with the Admiralty, with scientists, and with suppliers of philosophical and navigating instruments. He requested and ordered telescopes, a dip needle, a transit instrument, barometers, artificial horizons, compasses, an instrument for determining diurnal variation – the explorer's kit as it had developed for Admiralty expeditions. For Richardson and Drummond, he ordered a ream of paper for preserving plants.[78] Other scientific preparations included his and his officers' attendance at geological lectures given for them by William Fitton, President of the Geological Society. These lectures were to prove especially valuable for Richardson, and also furnished the occasion for the first meeting between Franklin, Richardson, and Back with Roderick Impey Murchison, who would later become the spokesman of British imperial geology and a statesman of science.[79]

Franklin necessarily combined science with politics, diplomacy, and naval administration. To George Simpson, governor of the Northern Department of the HBC's Territories, he wrote to explain the aims of his expedition, and to engage the company's support. He described the coordinated attempt that would be made by his expedition, and by Parry's and Beechey's. It was hoped that this would resolve the issue of the Northwest Passage, and in the process would map and secure territory for Britain and, Franklin noted, for its fur trading company. Franklin had in mind, and knew that Simpson did too, the rival exertions of Russian explorers. The Russian card was as effective with Simpson as it had been with Barrow. They all thought that they were in a race with Otto von Kotzebue, the Russian navigator whose first arctic expedition had set out in 1815, and had led to the discovery of Kotzebue Sound and Krusenstern Cape on the North American arctic shore.[80] In 1823, he had set out with two ships of war to

[77]   Eleanor Anne Franklin to Miss [Betsey] Franklin, 5 Feb. 1824, RGS MS.
[78]   Franklin to Colonial Office, 12 Feb. 1824, PRO MS CO 6 16 p. 8; PRO MS CO 6 17 f. 45; Franklin to Wilmot Horton, Undersecretary of State, 13 July 1824, PRO MS CO 6 16 f. 26.
[79]   R. A. Stafford, *Scientist of Empire: Sir Roderick Murchison, Scientific Exploration and Victorian Imperialism* (Cambridge, 1989) pp. 7, 68–9; John Warkentin, ed., "Geological Lectures by Dr. John Richardson," *Syllogeus* no. 22 (Ottawa, National Museum of Natural Science, 1979).
[80]   O. von Kotzebue, *A Voyage of Discovery into the South Sea and Beering's Straits for the Purpose of Exploring a North-East Passage, Undertaken in the Years 1815–1818*, 3 vols. (London, 1821).

take reinforcements to Kamchatka; he also took a group of scientists who did important work in ethnography and natural history.[81] Franklin wrongly assumed that he was bound for the American Arctic, "but as I find he will have to stop a little time at Kamtschatka, and to reexamine the sound named after him, before he goes on the northern coast, I have little doubt that we shall anticipate his discoveries, provided my party can be forwarded expeditiously through the country from Montreal, and have the arrangements made in the interior . . ."[82] Simpson authorized logistical support, arranged for an Inuk interpreter, and sought to reinforce the natural history side of the expedition. "I have sent out circulars," he told Franklin, "to all parts of the Country requesting that specimens of Quadrupeds, Birds, and Plants, be collected, with the necessary directions for preparing and preserving them: but few of our gentlemen have a turn for Natural History. I shall therefore not hold out the prospect of much success in that way. . . ."[83]

Another necessary preparation, given the rivalry and unresolved negotiations with Russia, was to obtain a Russian document serving as passport in case of any encounters with Russians in the North; the ukase had exempted "vessels sent by friendly Powers on voyages of discovery," as long as they carried such a document.[84] The Russian ambassador in London obliged with a letter for Franklin as an officer of an allied power.[85] Happily, relations between the rival powers were about to improve. Russia had already drawn up a treaty with the United States, and soon after Franklin's departure in 1825 Russia matched this with a treaty with Great Britain. The main part of Alaska and its islands west of 141° W were Russian, but Franklin, with diplomatic protection, saw his way clear.[86]

With Richardson, Drummond, Back, and Ernest Kendall, mate, Franklin embarked in February 1825 for New York, and traveled thence to York, now Toronto, and northward, arriving at Cumberland House in June.

[81]  Kotzebue, *A New Voyage round the World in the Years 1823–1826* (1830; reprinted Amsterdam, 1967).

[82]  Franklin to Simpson, 27 Feb. 1824, copy in SPRI MS 248/281/1 pp. 96 et seq.

[83]  Simpson to Franklin, 8 Aug. 1824, copy SPRI MS 248/281/1 pp. 162–3. See also PRO MS CO 6 16 ff. 143–56.

[84]  *New Times*, 7 October 1825; cutting in HBC archives A 71/1 f. 165, microfilm PAC HBC M9.

[85]  Franklin to R. W. Horton, 7 Dec. [1824], PRO MS CO 6 16 f. 32; Russian Ambassador to [Foreign Office] 12 Dec. 1824, PRO CO 6 16 f. 131.

[86]  N. L. Nicholson, *The Boundaries of Canada, Its Provinces and Territories* (Ottawa, 1954) pp. 27–8. S. R. Tompkins, "Drawing the Alaska Boundary," *Canadian Historical Review* 26 (1945) 1–24; *Chronicle*, 17 May 1825, "Convention between His Majesty and the Emperor of Russia."

There Drummond left them to botanize in the Rocky Mountains. Before he left, Franklin gave him instructions to make collections "of Plants, Insects, Birds, rare Quadrupeds and Minerals, and to avail himself of the aid kindly promised by the Company's Officers."[87] Drummond began as he was to go on, with enormous energy and dedication. Richardson had written from Fort William on Lake Superior to his wife that "Drummond is the best disposed and most indefatigable collector of Natural History I have ever seen. I expect to find him on my return to Cumberland House with three canoe loads of plants & skins."[88] The rest of the party went on to winter quarters at newly renamed Fort Franklin on Great Bear Lake.

While the others worked on their winter home, Franklin and Kendall descended the Mackenzie to check on sea ice conditions at its mouth – it was ice free – and Richardson surveyed the north shore of Great Bear Lake, "the finest lake in the world."[89] Franklin wrote enthusiastically to Murchison about the strata along the Mackenzie: coal formations, sandstones and limestones, the latter rich in fossils, and altogether worth a geological excursion.[90] Coal in the North indicated a warmer climate in the Paleozoic era, and such evidence contributed to controversy about climatic change, a controversy that can be followed in successive editions of Murchison's *Siluria: A History of the Oldest Rocks . . .*[91] Coal was of practical significance: British control of arctic coal measures would help to secure the region for British warships, rather than American or Russian ones.

Franklin also wrote to the Colonial Office, announcing that he differed from Alexander Mackenzie, first European navigator of the river named after him,[92] on several points of longitude, "which I attribute to his having laid them down by compass bearings, and his not having had the means of detecting its changes in variation, which we have found to be

---

[87]  Franklin, "Memo. given to Mr. Drummond at Fort William," 12 May 1825, SPRI MS 248/281/1 pp. 274–5.

[88]  J. Richardson to Mary Richardson, 12 May 1825, SPRI MS 1503/6/2.

[89]  Richardson to Robert McVicar, 7 Sept. 1825, McCord Museum MS M2740.

[90]  Franklin to Murchison, 4 Nov. 1825, SPRI MS 248/225/1–3.

[91]  Stafford, *Scientist of Empire*, p. 68. The first edition of *Siluria* appeared in 1854, the fourth in 1867. See also R. I. Murchison, *The Silurian System, Founded on Geological Researches in the Counties of Salop, Hereford. Radnor, Montgomery, Carmarthen, Brecon, Pembroke, Monmouth, Gloucester, Worcester, and Stafford: With Descriptions of the Coal-Fields and Overlying Formations* (London, 1839).

[92]  A. Mackenzie, *Voyages from Montreal on the River St. Laurence through the Continent of North America to the Frozen and Pacific Oceans in the years 1789 and 1793, with a Preliminary Account of the Rise, Progress, and Present State of the Fur Trade of that Country* (London, 1801; reprinted Rutland, Vermont, and Tokyo, 1971).

considerable . . ."[93] Franklin and Kendall made measurements of magnetic dip and variation, and found that there had been an increase in variation since Mackenzie's epic voyage some thirty years previously. They agreed very closely with Mackenzie in fixing the position of the mouth of the river. Franklin, who had superior instruments, including decent chronometers, patronizingly observed that Mackenzie's survey was surprisingly accurate, given his imperfect instruments and inexperience.

On their return to Fort Franklin, they formed part of a group of fifty, including Inuit, Indian hunters, voyageurs and seamen, who needed not only food but regular occupation. The officers started a school, teaching reading, writing, and arithmetic. There were games, Sunday services, and magnetic, meteorological, and thermometrical observations. Richardson undertook studies on solar radiation, Kendall made observations on the velocity of sound, and there were geological lectures.[94] Fitton of the Geological Society had not only instructed the officers in geology, but had furnished them with a portable reference collection; at Fort Franklin, Richardson explained the specimens to his colleagues.[95]

Richardson's main sources for his lectures were Robert Jameson's *Mineralogy* and Conybeare and Phillips' *Outlines of the Geology of England and Wales*.[96] Jameson's work was based on an essentially Wernerian taxonomy, whereas Conybeare and Phillips made use of William Smith's account of the use of fossils in identifying and classifying geological strata.[97] Smith was an engineer who, in observing the cuttings made in the development of railroads and canals in the early years of the nineteenth century,

[93]  Franklin to Horton, 8 November 1825, PRO CO 6 17 p. 12.
[94]  *Narrative of a Second Expedition to the Shores of the Polar Sea in the Years 1825, 1826, and 1827 by John Franklin, Captain R.N., F.R.S. and Commander of the Expedition: Including an Account of the Progress of a Detachment to the Eastward by John Richardson, M.D., F.R.S., F.L.S. Surgeon and Naturalist to the Expedition* (London 1828; reprinted Rutland, Vermont, 1971) pp. 54 et seq. See also Warkentin, "Geological Lectures."
[95]  Back took systematic notes during these sessions, and his notebook survives: Warkentin, "Geological Lectures" and Back, "Notes made by Lieutenant George Back during a series of lectures on geology given at Fort Franklin on Great Bear Lake by Dr. John Richardson from October 7th, 1825 to January 13th, 1826," SPRI MS 395/1.
[96]  Jameson, *Manual of Mineralogy: Containing an Account of Simple Minerals, and also a Description and Arrangement of Mountain Rocks* (Edinburgh, 1821); W. D. Conybeare and W. Phillips, *Outlines of the Geology of England and Wales* (London, 1822).
[97]  W. Smith, *A Memoir to the Map and Delineation of the Strata of England and Wales, with Parts of Scotland* (London, 1815); and *Stratigraphical System of Organized Fossils with Reference to the Specimens of the Original Geological Collection in the British Museum* (London, 1817). For a fuller account, see Warkentin "Geological Lectures." The principal Wernerian groups of rocks and strata, and their chronological and stratigraphic sequence, are briefly discussed in chapter 2.

hit on paleontology as the clue to the interpretation of stratigraphy, and produced the first accurate stratigraphic map of Britain. Now Richardson was to apply Smith's method, first in his lectures, and subsequently, though less coherently, in arctic fieldwork. Franklin described Richardson's lectures that winter as proving "a most agreeable and useful recreation to us all."[98] To Murchison, Franklin wrote ambiguously of Richardson's lectures:

Through [Richardson] we endeavour to keep up the information which Dr. Fitton first imparted. We have got Conybeare & Phillips . . . and Humboldt on the superposition of Rocks,[99] but to the inexperienced one lecture from a person conversant with the science is more profitable than many hours reading on subjects difficult to be comprehended. It is evident too, on the slightest enquiry into Geology, that a comparative knowledge of other sciences is requisite – Mineralogy & Chemistry for instance. . . .[100]

After leaving the fort in the spring of 1826, they proceeded down the Mackenzie River to the Arctic Ocean. Richardson and Kendall went eastward, exploring the coast between the Mackenzie and Coppermine Rivers; their discoveries included part of Victoria Island, which they called Wollaston Land. They corrected Hearne's latitude at the mouth of the Coppermine, and then returned to Fort Franklin, by way of the Coppermine and Dease rivers, and by overland travel, rejoining the rest of the party ahead of Franklin on 1 September.[101] Early in the expedition, Franklin had told his officers that they would make "a variety of Astronomical, Atmospherical & Magnetical observations," when the speed of their travel allowed.[102] They traveled hard: Richardson and Kendall covered 1,980 statute miles in 71 days. Observations en route were primarily for accurate mapping and navigation, departments in which Kendall played the principal role, as Richardson warmly recorded:

and as proof of his correctness I may state, that, unaided as he was by a chronometer, the longitude by the Dead (Estimated) Reckoning corrected from time to time by Lunar Observations, differed on reaching the mouth of the Coppermine River, only five minutes (or 2 miles of distance) from the position of that place as formerly established by you from the mean of three chronometers.[103]

[98] Franklin, *Narrative*, p. 56.
[99] A. von Humboldt, *A Geognostical Essay on the Superposition of Rocks in Both Hemispheres* (London, 1823).
[100] Franklin to Murchison, 4 November 1825, SPRI MS 248/225/1–3.
[101] This summary comes largely from Cooke and Holland, *The Exploration of Northern Canada*, p. 152.
[102] Franklin, "General Memorandum Issued to the Officers," SPRI MS 248/281/1 pp. 232–5.
[103] Richardson to Franklin, Fort Franklin, 4 Sept. 1826, PRO CO 6 17 f. 60.

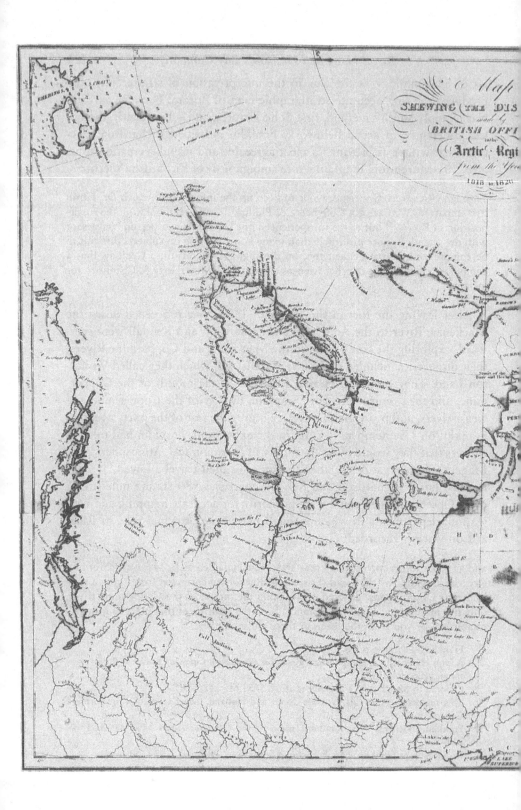

Figure 17. "Map Shewing the Discoveries made by British Officers in the Arctic Regions from the year 1818 to 1826," from John Franklin, *Narrative of a Second Expedition to the Shores of the Polar Sea, in the Years 1825, 1826, and 1827* (1828). Fisher Library, University of Toronto.

They also recorded weather and tides, and Richardson managed to collect a variety of animal, vegetable, and mineral natural history specimens. His coastal observations recorded about 175 flowering plants, "being one-fifth of the number of species which exist fifteen degrees of latitude further to the southward."[104]

Meanwhile, Franklin and Back went westward, in hopes of meeting Beechey on the *Blossom* at Icy Cape. After six weeks, they were only half way there, and turned back to avoid being caught by winter. In spite of the rigors of the journey, they managed to combine precise navigation with natural history observations, and made a good collection of plants.[105] They got back to Fort Franklin on 21 September 1826. Richardson wrote to his wife from Great Slave Lake, delighted at Franklin's safe return

from a voyage full of peril and difficulties, but more successful than could have been hoped under such circumstances. . . . [A]lthough he has not completed the Northwest passage, yet he has left so small a portion of the coast unsurveyed, that if Cap[n.] Beechey gets round Icy Cape he can scarcely fail in completing it. The search after the passage has employed three centuries but now that it may be considered as completed, the discovery will I suppose be committed like Juliet to the tomb of the Capulets, unless something more powerful than steam can render it available for the purpose of mercantile gain.[106]

Geographical and scientific curiosity were not so easily satisfied. The gaps in the passage were larger than Richardson suggested. Pressures from the whaling fleet, and especially the American fleet, also encouraged northern exploration.[107] The Royal Society and the Royal Geographical Society kept badgering the British government to extend geographical and other scientific researches. Completion of the mapping of the passage came only during the searches for Franklin and his crew after Franklin's next and fatal voyage. But for the present, they were in the north country, and well, and eager to add to knowledge of the North. Richardson set out in the winter of 1826 to join Drummond on the Saskatchewan River, to collect birds and animals, and to take advantage of an early spring to collect plants. Numerous type specimens of birds were collected there, by both Drummond and Richardson,[108] who met at Carlton House. Richardson rejoined Franklin at

[104]   Richardson in Franklin, *Narrative*, p. 264.
[105]   Richardson to Back, Fort Resolution, 21 Dec. 1826, SPRI MS 395/60/2.
[106]   Richardson to his wife Mary, 10 Nov. 1826, MS in McCord Museum, McGill University, Richardson papers folder 2.
[107]   J. R. Bockstoce, *Whales, Ice, and Men: The History of Whaling in the Western Arctic* (Seattle and London, 1986) illustrates the interplay of whaling and exploration.
[108]   C. S. Houston and M. G. Street, *The Birds of the Saskatchewan River Carlton to Cum-*

Cumberland House, and they headed homeward. The expedition returned to Britain in the fall of 1827.

Once home, Richardson presented his geological and topographical observations to the Geological Society, in whose museum he deposited the rock and mineral specimens from the expedition. His account mirrored the stance of his geological lectures at Fort Franklin, combining a broadly Wernerian scheme with attention to the succession of strata and to the fossils they contained. His narrative could have furnished him with the basis for a broad rather than detailed geological map of the routes and territories covered by the expedition. Other scientific appendixes to Franklin's *Narrative* presented meteorological data from Fort Franklin, including observations on parhelia and paraselenae; observations on solar radiation; Kendall's observations on the velocity of sound at different temperatures; observations on the aurora;[109] and measurements of magnetic dip and variation. "Accidental circumstances" prevented them from obtaining satisfactory observations on the magnetic force. Franklin's first expedition had suggested that aurora affected magnetic variation. Observations from his latest expedition confirmed this. The necessary observations were the result of teamwork; Kendall and Franklin read off the arc at each end of the needle, while Back and Richardson stayed outside the observatory to record changes in the auroral displays.[110] Franklin was looking for a wider set of correlations. As he wrote to Beaufort, "[i]t would appear there are many Laws and effects of Magnetism to be developed. Our observations lead us to suppose that the motions of the needle often depend on Atmospheric changes – on three occasions for instance, an Easterly Gale has produced an increase of variation."[111]

*Thomas Drummond, field botanist.* Franklin's magnetic and meteorological data were useful, but the principal scientific results of his expedition were in natural history. These results contributed substantially to Hooker's *Flora* and Richardson's *Fauna*.[112] Richardson was surgeon, explorer, and naturalist, and his skills in all three departments had been much in demand. In November 1825 he wrote to Hooker that Drummond "is my main stay in the botanical and entomological departments, my attention being much

   *berland*, Special Publication No. 2, Saskatchewan Natural History Society (Regina, 1959).

[109] George Back, "A Journal of the Proceedings of the Land Arctic Expedition under Command of John Franklin . . . ," Dec. 1824–June 1826, SPRI MS 395/6 pp. 196–204.

[110] Franklin, *Narrative*, appendixes.

[111] Franklin to Beaufort, Fort Franklin, 6 Feb. 1826, SPRI MS 1162/2.

[112] See the next section, "Interlude: *Flora* and *Fauna*."

directed to other objects."[113] Drummond's botanizing, indeed, had heroic qualities.

He observed and collected vigorously around Cumberland House until boats of the Hudson's Bay Company arrived in August 1825, and left with them along the Saskatchewan River for Carlton House. There, he noted, "the Indians were found to be in so unsettled a state, that it would have been very unsafe to make excursions in that neighbourhood, without the protection of a strong party; and I therefore decided upon proceeding with the brigade, until I should find a place better suited to my purpose."[114]

They went from Carlton House to Edmonton across the prairies, at a rate of twenty miles a day, with Drummond botanizing each day over fifteen or more miles, and complaining at the monotony of the flora: "indeed, I did not find a single plant that I had not seen within ten miles of Carlton House, although I had the opportunity of examining the country carefully, having performed the greater part of the journey on foot."[115] Most common were the so-called *papilionaceae*, leguminous plants whose flowers bear a fancied resemblance to butterflies; also common were the genera *phlox*, *liatris* (blazing-star and button-snakeroot), *malva* (mallow), and *erigonum* (wild buckwheat). Each day, Drummond stayed with the boats until they stopped to breakfast, then went on shore with his vasculum, following the river bank and from time to time making excursions inland, generally managing to rejoin the boats at their next camp. There, after supper, "I commenced laying down the plants gathered in the day's excursion, changed and dried the papers of those collected previously; which operation generally occupied me till daybreak, when the boats started. I then went on board and slept [for perhaps four hours] till the breakfast hour, when I landed and proceeded as before."[116] Apart from the plants, which were grist for Hooker's *Flora*, there were numerous animals and birds, most of which were described previously by Richardson or by Lewis and Clark.[117]

---

[113]   Quoted in *Edin. J. Science* 6 (1827) 107–10.
[114]   Drummond, "Sketch of a Journey to the Rocky Mountains and to the Columbia River in North America," Hooker's *Botanical Miscellany* 1 (1829–30) 178–219 at 182.
[115]   Drummond in Franklin, *Narrative*, pp. 308–9.
[116]   Drummond, "Sketch," p. 183.
[117]   We have encountered Richardson's contributions to natural history appendixes to narratives of previous expeditions. For Lewis and Clark, see M. Lewis and W. Clark, *The Journals of the Expedition under the Command of Capts. Lewis and Clark, to the Sources of the Missouri, thence across the Rocky Mountains and down the river Columbia to the Pacific Ocean, Performed during the Years 1804–5–6 by Order of the Government of the United States* (reprinted New York, 1962); Penguin Books have recently brought out an abridged version, *The Journals of Lewis and Clark*, ed. and introduced by F. Bergon (New York, 1989). See also R. D. Burrough, *Natural History of the Lewis and Clark Expedi-*

Drummond was particularly taken with the ruddy duck, "remarkable for the brilliant blue colour of the bill of the male, and the singular way in which, when courting . . . , it carries its tail . . . perfectly upright, giving the bird, at a little distance, the appearance of having two heads."[118] He singled out a flycatcher, the eastern kingbird,

for the courage with which it attacks all others that venture near its residence; . . . it is truly amusing to see it assault the *Falco borealis,* or any large bird. It soars above them, then darting down on the back of its opponent, applies its beak, with all the strength that it possesses, to its head, sometimes remaining in this position for a minute or more, and then it returns in triumph to its station. . . .[119]

From Edmonton, the brigade, accompanied by Drummond, made the hundred-mile portage to the Red Deer (now the Athabaska) River, where they determined that some of the party should go by land, because of the state of the river and the heavy load of the canoes. Drummond was of the land party; soon after their departure, a heavy snowfall put an end to systematic collecting for the season, and made the journey a hard one. Drummond nonetheless made the most of his opportunities: "The first indication which the vegetation afforded of our approach to the mountains, was the *Arbutus alpina* and *Dryas Drummondii;* the latter, with a beautiful yellow flower, was growing upon the gravelly battures formed by one of the mountain rivulets; *Dryas tenella* was also there, and an *Eriogonum* of considerable beauty."[120]

As he went, he collected where the snow allowed. He collected mosses for Richardson, and also sets for an exsiccati series (i. e., a set of dried specimens) to be called *Musci Americani.*[121] The party followed the river to Jasper House, near today's Jasper. They then went back to the junction of the Assinaboyne (Snake) River with the Athabaska River. There they halted to arrange the luggage before crossing the mountains. "The very great difficulty with which this process was attended, compelled me to give up the resolution I had formed of going for the winter to the Columbia River, and decided me upon remaining among the rocky mountains."[122] With an Iroquois hunter as guide, he left the mountains to camp for the winter on a

    *tion* (Ann Arbor, Michigan, 1961) and P. R. Cutright, *Lewis and Clark: Pioneering Naturalists* (Urbana, Illinois, 1969).

[118] Drummond, "Sketch," pp. 185–6.

[119] Ibid., p. 186. See also W. Swainson and J. Richardson, *Fauna Boreali-Americana* pt. 2, *The Birds* (London, 1831) p. 138.

[120] Drummond, "Sketch," p. 190.

[121] C. D. Bird, "Mosses Collected by Thomas Drummond"; Drummond, *Musci Americani* (Glasgow, 1828).

[122] Drummond, "Sketch," p. 192.

tributary of the Athabaska. In April, they snow-shoed the 175 miles back to Jasper in six days, and there Drummond received supplies that he had lacked through the winter – his tent, some tea and sugar, and a fresh supply of drying paper. Botanizing the next month, he was repeatedly charged by a sow grizzly bear, with two cubs that ran off. He found his powder was damp, and his gun would not fire. "My only resource was to plant myself firm and stationary, in the hope of disabling the bear by a blow on her head with the butt end of my gun." The return of the brigade scared off the bear. Drummond resolved for the future to keep his gun in better order, but found, "by future experience, that the best mode of getting rid of the bears when attacked by them, was to rattle my vasculum, or specimen box, when they immediately decamp."[123] This particular adventure did not, however, prevent Drummond from collecting the moss he was after. He botanized in the mountains for the next two months, enjoying the spring glories of the alpine meadows and the orchids in the woods. In spite of the hardships he encountered, or perhaps in part because of them, Drummond was, as he wrote to Hooker, grateful for the opportunity of exploring "scenes so congenial to my inclination."[124] Apart from botanizing, he continued to observe and collect birds and animals. In the neighborhood of Jasper, he collected two species of birds new to science, the white-tailed ptarmigan and the black-backed three-toed woodpecker. He also collected a dipper that is now the type for the subspecies *cinclus mexicanus unicolor* Bonaparte. He collected at least three other type specimens: Forster's tern, collected upriver from Cumberland House; a subspecies of the loggerhead shrike, collected at Carlton; and the arctic subspecies of the rufous-sided towhee, also collected at Carlton.[125]

In summer he crossed the Great Divide, taking much satisfaction in the confirmation of his idea that the vegetation would differ considerably on the other side. He recrossed the divide and returned to Jasper House in October, a strenuous journey in deep snow. He was back at Edmonton in December, and there learned that Richardson's part of the expedition had been an entire success, and that he was to join Richardson at Carlton House. "Accordingly, I quitted Edmonton House in the middle of March, taking with me a single specimen of every plant gathered among the Rocky Mountains."[126] Richardson was delighted to see Drummond again, not least because of his collection of plants, which he described as "copious and

[123]  Ibid., p. 197.
[124]  Drummond to Hooker, Rocky Mountains, 24 April 1826, Kew Misc. Letters XLIV 72.
[125]  C. S. Houston and M. G. Street, *The Birds of the Saskatchewan River* (1959) p. 13.
[126]  Ibid., p. 207.

extremely interesting and . . . very valuable."[127] David Douglas was in the
meantime returning across the continent guided by Edward Ermatinger, a
fur trader with the Hudson's Bay Company, on a vigorously forced march
from Fort Vancouver to Hudson's Bay via Athabaska Pass. In July 1827 he
and Drummond met at Carlton House,[128] and were to meet again and botanize together on the shores of Hudson Bay. In September, visiting the ship
in a small boat with Back, Kendall, and nine others, they were swept out to
sea in a storm that carried their masts away, and took two punishing days
and nights to get back. Kendall and Douglas suffered particularly from exposure and exhaustion, and needed the homeward voyage to get back on
their feet; Drummond, however, was out the next day, botanizing through
muskeg and swamp, "and the only plant which recompensed me for all my
labour was *Polytrichum formosum*."[129] But he stayed on his feet, and no
doubt he was immune from seasickness on the return to Britain. He became
for a while curator of the new Botanic Gardens at Belfast, but as Hooker
remarked, he was not suited for "stay-at-home employment";[130] what he
wanted was another collecting expedition, preferably to somewhere with a
cold climate. But Drummond's later expeditions were more and more
southerly, and he died in Havana in 1835.

### INTERLUDE: *FLORA* AND *FAUNA*

Back in England, Richardson set about obtaining proper publication of his
and Drummond's work. As he told Franklin, the collections "are, particularly in the branches of Ornithology and Botany, by much the most copious

[127] Richardson to Back, Cumberland House, 14 June 1827, SPRI MS 395/60/3. Among
the plants named for Drummond in Hooker's North American flora were *Drummondia
mitelloides, Agrostis drummondii, Astragalus drummondii, Dryas drummondii, Litho-
spermum drummondii, Potentilla drummondii, Salix drummondiana,* and *Silene drummondii.* For a broader based list, see J. and E. D. Ewan, *Biological Dictionary of Rocky
Mountain Naturalists: a Guide to the Writings and Collections of Botanists, Zoologists,
Geologists, Artists and Photographers 1682–1932* (Utrecht, 1981); under DRUM-
MOND, they list *Drummondia,* W. J. Hooker; *Phlox drummondii* Hook; *Bruchia drummondii* James; *Cylindrothecicum drummondii* Bry. Eur.; *Grimmia drummondii* Hook. f.
and Wils.; *Entosthodona drummondii* Sullivant; *Webera drummondii* Lesq. and James;
*Dicranium drummondii* C.Meill.

[128] Ewan, *Dictionary of Rocky Mountain Naturalists,* articles under DRUMMOND,
DOUGLAS, and ERMATINGER. See also Samuel Wood Geiser, *Naturalists of the
Frontier* (Dallas, 1937; 2nd ed. 1948).

[129] Drummond, *Sketch,* p. 218.

[130] Hooker to Richardson, 23 May 1828, Kew MSS Hooker to Richardson, 1819–43
ff. 65–6.

in specimens that have ever been formed at any one period, either by Public agents or Private collectors, in British North America." A properly illustrated work would not be a commercial proposition for a publisher, and could only be undertaken under the auspices of government – a circumstance for which there were precedents. Hooker would describe the plants without charge, and had offered to publish them at his own risk; but figures and engravings would be needed for the rarer kinds, and plates were necessary for some of the zoological specimens. A subsidy of up to £1,000 pounds would be needed.[131] Franklin agreed with Richardson, and wrote to the Colonial Department supporting his request.[132]

Government support was made available, and work went briskly ahead. Hooker got his teeth into the *Flora,* describing the plants, arranging them according to the natural orders and genera, and putting them all in their proper habitats; Drummond provided assistance, especially on mosses. There were some 14,000 specimens from Richardson and Drummond's collection alone, and Hooker also solicited plants from other arctic explorers, and from Douglas. Hooker looked out for Drummond's interests, and his own and Richardson's, proposing to publish a notice of all new species to ensure their priority, before David Douglas published his accounts.[133] Richardson undertook the *Fauna,* with help from friends. William Kirby, clergyman and entomologist, set about the insects:

I perceive several new forms amongst them, & several good things – & the specimens are in better order in general than might be expected. The only ones that seem past all remedy are the Dragon flies except one or two of the smaller ones. . . .

I have named specimens after you [Richardson] & Captn. Franklin, if there are any that were your particular favourites, I should like so to designate them. I should likewise like to name one after the collector [Drummond]. . . .[134]

The reward system of naming species after their discoverer was in full swing. Richardson, however, needed help more than glory. Hooker introduced him to the naturalist and author William Swainson when they met at Sotheby's auction room, and Swainson undertook to work on the birds with Richardson; his particular responsibility was taxonomic, and he also made a nicely advantageous arrangement for the plates.[135] Richardson had

[131]  Richardson to Franklin, London, 25 October 1827, PRO CO 6 17 p. 29.
[132]  29 Oct. 1827, PRO CO 6 17 f. 27.
[133]  Hooker to Richardson, 21 March 1828, Kew MSS Hooker to Richardson, 1819–43 ff. 62–4.
[134]  Kirby to Richardson, Barham, 4 Feb. 1828, SPRI Voss collection, MS 1503/7/2.
[135]  Hooker to Swainson, Glasgow, 6 Dec. 1827 and 3 Jan. 1831, MSS Linnean Society, Swainson correspondence; Hooker to Richardson, 23 Jan. and 23 May 1828, Kew MSS Hooker to Richardson, 1819–43 ff. 54–5, 65–6. For Swainson's taxonomy, see notes

Figure 18. "Arctic or White Horned Owl," by William Swainson, in Swainson and Richardson, *Fauna Boreali-Americana* vol. 2 (1831). Metropolitan Toronto Library.

to caution him against expecting too much: "Works of natural history very rarely in this century . . . bring money to the authors."[136]

Richardson wrote most of the volume on birds, and sent his manuscript to Swainson for inspection. Swainson wrote the preface, and a highly idiosyncratic introduction to a natural system of classification. He stated that there were two principal approaches to classification. One, best exemplified

138–40, and also D. M. Knight, "William Swainson: Types, Circles, and Affinities," in J. D. North and J. J. Roche, *The Light of Nature: Essays in the History and Philosophy of Science Presented to A. C. Crombie* (Dordrecht, 1985) pp. 83–94. See also Richardson to Swainson, Chatham, 2 Nov. 1830, MS Linnean Society, Swainson correspondence: "You informed me, however, at the outset that you could not undertake more than the synonyms and general remarks without a compensation for your time which I had no means of providing."

[136] Richardson to Swainson, Dumfries, 8 Sept. 1829, MS Linnean Society, Swainson correspondence. The unaccomplished but most costly publication for its author, arising from natural history collections during an expeditionary voyage, was Joseph Banks's *Florilegium,* which is only now being completed and published from materials in the British Museum, Natural History.

by the system of Linnaeus, was artificial, a way of organizing our knowledge of the natural world that stressed clarity and convenience. The other was natural, seeking to organize our knowledge according to relations truly existing in nature. Cuvier, for example, classified organisms both by the similarities and convergence of species, and by the principle of the subordination of characteristics.[137] Swainson was ambitious, being convinced "that a zoological system which aimed at illustrating the general laws of creation was that only which deserved to be called *Natural*."[138] He looked at the system of William Sharpe MacLeay, which was built upon two forms of relationship: analogy, in which similar features occurred in groups otherwise remote, for example whales and fishes, and affinity, based upon similarity of structural relations, for example between whales and other mammals. According to MacLeay, all the animals in any group form a circle based on affinities, with the closest relationships being those between neighboring members, while all the animals of any one group are analogous to those of all other groups. The circular groups themselves could be arranged in circles. MacLeay decided that each circular group contained five members, and hence his system was known as the circular and quinarian theory. Swainson argued that "the primary divisions of every natural Group, of whatsoever extent or value, are three, each of which forms its own circle."[139] Thus, for example, the vegetable kingdom was divided into three, monocotyledons, dicotyledons, and acotyledons, while another circle could be made of fishes, aquatic serpents, and aquatic quadrupeds. It is little wonder that Louis Agassiz, in his *Essay on Classification*, said that the only merit of MacLeay's system consisted in his having drawn attention to the distinction between affinity and analogy; it is worth noting that this is a most important distinction. Agassiz remarked that in order "that I may not appear to underrate the merits of this system, I will present it in the very words of its most zealous admirer and self-complacent expounder, the learned William Swainson."[140]

Fortunately, Swainson's introductory observations were largely distinct from the descriptions of individual species, which were almost entirely Richardson's. Swainson wanted to give Richardson's name to a variety of

---

[137] G. Cuvier, *Le Règne Animal Distribué d'après son Organisation* . . . , 4 vols. (Paris, 1817) vol. 1 p. xiv. For Linnaeus see J. L. Larson, *Reason and Experience: the Representation of Natural Order in the Work of Carl von Linne* (Berkeley, California, 1971). For a general account of classification, see D. M. Knight, *Ordering the World: A History of Classifying Man* (London, 1981). A valuable essay is Louis Agassiz, *Essay on Classification* (London, 1859; reprinted Cambridge, Mass., 1962, ed. E. Lurie).

[138] Swainson in Richardson and Swainson, *Fauna*, p. xlii.

[139] Ibid., p. xlviii.                        [140] Agassiz, *Essay on Classification*, p. 242.

species and subspecies, an inclination that Richardson urged him to curb, because he felt that his name had already received more than adequate exposure.[141] Richardson argued that Swainson's name should be attached to species that the latter had identified as new, and there were plenty of these.[142] Swainson was indeed inclined to find species where others were to find only subspecies at most. He proposed, for example, to name a specimen of the blue grouse, an inhabitant of the western mountains, *tetrao richardsonii*; Richardson noted that, in spite of minor differences, he believed that it was the species that Bonaparte had already called *tetrao obscurus*, the dusky grouse.[143] Richardson was right, as he was in several other cases. One that did slip through the net was *lestris richardsonii*, Richardson's jaeger, which so appears in the *Fauna*, but which is in fact a dark-phase parasitic jaeger.[144] Richardson had a hard time with the jaegers, and it is worth seeing just how thoroughly he took his task. On 2 May 1831 he wrote to Swainson:

I have also, to avoid delay, been under the necessity of sending the manuscripts of the *Lestri* to the printer without comparing *Lestris Richardsonii* again with the British Museum specimens. I found it necessary to make some observations on *L. parasitica,* its character coming nearer to *L. Buffonii* of Bonap[arte] than to his *parasitica* but approaching to *L. parasitica* of Temminck. . . . It does not, however, agree perfectly with either and there are Spitzbergen specimens of a Lestris which have some of the characters of *L. parasitica* of Bonap[arte] but want the others. The Spitzbergen bird is clearly distinct from the more common *L. parasitica* the transverse diameter of its bill, at the front being 7/12 of an inch, while that of *L. parasitica* is only 4/12. . . . In your remark on *L. Richardsonii* you say the feet & toes of Edwards bird are *all* yellow (pl. 149) but in plate 148 which is the one quoted by Bonaparte the tarsi are yellow & the *toes* black.[145]

Richardson may have left to Swainson the final word on the identification and naming of new species, but even here his own contribution was clearly major. Swainson's identifications were enthusiastic. He was no lumper of several species into one, but rather a splitter of species into more species, in a debate common to other branches of natural history. Hooker complained that

[141] Richardson to Swainson, Portland Place, 5 Oct. 1830, MS Linnean Society, Swainson correspondence.
[142] Ibid., Richardson to Swainson, London, 13 April 1831.
[143] Ibid., Richardson to Swainson, 24 Oct. 1830.
[144] Richardson and Swainson, *Fauna,* pt. 2 pp. 433–4.
[145] Linnean Society, Swainson correspondence. C. L. J. L. Bonaparte, *American Ornithology, or, the Natural History of Birds Inhabiting the United States,* 4 vols. (Philadelphia, 1825–33).

modern Botanists are carrying their ideas of division & subdivision to a most unwarrantable length, and that they are thereby doing an injury rather than a benefit to science & are deterring many from undertaking the study altogether. . . . How the *Ladies* are to comprehend such nice distinctions I cannot conceive. . . . You and I may set our faces as we will against such these species mongers; but tis to no purpose.[146]

Swainson, although responsible for less of the volume than his name on the title page suggests, took the problem of speciation seriously, comparing Richardson's specimens with those from earlier expeditions, and corresponding with other ornithologists, including John Gould,[147] about their American birds. But it was Richardson who most consistently noted discrepancies between his specimens and observations and those of earlier naturalists, especially the French.[148]

Richardson was right to be critically aware of the work of other ornithologists, both in the matter of accuracy, and by way of competition. This was the heroic age of American and British North American ornithology. Lucien Bonaparte had published three of the four volumes of his *American Ornithology;*[149] at the same time, John James Audubon had begun to issue his *Birds of America.* Richardson and Swainson's volume was published in 1831, the same year as a new edition of Alexander Wilson's *American Birds,* as well as the first volume of Audubon's *Ornithological Biography.*[150] Audubon, as Cuvier first remarked, was not a scientist, but an outstanding field observer and artist.[151] He knew a great deal, but was concerned both for his priority and for his sales. When Richardson, preparing a report on North American zoology for the 1836 meeting of the British Association for the Advancement of Science, wrote to him for information about bird distribution in the United States, Audubon declined to send any, since he was nearly ready to publish the results of more than thirty years'

[146]  Hooker to N. Winch, Glasgow, 28 Jan. 1831, MS Linnean Society, Winch correspondence. Hooker's comment on the ladies reflects the vogue for amateur botanizing.

[147]  Gould to Swainson, 12 Dec. 1830, Linnean Society, Swainson correspondence. Gould was first and foremost an outstanding illustrator, and he had a scientist's as well as an artist's eye; he would later confirm that Darwin's finches from the Galapagos were of different species, and not merely different varieties.

[148]  E. g., Richardson to Swainson, 13 March and 22 March 1831, Linnean Society, Swainson correspondence.

[149]  See note 142.

[150]  J. J. Audubon, *Birds of America* (London, 1828 et seq.); *Ornithological Biography,* W. Macgillivray, ed., 5 vols. (Edinburgh, 1831–9). Alexander Wilson, *American Birds; Or, the Natural History of the Birds of the United States,* 4 vols. (Edinburgh, 1831).

[151]  G. Cuvier, *Journal des Savants* (1833) 706, quoted in P. L. Farber, *The Emergence of Ornithology as a Scientific Discipline: 1760–1830* (Dordrecht, 1982) p. 106.

work.[152] Jameson wrote encouragingly to Richardson, urging him to send sheets in advance of publication, so that he could give proper notice of his and Swainson's work, alongside that of Wilson and Audubon.[153] In the same month, May 1836, James Wilson, a Scottish naturalist, wrote a review of the first volume of Richardson's *Fauna* in the *Edinburgh Review*, speaking handsomely of it, but giving little information beyond what could be found in the introduction to the volume.[154]

With the birds out of the way, Richardson moved on to the fishes for the third volume of the *Fauna*, while Kirby went ahead with the insects. Swainson wanted Richardson to follow *his* arrangement of fishes, but Richardson, by now more than a little tired of Swainson's self-important enthusiasm for a taxonomy that did not meet with universal acclaim, made his excuses to Swainson, explained that it was not possible at that stage, and bade him farewell.[155]

Swainson's theoretical extravagance contrasted with the modest thoroughness of Hooker in his work on the *Flora*. The work, when it appeared in 1840, was dedicated to Franklin and Richardson. The English part of the title, *The Botany of the Northern Parts of British North America*, was adopted to match that of the *Fauna*, but it was misleading, since Hooker had covered a wider canvas by presenting "the vegetation of all that portion of North America Proper, which, commencing with the extreme Arctic Islands, stretches south to the boundary, so far as it has been ascertained, of the United States and California"; it thus encompassed territories belonging to the Russian Empire, or claimed by the United States.[156] Where

[152] Audubon to Richardson, London, 31 May 1836, SPRI MS 1503/14/9, Voss collection. For Richardson's report to the BAAS, see note 155.

[153] Richardson to Swainson, [20] June 1831, Linnean Society, Swainson correspondence.

[154] [James Wilson], "Geographical Distribution of Animals," *Edinburgh Review* 52 (1831) 328–60; Richardson to Swainson, 29 June 1831, Linnean Society, Swainson correspondence.

[155] Richardson to Swainson, Gosport, 15 Feb. 1840, Linnean Society, Swainson correspondence. Richardson was an informed conservative in taxonomy, and, with Charles Darwin and others, was a member of the committee set up by the British Association for the Advancement of Science in 1842 "to draw up a series of rules with a view to establishing the nomenclature of zoology on a uniform and permanent basis." [Quoted in W. Jardine, *Memoirs of Hugh Edwin Strickland* (1858) p. cxciii.] The report, published in 1842, defeated the nomenclature radicals, of whom Swainson was one, and checked the constant changing of names in zoology. I am grateful to Gordon McOuat for his paper "Ownership of Names: A Fight over Zoological Nomenclature in the Early Nineteenth Century," read to the Canadian Society for the History and Philosophy of Science at Kingston, Ontario, 1991: he alerted me to this dimension of Richardson's exchange with Swainson.

[156] Hooker, *Flora*, p. iv.

Richardson or Drummond had been, the plant life was as well known as anywhere in British North America. Hooker found the vegetation along the southern reaches of their travels especially interesting, since it had many plants in common with the regions of the Mississippi and Missouri rivers. The result was a remarkably comprehensive volume, excluding, for reasons of space and time, ferns and cryptogams, but giving a splendid account of the nature and distribution of arctic plants, from the forests to the tree line where willows and alders yield to dwarf birches and shrubby willows only inches high. Sedges, grasses, and heaths predominate on the tundra, and showier flowers of the far North include arctic avens, saxifrages, buttercups, and poppies. Hooker's volume describes some 5,000 plants in an ambitious, encyclopedic, and eminently successful work. Richardson and Drummond, as well as Douglas and the naturalists on both earlier and later expeditions, had furnished the material, and Hooker had fully justified their collecting, as he said in his dedication, "under circumstances of singular difficulty, hardship, and danger."

The gains to natural history from the recent expeditions had been substantial. But that was hardly the navy's first priority. As Richardson had pointed out, the Northwest Passage had been almost although not quite completed, and shown to be of no immediate economic significance. The treaty with Russia meant that the arctic coast was militarily less sensitive than it had been a few years earlier. All in all, the arguments for naval involvement were no longer urgent. In 1830 Sir James Graham, a reform politician with no naval experience, was made first sea lord, and support for arctic science seemed less rather than more likely to be forthcoming. Richardson, who since returning with Franklin had spent a good deal of time inventing proposals for further natural history expeditions to the far North, was thoroughly discouraged. "I have given up every intention of going out to Hudson's Bay again," he wrote to Robert Jameson in Edinburgh,

& I do not think that the present ministry or at least the Admiralty will encourage scientific expeditions to any quarter. Whenever we have had a purely naval admiralty they have always been noted for a want of attention to science and Sir James Graham in all practical points yields entirely to the opinions of the Naval Lords. The head of the Colonial office is very differently disposed but the want of money and still more the dread of what the people will say of them deters them from promoting any enterprise that is likely to cost them any thing.[157]

---

[157]   Richardson to Jameson, Chatham, 29 Aug. 1832, Edinburgh University Library MS Gen. 129 f.192.

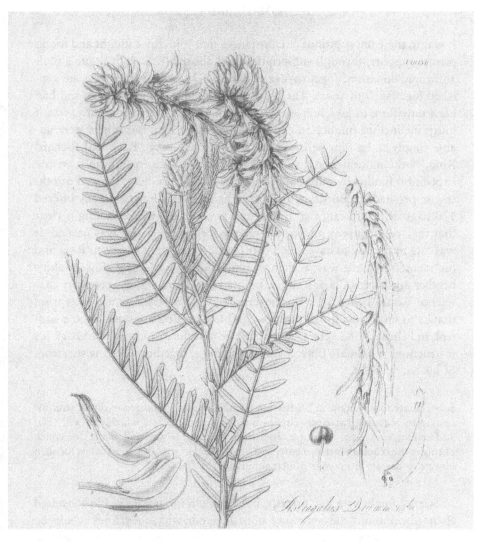

Figure 19. *Astragalus Drummondii,* in W. J. Hooker (1840). Fisher Library. University of Toronto.

Parsimony was one principle of policy. Besides, there were always other theaters for exploration or military action; the Arctic was only one theater in Britain's expanding world role, and active service brought faster promotion than exploration. The Admiralty was willing to ease up in the Arctic.

## GEORGE BACK

It was in these unpropitious circumstances that John Ross sought and found private support, through subscription and above all through Booth's philanthropic donation, for a voyage north in 1829. As we have seen, he vanished for a spell of years. The navy had little love for John Ross, and had been unwilling to give him an Arctic commission; nor, apparently, was it much inclined to finance an expedition to rescue him. But others were unable simply to let him perish. After four years, George Back and Richard King,[158] the naturalist and surgeon, set out to search for him in a private expedition funded by public subscription. The government, shamed or otherwise persuaded, contributed £2,000; the Hudson's Bay Company offered 120 bags of pemmican, two boats, and two canoes, while making it clear that they could supply and would charge for whatever else was needed. It was late in the day to have much confidence in bringing succor to Ross and his men; but there was a chance, and besides, as George Ross, John's brother and James Clark's father, pointed out, there was the certainty that science would benefit.[159] The government had a stake in the enterprise, thanks to its financial contribution, and felt entitled to exercise some control. In February 1833 the Colonial Office (not the Admiralty) sent Back his instructions. The main object was to find Ross, together with any survivors of his party. Then,

Subordinate to your object [of] finding Captain Ross . . . you are to direct your attention to Mapping what yet remains unknown of the coasts which you will visit, and making such other scientific observations as your leisure will admit; for which purposes the requisite Instru[ments] will be supplied to you. But you are not for such objects to deviate from your principal pursuit. . . .[160]

Science might be a low priority, but Back did not ignore it. He consulted Richardson about the necessary impedimenta: what medicines would be needed "for three years for 20 [men]," what about powder for preserving

---

[158]   Alan Cooke, "RICHARD KING," *Dictionary of Canadian Biography vol. X 1871 to 1880*, M. la Terreur, ed. (Toronto and Quebec, 1972) pp. 406–8; H. N. Wallace, *The Navy, the Company, and Richard King: British Exploration in the Canadian Arctic, 1829–1860* (Montreal, 1980).

[159]   George Ross to Richardson, London, 23 April 1833, Voss Collection, SPRI MS 1503/10/1.

[160]   Instructions from Colonial Office to Back, 4 Feb. 1833, SPRI MS 395/79/1.

animal and bird skins, what sort of geological hammer did Richardson recommend? Instruments and guns were already ordered by December 1832.[161]

Back and King sailed to New York, made their way to Great Slave Lake, where Fort Reliance was built as winter quarters, and where in the following spring they learned of Ross's return to England. Nonetheless, they decided to stay on, descended the Great Fish (now Back) River to its mouth, explored the arctic coast for three weeks, and returned to Fort Reliance where they spent a second winter. Back returned via New York in 1835, King followed later with the crew and gear, having, as expedition naturalist, made useful observations, especially in zoology and meteorology.[162] Back observed that King deserved all the more credit in view of the difficulties under which he had labored: "We were without the kind of shot calculated for killing small birds, inconvenienced by want of room in our single boat, and assailed by almost constant rain, while the barren grounds afforded little beyond moss for fuel."[163] Richardson wrote the "Zoological Remarks," with descriptions of species and thoughtful remarks about geographical distribution. The new ground explored by Back was composed for the most part of primitive (Precambrian) rock. The geological specimens and notes that he brought back were extensive and valuable, and were incorporated often verbatim into Fitton's "Geological Notice."[164]

Back took the main responsibility for auroral and magnetic observations. His auroral observations were the most detailed that I have seen to that date.[165] On 7 December 1833, he wrote to Beaufort from Fort Reliance that "I have a very superior observatory and the needle is performing with diurnal regularity" – a somewhat unusual state of affairs.[166] Back's work was a valuable addition to scientific knowledge of the Arctic, in what was now a well-established mold.

[161]  Back to Richardson, 19 Dec. 1832, SPRI MS 1503/9/24.
[162]  G. Back, *Narrative of the Arctic Land Expedition to the Mouth of the Great Fish River, and along the shores of the Arctic Ocean, in the Years 1833, 1834, and 1835* (London, 1836), with botanical and meteorological appendixes by King; Cooke and Holland, *The Exploration of Northern Canada*, p. 160; H. N. Wallace, *The Navy, the Company, and Richard King*, pp. 15–30.
[163]  Back, *Narrative*, p. 475.
[164]  W. H. Fitton to Richardson, 26 March 1836, SPRI MS 1503/14/6. Back's "Geological Notices," misdated, are in SPRI MS 395/94.
[165]  Back, Observations on the Aurora Borealis 1833–5, McCord Museum, McGill University, MS M2634, 40 leaves.
[166]  Back's Letterbook 12 April 1833–28, April 1834, McCord Museum MS M2635.

Following Back's return, there was the by now habitual lobbying of the Admiralty for yet another arctic expedition. Richardson presented Beaufort with reasons and plans for completion of the Northwest Passage by boats, in an expedition that he offered to lead.[167] The Royal Geographical Society made its recommendations to the Admiralty for further arctic exploration, using Franklin's arguments. There were really no significant new arguments, and the Admiralty at first said no;[168] they then reserved their decision, perhaps uncharacteristically wilting under the weight of lobbying, and the expedition that Franklin had proposed was finally approved, with Back, Franklin's choice as leader, in command.

The expedition was intended to complete the survey of the north coast of America, between the west end of Fury and Hecla Strait and Turnagain Point. Geographical exploration came first. "Nevertheless," Back was instructed,

on the return of the parties when they can estimate what time they have to spare, and at every nightly station, every adverse gale or impervious fog will afford opportunities for observing the magnetic dip and intensity and for encouraging a variety of valuable researches in other branches of science, . . . the opportunities for effecting which you will no doubt discreetly employ.[169]

The expedition was meant to go out, and if possible to return before the winter. Only in the event of an enforced over-wintering would science come to the fore; in that case, "you will continually and sedulously employ all the scientific means at your disposal in rendering your long winter as beneficial as circumstances will permit."[170] Back *was* caught in the ice, and spent the winter drifting until released in the following July near the west end of Hudson Strait. With a badly damaged boat, he headed for home.[171]

Here, if the Admiralty needed it, was definitive discouragement from the support of further arctic exploration: cost and risk were disproportionate to geographical and other scientific gain, and the military advantage was invisible. Sir John Franklin, senior arctic explorer, F.R.S., and persuasive advocate for northern exploration, had been sent symbolically south in the summer of 1836, while Back was heading north. Franklin became lieutenant governor of Van Diemen's Land, and there established the Royal Society of Hobart Town. He interested himself in antarctic voyages, including that of Dumont d'Urville, and entertained his old friend James Clark Ross on

[167]   Richardson to Beaufort, 6 Feb. 1836, SPRI MS 1020/3.
[168]   PRO CO 6/18 ff. 170–7, 180.
[169]   Instructions to Back, accompanying letter dated 12 June 1836, SPRI MS 395/86.
[170]   Ibid.
[171]   Cooke and Holland, *The Exploration of Northern Canada*, p. 163.

his way to the Antarctic.[172] Franklin remained away until 1844, in constant scientific correspondence with John Richardson, his friend and companion in exploration.[173]

In Franklin's absence, the Royal Navy was inactive in northern exploration. The Hudson's Bay Company was the principal agent in such work, and, although they contributed much to geographical knowledge, they came only gradually to support the broader advancement of science.[174]

Meanwhile, the Admiralty had achieved a clear notion of the kind of science it could support, and the Royal Society had an equally clear and complementary notion of what it could expect from the navy, including a willingness to allow naval officers to undertake supplementary scientific observations, and a reluctance to give much space to civilian observers. The Admiralty and the Royal Society between them defined naval science, and within it, arctic science. Their cooperation in that definition, and the scientific context in which it developed, is the subject of the next chapter.

[172] Franklin to J. C. Ross, 18 October 1840, SPRI MS 248/316/2.
[173] *Dictionary of National Biography*, art. John Franklin; A. H. Markham, *Life of Sir John Franklin and the North-West Passage* (London, 1891); Richardson–Franklin letters, Royal Botanic Gardens, Kew.
[174] S. Zeller, "The Spirit of Bacon: Science and Self-Perception in the Hudson's Bay Company, 1830–1870," *Scientia Canadensis* 13 (1989) 79–101. The only major scientific expedition to the north carried out during Franklin's antipodean absence was John Henry Lefroy's magnetic survey: see chapter 4.

# 4

Mid-century: The *Admiralty Manual* and the State
of Arctic Science

By the time that Franklin sailed to Australia, it had become clear to any
rational observer that the principal arguments in favor of continued arctic
exploration were scientific; that military and political considerations had
less weight, now that the Russian question appeared to have been resolved,
and the American question had not yet arisen in the Arctic; and that commercial arguments were unconvincing. A viable route to the Far East was
not conceivable until the most recent developments in icebreaker technology allowed passage. Commercial advantage did come sooner, as the Americans were to show, through the short-lived expansion of the whale fishery,
both in the western Arctic and northward to the west of Greenland. Benefits to the sciences, however, could accrue from the outset, as Sabine and
Foster demonstrated on Parry's first and third expeditions respectively.
Sabine, then a captain in the Royal Artillery, received the Copley Medal of
the Royal Society – its highest award – for his geophysical work, and Foster, a naval man, received the same award for his astronomical work. The
close connection between science and the navy was disapprovingly noted by
Charles Babbage in his polemic, *On the Decline of Science in England, and
on its Causes* (London, 1830), in which he deplored the lack of professionalism in science, and the undue recognition given to descriptive work performed by relative amateurs.[1] Richardson's natural history, although
disciplined and informed, was the kind of achievement that Babbage ranked
as a decided second to the experimental and theoretical work of natural

---

[1] This, and the following paragraphs, rehearse the argument David Knight gave in his introduction to the *Admiralty Manual of Scientific Enquiry: Prepared for the Use of Officers
in Her Majesty's Navy; and Travellers in General*, J. F. W. Herschel, ed., 2nd. ed. (London,
1851; reprinted by Dawson, Folkestone, 1974), from which otherwise unacknowledged
biographical information about contributors to the *Manual* is taken.

philosophers like Michael Faraday or John Herschel. Faraday was rapidly making the fields of electricity and magnetism his own. Herschel was the author of an authoritative study on scientific method, a versatile experimentalist, and the scientists' rather than the socialites' candidate for the presidency of the Royal Society. But as we have seen, Richardson's work, for example on lichens, was disciplined, informed, and accurate; we could anachronistically but well call it professional. Babbage clearly valued the physical sciences above natural history, a partial and unreasonable judgment. Geography, the branch of the sciences to which naval officers contributed most, was more akin to natural history than to physics. Babbage could scarcely have been delighted at the stream of arctic and other naval officers elected to the Royal Society, and he must have deplored Franklin's election to the council of that body.[2]

In spite of Babbage, the connection between the sciences and the navy remained strong, and the Royal Society regularly advised the Admiralty on scientific aspects of naval expeditions. A similar connection existed between the sciences and the ordnance of the army, the Royal Artillery and the Royal Engineers. At Woolwich, scientific instruction was more systematic than in the navy, where apprenticeship and learning on the job, or rather on the voyage, were the order of the day. As David Knight has pointed out, "Flinders had sailed with Cook; Franklin had sailed with Flinders; and Beechey, a contributor to the *Admiralty Manual,* had learnt his surveying on a voyage with Franklin."[3] Geophysical and astronomical work required teamwork, regular observations, sophisticated apparatus, and often an observatory. Even natural history, in the hands of an energetic naturalist who was often also the ship's surgeon, was enriched by the collections made by other officers and crew members. Given the opportunities of collecting specimens in remote seas and lands, it is unsurprising that most of the best naturalists in nineteenth-century Britain had sailed on an expedition,[4] either as naval officers or under naval direction. This was true of Thomas Henry Huxley and Joseph Dalton Hooker. Huxley, Darwin's bulldog, was

---

[2]  On the reform movement in the Royal Society, with Herschel at its center, see M. B. Hall, *All Scientists Now* (Cambridge, 1984) pp. 63–91; for the institutional background in science in the 1830s, see J. Morrell and A. Thackray, *Gentlemen of Science: Early Years of the British Association for the Advancement of Science* (Oxford and New York, 1971). A broader survey for Britain in the nineteenth century is D. S. L. Cardwell, *The Organisation of Science in England* revised ed. (London, 1972). Herschel's book on scientific method was his *A Preliminary Discourse on the Study of Natural Philosophy* (London, 1831; reprinted Chicago, 1987).

[3]  Ibid., Introduction p. xiii.

[4]  D. M. Knight, *The Nature of Science: The History of Science in Western Culture since 1600* (London, 1976) p. 147.

a pupil of John Richardson during the latter's tenure as physician of the Royal Hospital at Haslar; Hooker, botanist, Darwin's ally and future president of the Royal Society, was another of Richardson's protégés. Even Charles Darwin got his start on a naval surveying voyage to South America.

Naval officers were not expected by the Admiralty to make theoretical discoveries, but they were expected to be accurate observers and reporters, and to be familiar with the instruments and calculations needed for navigation. With such skills, they could be instructed as observers in new branches of science, such as geomagnetism. The Lords Commissioners of the Admiralty were persuaded "that it would be to the honour and advantage of the Navy, and conduce to the general interests of Science, if new facilities and encouragement were given to the collection of information upon scientific subjects by the officers, and more particularly by the medical officers, of Her Majesty's Navy, when upon foreign service."[5] The emphasis was upon the inventory sciences, based upon the collection of specimens and the tabulation of data. The Admiralty clearly believed that such an enterprise was of intrinsic scientific worth; so did the Royal Society, judging by the number of naval officers it elected as fellows. This agreement between the two institutions was the foundation and inspiration for the *Admiralty Manual*, and so reading it tells us something about the state of mid-nineteenth-century science and about the Royal Society, as well as about the navy.

Shortly before the first edition of the *Manual* was published in 1849, James Clark Ross had sailed for the Antarctic, with detailed instructions based upon his and others' arctic experience.[6] That experience, as well as the scientific expertise of Herschel and his coauthors, helped to shape the *Manual*. The aim was to provide a guide for nonspecialists, without invoking elaborate training or apparatus. The navy had long shown a reluctance to entrust major responsibilities to civilians, and did not plan regularly to carry civilian scientists on voyages. They wanted a volume that would be useful to naval officers in different regions of the globe, that would meet the needs of the Admiralty and also be useful to scientists. They and the authors of the *Manual* were effectively defining the relationship between the navy and the sciences, so that both could benefit. The sciences could be appended to such naval work as search and geographical discovery, or rather, geographical science could be amplified by other branches of natural science.

---

[5]　*Admiralty Manual*, Memorandum p. iii.

[6]　J. C. Ross, *A Voyage of Discovery and Research in the Southern and Antarctic Regions, During the Years 1839–43*, 2 vols. (London, 1847; reprinted Newton Abbot, 1969); M. J. Ross, *Ross in the Antarctic* (Whitby, 1982).

Besides the exact sciences, their Lordships noted that they would look, "in many instances, for Reports upon National Character and Customs, Religious Ceremonies, Agriculture and Mechanical Arts, Language, Navigation, Medicine, Tokens of value, and other subjects; but for these only very general instructions can be given . . . "[7] This was not vagueness, but a recognition of a lack of a generally accepted methodological coherence in the emergent disciplines of ethnology and anthropology.

The *Manual* that Herschel produced for the Admiralty was organized in a sequence, suggesting a hierarchy that began with astronomy and geophysics, sciences crucial for navigation, and moved on to sciences of the sea (hydrography and oceanography), which were still very much the navy's preserve. Then came the natural history sciences, including geology, which Banks had dominated on Cook's first expedition, but which in the nineteenth century were often entrusted to the ship's surgeon.[8] Finally came the human sciences, least developed and very much emergent through the nineteenth century. In the following pages, we shall adopt that sequence, while considering only those aspects of the *Manual* that had a bearing on the navy's practice of science in the Arctic. But the sciences pursued in the Arctic inevitably went beyond the *Manual,* so we must do so too. This is especially although not exclusively true of magnetism, natural history (including geology), and zoology. Most of the sciences presented in the *Manual* had been enriched by the experiences of almost a quarter of a century's exploration. Drawing on those experiences, the volume would serve as a guide and foundation for the next half-century's arctic work. This chapter, like the book and the sciences it describes, thus has a pivotal role in the overall narrative. It incorporates discoveries, and sets out directions for future research; it takes stock and looks forward. Alongside summary and description, it offers comment, establishing context and exploring implications.

### ASTRONOMICAL AND GEOPHYSICAL SCIENCES

*Astronomy.* Astronomy, queen of the sciences, came first in the volume, and first in prestige in Britain, with its Royal Observatory dating from the

[7] *Admiralty Manual,* p. iv.
[8] Charles Darwin was *a* naturalist on H.M.S. *Beagle* under Captain Robert Fitzroy, but the surgeon on that voyage, Robert McCormick, who had sailed with James Ross to the Antarctic, regarded himself as *the* naturalist. McCormick's diaries are in the Wellcome Institute, London: he sailed in 1852–4 in H.M.S. *North Star* in search of Franklin. See his *Voyages of Discovery in the Arctic and Antarctic Seas and Round the World* (London, 1884).

seventeenth century, and the association with Isaac Newton's *Principia* to give it authority. The science was presented in the *Manual* by Sir George Airy, Astronomer Royal since 1835. Because he remained in this position until 1881, he became a major figure in the patronage and politics of nineteenth-century science. Airy's first consideration was the advancement of general astronomy: "observation of the places of comets or other extraordinary bodies," using a sextant to locate the comet in relation to three conspicuous stars; observing eclipses of the sun from harbors of known latitude and longitude; occultations of stars by the moon and eclipses of the satellites of Jupiter; investigating the effects of humidity on atmospheric refraction; observations of the brightness of stars of variable magnitude in the southern hemisphere; and other observations that might be favored by time and place. In all these he stressed that "a bad observation, or an observation which is given without the means of verification, is worse than no observation at all" (*Manual,* pp. 1–2). This was a warning of very general validity, and its neglect vitiated much otherwise disciplined observation.[9]

Nautical astronomy he dismissed relatively briefly. "So much attention has been given to every detail of Nautical Astronomy, that it is very difficult to fix upon any part of it to which the attention of navigators should be especially directed with a view to its improvement" (pp. 7–8). But he did fix on the problem of using a sextant to observe the altitude of stars at night, and of the sun and moon when the horizon was ill-defined. It was important to learn the use of a good artificial horizon such as Becker's.[10] He also stressed the importance of learning to observe occultations of stars by the moon, and eclipses of Jupiter's satellites when at sea:

Occultations occur rarely, but the result which they give for longitude is usually so much more accurate than that given by lunar distances, that, in long voyages where little dependence can be placed on the chronometer, the observation of an occultation must be extremely valuable. The eclipses of Jupiter's satellites afford less accurate determinations of longitude, but they occur very much more frequently, and may be very useful where chronometers cannot be trusted.

---

9   This was a problem in other areas of science, for example in magnetic observations when a Fox dip circle was broken and repaired by the ship's armorer without being recalibrated during the British Arctic Expedition of 1875–6: see chap. 7.

10  Christopher Becker was working in Arnhem around the time of the *Manual*'s publication; he later founded Becker and Sons in Brooklyn, New York (Turner, *Nineteenth-Century Scientific Instruments,* [Berkeley and Los Angeles, 1984] p. 251). Natural horizons were often hazy, making surveying and navigation difficult. A mercury trough provided one of the simplest kinds of artificial horizons (thanks to the horizontal plane of the surface of the mercury), and there were numerous more complicated devices.

Airy seems to have had an excessive suspicion of chronometers, fostered by Greenwich tradition, which favored the observation of lunar distances and the use of the *Nautical Almanack*.[11] By the mid-nineteenth century, however, chronometers were pretty dependable, and had indeed been so since Captain Cook's day. The difference between time by the chronometer, and local time as observed by the sun or other celestial bodies, corresponded to the difference in longitude between the place where the chronometers had been set[12] and the place whose longitude was being determined. One hour's difference meant fifteen degrees' difference in longitude, and in the nineteenth-century Northwest Passage expeditions, positions on land were reliably noted to less than one minute of arc, that is, to less than one-sixtieth part of a degree. The slight errors that might build up after a journey of months over thousands of miles could be rectified by careful astronomical observations on land, with a transit instrument to record the moment at which a given celestial body reached the meridian. The best chronometers, for example those made by Parkinson and Frodsham, could be trusted, and this was doubly ensured by the practice of taking several chronometers on expeditions, and of checking their accuracy against stellar tables.

The more complicated method of "lunars" for finding longitude at sea had been developed by Tobias Mayer, who received less credit and less reward for it than he deserved.[13] It depended for its accuracy upon the reliability of tables of the moon's position, determined by the angular distance apart of the moon from other heavenly bodies.[14] Mayer's method was not the best for use at sea, but it worked well on shore. His observations had been worked up by the splendid mathematician Euler, and incorporated in the tables of the *Nautical Almanack* by Maskelyne.[15] Airy instructed his naval readership:

[11] See chap. 2 notes 68 and 124 above, and D. Howse, *Nevil Maskelyne: The Seaman's Astronomer* (Cambridge, 1989), chap. 9.

[12] For the British, since the late seventeenth century, this was the Greenwich Observatory; the French, for about the same length of time, had the Paris Observatory.

[13] For information on Mayer, see the writings of Eric Forbes, including "Tobias Mayer's lunar tables [1755]," *Annals of Science* 22 (1966) 105–16, and *Tobias Mayer (1723–62), Pioneer of Enlightened Science in Germany* (Göttingen, 1980).

[14] The moon's orbit had given Newton a headache in his work on the *Principia*. In fact, it was supposedly his biggest problem in preparing his *Mathematical Principles of Natural Philosophy*, and it was one of the reasons necessitating a second edition.

[15] See chap. 1. E. G. R. Taylor tells us (in *The Mathematical Practitioners of Hanoverian England* [Cambridge, 1966] p. 268) that in "the first issue of the *Nautical Almanac* Maskelyne wrote: 'The Tables of the Moon had been brought by the late Professor Mayer of Gottingen to a sufficient exactness to determine the Longitude at Sea within a Degree, as appeared by the Trials of several Persons who made use of them. The Difficulty and

One or two stars at least, as near the pole as possible, should be observed every night, in addition to the Nautical Almanac stars necessary for chronometer error, and the moon-culminating stars[16] which are observed with the moon. The instrument should be reversed[17] on alternate nights; and, if possible, as many transits of the moon should be taken after the full moon as before the full moon. (p. 10)

The importance of accurately pinpointing position is obvious. In the Arctic, navigation by the sun, or in winter by the moon and stars, was all the more crucial because of the unpredictability of the compass needle, which was ever less reliable as one neared the magnetic pole.

*Terrestrial magnetism.* The very real problems associated with navigation by the compass, reinforced by the theoretical and potential practical importance of polar magnetic studies, had made such studies a regular feature of naval arctic expeditions, from the earliest work of Sabine and James Clark Ross. But although Britain took the initial lead in magnetic investigations in the Canadian Arctic, it did not function as the scientific leader in the field. That role was taken by German physicists, mathematicians, ˜ d instrument makers.

In 1834, C. F. Gauss and the physicist Wilhelm Weber established their *Magnetische Verein;* by the following year, there was a network of eighteen geomagnetic observatories in northern Europe and Russia, and, with the participation of French and British geophysical observers, the number had grown to fifty by mid-century.[18] The model for all these observatories, in organization, theory, and instrumentation, was based on the example of Gauss and Weber.[19] British participation owed much to the energy and enterprise of Edward Sabine, working with John Herschel, James Clark Ross, and Professor Humphrey Lloyd of Dublin, of whom we will learn more in this chapter.[20] Given the theoretical and practical significance of the field,

---

Length of the necessary Calculations seemed the only Obstacles to hinder them from becoming of general Use.' " See also Howse, *Nevil Maskelyne.*

[16]  I.e., stars that reach their greatest elevation with the moon.

[17]  Errors in measurement arising from asymmetry in the instrument, e. g., of the telescope in relation to its axis or support, or, in a magnetic instrument, in the needle or its support, could be largely eliminated by reversing the instrument between observations, and taking the mean.

[18]  J. Cawood, "Terrestrial Magnetism and the Development of International Collaboration in the Early Nineteenth Century," *Annals of Science 34* (1977) 551–87.

[19]  See, e. g., J. G. O'Hara, "Gauss and the Royal Society: the Reception of his Ideas on Magnetism in Britain (1832–1842)," *Notes and Records of the Royal Society of London 38* (1983) 17–78, at pp. 45–50.

[20]  Cawood, "The Magnetic Crusade: Science and Politics in Early Victorian Britain," *Isis 70* (1979) 493–518.

and because the key observations could be carried out by minimally trained but highly disciplined naval officers, here was a science tailor-made for northern naval expeditions. Arctic and then antarctic naval commanders were left in no doubt that magnetic studies were scientifically the most important part of their mandate. The balance of this chapter reflects that emphasis.

In 1833, the year before the inception of the *Magnetische Verein*, the physicist Samuel Hunter Christie addressed the young British Association for the Advancement of Science on "the State of our Knowledge respecting the Magnetism of the Earth."[21] He pointed out that dip and variation were both subject to change, both daily and over a longer term. The daily changes in dip were too small to be measured "with the imperfect instruments which we possess."[22] Edmond Halley had suggested that the different direction of the magnetic needle at different places on the earth's surface might be explained by assuming that there were four magnetic poles, two of which were fixed and two moving.[23] More recently, the Norwegian physicist, astronomer, and mathematician Christian Hansteen had revived the four-pole theory, but assumed that all four poles were in motion.[24] Results calculated on the basis of Hansteen's hypothesis were in fair agreement with observation, but "as considerable uncertainty attends magnetical observations, excepting those of the variation made at fixed observatories," it was too soon to be confident about Hansteen's theory.

There was a plethora of hypotheses about the causes of geomagnetism and its changes, involving heat, electricity, chemical action, or the contact of dissimilar metals. The influence of the aurora borealis on geomagnetism was also a subject for speculation and further investigation.[25] But, independent of any causal theory, there was a need to record and map the magnetic vectors. A very recent effort had:

represented the lines of equal variation on a globe, from a great mass of the most recent documents connected with the variation, furnished to him by the Admiralty, the East India Company, and from other sources. If to the lines of equal variation were added the magnetic meridians and their normals, the isodynamic lines, with

21   Christie, *Report of the Third Meeting of the British Association for the Advancement of Science; held at Cambridge in 1833* (1834) 105–30.
22   Ibid., p. 110.      23   Ibid., p. 111; Halley, *Phil. Trans.* 13 (1683) 208.
24   C. Hansteen, *Untersuchungen über den Magnetismus der Erde* ([Oslo], 1819).
25   Using long-term observations, Sabine was able to report by 1852 that years of greatest and least magnetic disturbances coincided with years of greatest and least sunspot activity, and gradually a correlation, though less clear cut, emerged between auroral displays and sunspot activity; see A. J. Meadows and J. E. Kennedy, "The Origin of Solar-Terrestrial Studies," *Vistas in Astronomy* 25 (1982) 419–26.

those of equal dip, such a globe would form the most complete representation of facts connected with terrestrial magnetism that has ever been exhibited, and might indicate relations which have hitherto been overlooked.[26]

The production of maps of isodynamic lines was to be a major preoccupation of Sabine and his fellows, and the creation of permanent magnetic observatories supported by winter-long temporary observatories was materially to aid this enterprise. The search for a mathematical expression having the force of a physical law was more frustrating. Sabine took two formulas derived on the assumption of two magnetic poles near the center of the earth; one of them, relating force to dip, was at variance with his data, whereas the other matched those data. Christie found it difficult to conceive how "the same set of observations should be in remarkable accordance with the one formula and at variance with the other, when these formulae are dependent on each other."[27] The complexity of the phenomena had for the time being outrun the capacity of theory to handle them.

The final section of Christie's report was devoted to the problems of measuring magnetic intensity. Sabine and others had observed the vibrations of a dip needle in the plane of the magnetic meridian, but friction on the axis of the needle made it difficult to obtain enough vibrations to ensure accuracy. Gauss had a partial solution to the problem – but it was only partial, because it was limited to the horizontal intensity. He first observed the vibrations of a magnetized bar. Such bars, when suspended by a single thread, were referred to as unifilar instruments, declinometers, or magnetometers. They became standard instruments.[28] Gauss, having observed the first suspended bar, then introduced a second one, and observed at different distances the joint effects on it of the first bar and of terrestrial magnetism. By a comparison of the first with the second set of observations, he was able to obtain an absolute measure of the horizontal component of terrestrial magnetic intensity. When observatories were set up in Britain, they used Gaussian instruments.[29]

---

[26]   Christie, *Report of the Third Meeting of the BAAS*, pp. 116–17.
[27]   Ibid., p. 121.
[28]   The one developed in the observatory at Kew was known as the Kew instrument or the Kew unifilar.
[29]   J. G. O'Hara, "Gauss's Method for Measuring the Terrestrial Magnetic Force in Absolute Measure: Its Invention and Introduction in Geomagnetic Research," *Centaurus* 27 (1984) 121–47, and "Gauss and the Royal Society"; C. F. Gauss and W. Weber, "Remarks on the Construction of Magnetic Observatories and the Instruments Which They Should Contain," *Annals of Electricity* 3 (1839) 92–108, trans. from the German of 1833. Gauss's contributions were theoretical as well as practical; for example, in his *Allgemeine Theorie des Erdmagnetismus* (1838) he made use of his invention of spherical harmonic analysis for representing the distribution of a function over a spherical body – the earth.

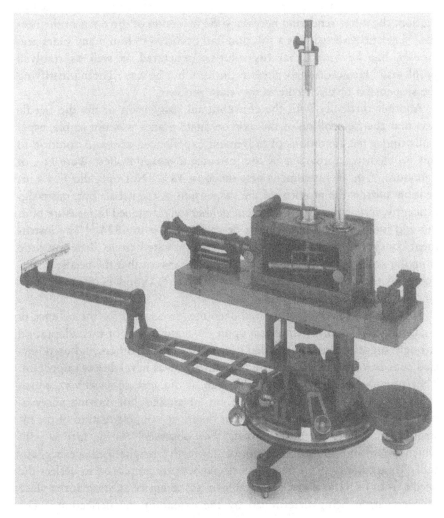

Figure 20. Kew pattern unifilar magnetometer by Thomas Jones ca. 1836. Science Museum London 1915–144, neg. no. 56/79.

There remained the problem of the vertical intensity, connected to the dip vector. Dip circles provided by the navy for the early Northwest Passage expeditions were generally of inferior construction, as Sabine had complained.[30] Christie had identified the friction of the axis as a primary cause of uncertainty in dip observations. A lesser problem arose because of the "non-coincidence of the centre of gravity of the needle with the axis of motion, the latter rendering necessary the inversion of the poles of the needle." Christie asserted that a solution had occurred to him many years previously, but he found that his solution generated as well as resolved problems.[31] He accordingly gave up the idea, but he was right in identifying the suspension of the needle as the main problem.

Another difficulty with the conventional suspension of the dip needle was that the free rolling of the axis on agate planes was not stable, especially under the conditions of movement experienced at sea. A solution to this problem was provided by the instrument maker Robert Were Fox of Falmouth,[32] in an instrument first made in 1832. Not only did Fox's invention address the problem of the suspension of the needle but, more significantly, it was the first instrument deliberately designed to measure both dip and intensity. Fox published a description of it in 1834.[33] The instrument consisted "of a dipping needle and graduated circle, differing little from the accustomed form of an Inclinometer, except that the needle is supported by the ends of the axle, which terminate in cylinders of small diameter working in jewelled holes."[34]

Fox's model for this feature had been the jewels used for the balance of a chronometer, and his emphasis upon the significance of this adaptation strongly suggests that he regarded it as novel. In fact, others had used jeweled cups before him.[35] Fox's use of such cups was nevertheless important. They provided greater stability, although at the cost of sensitivity, which was on balance, as Fox pointed out, an advantage "for moving observations" such as those taken at sea. "But the most valuable feature of the instrument, because the most wanted," Fox observed, "is the facility with which it will indicate the intensity of the earth's magnetism in every latitude. To accomplish this object, steel magnets are employed to deflect the needle from its natural dip, the greater or less intensity existing at the place of observation being determined by the extent of the deflection." The

[31]   Christie, *Phil. Trans. 123* (1833) 343–58 at 343.
[32]   Fuller information regarding Fox and Arctic science may be found in Levere, "Magnetic Instruments in the Canadian Arctic Expeditions of Franklin, Lefroy, and Nares," *Annals of Science 43* (1986) 57–76, from which passages are here reproduced by permission.
[33]   Fox, "Notice of an Instrument for Ascertaining Various Properties of Terrestrial Magnetism, and Affording a Permanent Standard Measure of its Intensity in Every Latitude," *Philosophical Magazine* 3rd series, 4 (1834) 81–8.
[34]   Sabine in *Admiralty Manual*, 2nd ed. p. 20.
[35]   A. McConnell, "Nineteenth-Century Geomagnetic Instruments and their Makers," in *Nineteenth-Century Scientific Instruments and their Makers: Papers presented at the Fourth Scientific Instrument Symposium, Amsterdam 23–26 October 1984* (Leiden and Amsterdam, 1985) pp. 29–52 at pp. 38–9.

Figure 21. Fox dip instrument by W. George. Science Museum London 1908–97, neg. no. 113/69.

instrument described and recommended by Sabine in the *Admiralty Manual* had a small grooved wheel carried on the axle, which received

a thread of unspun silk, furnished at each extremity with hooks to which small weights may be attached, for the purpose of deflecting the needle from its position of rest in the magnetic direction, in which it is in equilibrium between the opposing forces of the Earth's magnetism and the deflecting weight. The weight being constant, and the magnetism of the needle assumed to be so, the intensity of the Earth's magnetic force in different localities is inversely as the sine of the angles of deflection. (p. 20)

Using the weight meant opening the box, a procedure inadvisable under conditions of rain, wind, or snow. On days when conditions did not allow the use of deflector weights, deflector magnets were used.[36]

John Franklin knew Fox before his instrument was generally adopted. On 22 November 1834 he wrote to him[37] that he had just breakfasted with Captain Chesney of the Royal Artillery, who was about to leave on an expedition to survey the Euphrates.[38] He was to be accompanied by officers responsible for scientific observations. "This appeared to me," wrote Franklin,

such a favourable opportunity of getting careful observations made with your Instrument and of bringing it into notice that I did not hesitate in mentioning it to Capt. Chesney. He said that he should be delighted to have the opportunity of making experiments with the instrument, and would gladly take it with him for that purpose, if you approved. . . . As the instrument has not yet been in general use I fear the Government, in these economical times, would not sanction its been [*sic*] purchased at once for the use of the Expedition though they might perhaps sanction its purpose after it had been repeatedly tried – if the report were favourable.

Chesney and Fox were forthwith put in touch.

Meanwhile, Franklin was also working directly on the Admiralty through Francis Beaufort, who was now the hydrographer. On 6 December 1834, he informed Fox that "I shewed your notice of the Instrument to Capt. Beaufort – who . . . thought so highly of what he had heard & read of it, as to make him desire much to see the Instrument. He thinks the Astronomical Society a very proper place for its being sent."

There is then a gap in the correspondence; Franklin wrote to Fox again on 26 March 1836, when he informed him that the proposal he had put before the Royal Geographical Society was the one selected for submission to the government, to urge the desirability of another expedition to com-

---

[36]   For this and other portable magnetic instruments, see also A. McConnell, *Geomagnetic Instruments before 1900.*

[37]   Public Archives of Canada (PAC) MG 24 H67, containing all Franklin–Fox letters quoted or cited.

[38]   This book is about the Canadian Arctic, but it is worth bearing in mind what Chesney's job suggests: the very far flung nature of Britain's naval activity throughout the nineteenth century, which literally spanned the globe.

plete the survey of the north coast of North America. He hoped to persuade Beaufort to purchase one of Fox's instruments. He went on:

I was indeed pleased with the result of your observations in Ireland – and with the improvements you have made in the needle which I have no doubt will give more equable results. I am more convinced that the Instrument must be adopted when its comprehensive merits & uses are known, and I sincerely trust if any Expedition does go out to the North that Beaufort will yield to my solicitation and allow it to be taken. I think Capt. Back will command the expedition.

Beaufort's diary for 26 April records a session in which a party including James Clark Ross and Sabine spent "3 hours in Admy Garden trying Friend Fox's new dip$^{g.}$ needle."[39] The results must have been satisfactory, for when Back, Franklin's protégé, took command of the expedition that Franklin had lobbied for, he had a Fox circle with him on H.M.S. *Terror.* Years later in his autobiography John Henry Lefroy, then a general with many honors,[40] but in 1837 a lieutenant in the Royal Artillery, recalled the return of the *Terror* into the Chatham dockyard in November 1837. "I had some pleasant intercourse with her officers, particularly with the second lieutenant, Owen Stanley, who introduced me to an instrument I was afterwards destined to make extensive use of Fox's dip circle. He had been the first observer to employ it in Arctic regions, or indeed on any distant voyage."[41] In summer, on the outward voyage, Stanley had referred to "Mr. Fox's Admirable dipping needle," but in the winter he was to complain that "the dipping needle gave me the most trouble the weights used in ascertaining the intensity being so small & delicate as not to be easily handled with cold fingers."[42]

Christie had taken care of the dispatch of magnetic instruments[43] and subsequently reduced and commented on Back's magnetic observations.[44] It is not clear whether the Admiralty provided the Fox dip circle that Back took with him, but that instrument became a standard part of the

[39] Huntington MS FB 21.

[40] His greatest service to science was his work at Toronto's colonial magnetic observatory, and his completion of a heroic magnetic survey in the Canadian North: see note 60 below.

[41] *Autobiography of General Sir John Henry Lefroy, C.B., K.C.M.G., F.R.S., etc.,* Colonel Commandant Royal Artillery, Lady Lefroy, ed. (printed for private circulation only, n.p., n.d.) p. 26.

[42] Owen Stanley, "Journal of a Voyage for the discovery of the N.W. Passage in H.M.S. *Terror* Captn. Geo Back . . . in the Years 1836, 1837," MS in Fisher Rare Book Library, University of Toronto.

[43] Christie to Back from the Royal Military Academy, 10 June 1836, SPRI MS 395/87: [My son James] "is just returned from seeing your Instruments safe deposited at the Victualling Office Deptford: viz. Dollond's Instrument, belonging to the Admiralty, your own dipping Instr. and Hansteen's; and I trust you will receive them all safe."

[44] *Philosophical Magazine,* 3rd series, 9 (1836) 523–5, 529–30.

equipment of Royal Naval polar voyages. James Clark Ross took one on each of his ships to the Antarctic in his voyage of 1839–43, together with instructions from the Royal Society specifying that the Fox circle be taken as part of the apparatus, and stressing that geomagnetism "must be considered as, in an emphatic manner, the great scientific object of the Expedition."[45] When it came to the section on magnetic observations in the *Admiralty Manual*, Sabine not only specified the Fox as the instrument for making observations at sea to provide relative measurements of magnetic force, but stated that it had "contributed more to a knowledge of the geographical distribution of terrestrial magnetism than any other recent invention."[46]

Fox's instrument offered one solution to the problem of measuring the total magnetic intensity at any point, and thus of allowing for the vertical component of the magnetic force. It was undoubtedly the best solution on board moving vessels. A more refined solution, requiring a stable observing base – that is, an observatory on shore – was provided, on Gauss's model,[47] by Humphrey Lloyd, who had the direction of Dublin's magnetic observatory.[48] Lloyd, like Fox, had to provide a more delicate suspension for the needle. He did so by incorporating knife edges in the needle, and resting them on planes of agate; the action of weights moved by screws then brought the needle from its inclined position to a horizontal one. Lloyd reckoned that with this instrument, he could detect a change of one part in 40,000 of the total magnetic intensity.[49]

Lloyd's work in Dublin was possible because the magnetic crusade in Britain had effectively found support in the Royal Society, the British Association for the Advancement of Science, and the Admiralty. In 1838 a committee of the BAAS was appointed "to conduct the co-operation of the British Association in the system of simultaneous magnetical and meteorological observations," including the establishment of colonial observatories,

[45]   J. C. Ross, *A Voyage of Discovery* (1847) vol. I pp. xxx, xxxiii; "Instructions for the Scientific Expedition to the Antarctic Regions, prepared by the President and Council of the Royal Society," *Philosophical Magazine* 3rd series, 15 (1839) 177–241, including "Account of the Magnetical Instruments to be Employed, and of the Mode of Observation to be Adopted, in the Magnetical Observatories about to be Established by Her Majesty's Government," pp. 224–41.

[46]   *Manual*, p. 20.

[47]   See note 29 above and O'Hara, "Gauss and the Royal Society."

[48]   Lloyd, "On a New Magnetical Instrument for the Measurement of Inclination and its Changes," *Proceedings of the Royal Irish Academy* 2 (1840–4) 210–17, 226–32; see also Lloyd, *Account of the Magnetical Observatory of Dublin, and of the Instruments and Methods of Observation employed there* (Dublin, 1842).

[49]   Herschel, "Terrestrial Magnetism," *Quarterly Review* 66 (1840) 271–312 at 292.

one of which should be in Canada.[50] In 1839 they obtained government support for these observatories. In 1840, Herschel wrote an essay on terrestrial magnetism for the *Quarterly Review,* asserting that no currently flourishing branch of science was "more eminently practical in its bearings and applications."[51] Here was "a science of *observation*" requiring coordinated teamwork over long reaches of time and space. It was true that the original impetus for such work had come from Germany. England had not been the first nation to join the enterprise:

But that in the pursuit of great and worthy objects we are coldly to hold back, and wait till foreign nations shall have led the way and roused us by their example, is a doctrine which, as Englishmen, we must repudiate, and which, if acted on by all, would annihilate the principle of national support altogether. And in the case before us, we hold it by no means creditable to have allowed other nations . . . to precede us to the extent to which . . . they have done. But let that pass, since a better era is arrived.[52]

British government support for science and for arctic exploration was generally laggard throughout the nineteenth century, and the specter of foreign nations winning glory that should have been Britain's was repeatedly raised by men of science and by geographers to persuade or shame the government into supporting their aims. The magnetic crusade was here unusually successful; every point in the 1838 recommendations of the British Association was adopted. "Ships, buildings, instruments, and, what is of infinitely the most importance, officers and observers selected with care and imbued with the full spirit of their work, have been provided and appointed; while . . . every department of the public service . . . responded with alacrity to the call."[53] Sabine and Lloyd visited Göttingen and Berlin to confer with the leaders in geomagnetic observation, so as to set up schedules and methods of observation compatible with those already in place. One of the new observatories was in Toronto, and it was clear that a wide field of magnetic research existed throughout North America, "the deficiency of trustworthy magnetic observations in all that vast region being lamentable."[54]

[50] O'Hara, "Gauss and the Royal Society," 54; the members of the committee were John Herschel, Humphrey Lloyd (Professor of Natural Philosophy and superintendent of the magnetic observatory, Trinity College, Dublin), George Peacock (Cambridge mathematician and astronomer, who became Dean of Ely cathedral), Sabine, and the scientific pedagogue and polymath William Whewell.
[51] Herschel, "Terrestrial Magnetism," p. 271.     [52] Ibid., p. 295.     [53] Ibid., pp. 297–8.
[54] Ibid., p. 304; A. D. Thiessen, "The Founding of the Toronto Magnetic Observatory and the Canadian Meteorological Service," *Journal of the Royal Astronomical Society of Canada* 34 (1940) 308–48.

Sabine proposed that, besides the work in the Toronto observatory, there should be a magnetic expedition to the North-Western Territory. In 1839 he wrote to Lefroy about Canadian service; Lefroy was willing, and went briefly to Dublin to learn from Lloyd the methods of observation. He found that his stay "was not long enough for thorough mastery of our work. In fact, the visit was premature, because the Professor had not himself matured his plans; but it was in all respects agreeable."[55] After this inadequate training, Lefroy headed first for St. Helena, site of another British colonial magnetic observatory, and then, in 1842, to Canada. In 1843 he set out for the Hudson Bay territory, on a grueling magnetic survey of the Northwest. Sabine subsequently defined the motivation for the journey. Existing observations had revealed unexpectedly that

the highest isodynamic lines of the northern hemisphere were closed and irregularly elliptical curves, extending across the North American Continent nearly in a northwest and south-east direction, and having their central point of maximum of Force, approximately in 52° north latitude, and 270° east [90° W] longitude.

To confirm this previously unsuspected characteristic of the magnetic system of the globe, – to establish beyond a question so remarkable a fact in Physical Geography, – to fix within narrow limits the geographical situation of the point of maximum corresponding to a particular epoch, – to ascertain with the precision of modern instruments and methods the intensity of the magnetic Force at its point of maximum, – and to assign the form and geographical localities of the isodynamic curves adjacent to that point, – were objects which presented themselves amongst the most important desiderata. . . .[56]

Thus the main reason for Lefroy's journey was to confirm that the point of maximum total intensity was not coincident with the dip pole that James Ross had discovered, but well to the south of it.[57]

Carrying delicate instruments overland would have done them too much damage, so transportation in the canoes of the Hudson's Bay Company was the obvious solution, and Lefroy adopted it. His personal equipment was simple, even meager, but he took with him ten cases of instruments. He also had with him one assistant, a taciturn but dependable bombardier called William Henry. Together, they observed tirelessly, sometimes at five-minute intervals for twenty-four hours at a stretch. The data thus obtained, Lefroy later complained,

[55] Lefroy, *Autobiography*, p. 35. This discussion of Lefroy's work is largely taken from Levere "Magnetic Instruments," pp. 64–7.

[56] Sabine, "Contributions to Terrestrial Magnetism. – No.vii," *Phil. Trans. 136* (1846) 238.

[57] Cf. David G. Knapp, "Arctic Aspects of Geomagnetism," in V. Stefansson, ed., *Encyclopedia Arctica,* typescript in Dartmouth College Library, Stefansson Collection MS 96 (11). The work was not published.

were not of much interest, and only swelled the volume of wasted labour, for no-body that I am aware of has even tried to sift them or obtain comprehensive results. Working always has been ahead of thinking and accumulating data, of comparing and reducing them. I . . . tried at least to do my own share of this work, but it has never been noticed because the interest of the whole inquiry was largely factitious.[58]

Lefroy was bitter, partly because Sabine's envy of his subordinate's field-work led to friction between them, and the junior officer could only be the loser. He was also partly wrong; the interest was genuine enough, but the difficulties of reducing and interpreting the many volumes of data proved overwhelming throughout the nineteenth century.[59] Certainly, as Sabine admitted in the *Admiralty Manual,* the Royal Artillery's establishment at Woolwich, where the work of coordination and reduction was principally executed, was "very limited in comparison with the duties which it endeav-ours to perform"(p. 32).

In 1843, however, Lefroy was enthusiastic and dedicated. His letters home and his diary[60] present a detailed and fascinating account of a jour-ney that took him over some 6,000 miles (he thought it had been closer to 10,000) of lakes and rivers, in country still effectively owned by the Indians and the Hudson's Bay Company. From Toronto he went to Montreal, and there called on George Simpson, "the toughest looking old fellow I ever saw, built upon the Egyptian model, height two diameters. . . . He is a fel-low whom nothing will kill; I don't suppose myself able to do all that he calls practicable."[61] Simpson furnished him with a letter of introduction to the heads of the company's posts and districts, and Lefroy was to benefit from their logistical support and general cooperation, as had scientist-explorers before him. Bombardier Henry arrived soon afterward with the instruments, all in good shape except the Fox, which was "almost shaken to pieces." They joined one of the company's fur brigades at Lachine, went north and west to Lake Winnipeg, and reached Norway House, their base for northern observations. In July they were at York Factory on Hudson

---

[58] Lefroy, *Autobiography,* p. 73.

[59] As late as 1887, the British Association for the Advancement of Science was receiving re-ports on reduction: Balfour Stewart, "Third Report of the Committee . . . Appointed for the Purpose of Considering the Best Means of Comparing and Reducing Magnetic Ob-servations," Appendix III, BAAS *Annual Report* (1887) p. 327.

[60] Lefroy, *Diary of a Magnetic Survey of a Portion of the Dominion of Canada chiefly in the North-Western Territories Executed in the Years 1843–1844* (London, 1883); Lefroy's manuscript journal is in the Yale University Library, and I have used a microfilm in the PAC, M–2314.

[61] Lefroy to Younghusband, Montreal, 25 April 1843, in Lefroy, *In Search of the Magnetic North: A Soldier-Surveyor's Letters from the North-West 1843–44,* G. F. G. Stanley, ed. (Toronto, 1955) p. 6; the original manuscripts are in the PRO, Kew.

Bay, and October saw them at Fort Chipewyan on Lake Athabaska, where they celebrated Christmas and made extensive magnetic observations. From there they descended the Mackenzie as far as Fort Good Hope, at nearly the same latitude as the north shore of Great Bear Lake, which, as Lefroy noted, was just nudging the Arctic. They then retraced their route to Lower Fort Garry, and thence to the St. Lawrence in the fall.

Traveling was hard on sensitive and delicate apparatus. Light needles suffered particularly from inequalities of the axis; heavy ones were unwelcome at portages, and needles broke – Lefroy's diary for October 1843 records new Lloyd's needles – or their axles broke – the same entry records new Fox's axles. The magnetism of needles decayed, especially at first. Much attention was paid to the means of making artificial magnets, and some of the needles Lefroy used had been laid by for four years "in the hope that their magnetism might become steady."

Some instruments were more portable than others;[62] several of them were designed for observations on term-days, and for observations in case Lefroy was detained at any of the HBC posts. Term-days were days set aside for intensive observations, when the colonial observatories were all measuring dip, intensity, and declination in concert. Such days were in contrast to days of hard traveling, when few and brief observations were the most that could be accomplished. On 17 July 1843, for example, Lefroy noted in his journal (PAC M-2314):

To go on daily without observing at all, seems remiss, yet believe I am acting right in saving time, since I must return over the same ground; the loss is as to the *number* of obsns. If judged by that, every day should have one to shew, but my impression is that mere number is not the object and that it must be balanced with other observations.

There were to be many such days. Lefroy and Henry used all their instruments, breaking some, losing a couple, but returning with an impressive set of observations.

Lefroy set about reducing some of his observations. This was not usually the responsibility of field observers, whose job was simply the collection of data – the direction of the magnetic force, its absolute values measured on shore, and its relative values measured on board ship. The Fox was by mid-century the standard naval instrument for the latter observations, whereas the unifilar magnetometer served for the former. Absolute measurements were always to be preferred to relative ones, because of secular variation,

[62]    C. J. B. Riddell, *Magnetical Instructions for the Use of Portable Instruments* (London, 1840). Riddell was the first director of the Toronto magnetic observatory.

the continuous change of magnetic elements, "so that at no one spot on the surface of the globe can the intensity be assumed to remain constant, and thus to afford a secure unvarying basis" for relative measurements (p. 18).

The results of surveys completed and published included, by the time Sabine wrote his article for the *Manual,* those of most of the results of Ross's Antarctic survey, and maps of the inclination and force in British North America, based on Lefroy's observations. Surveys completed and "in progress of reduction but not yet published" included observations by Richardson and Lefroy in their Franklin search expedition.[63] Among observations now in progress, Sabine identified three to "the Arctic Polar Sea," one of them commanded by Sir John Franklin (pp. 30–1). Observations in the polar regions were most likely to illuminate and perhaps even resolve theoretical debate in geomagnetic science. But Sabine, who by now was running a magnetic empire, wanted it to be in both senses one on which the sun never set, and so he described the work very differently from Herschel or Gauss or von Humboldt as being "general and purely experimental" in character, "unconnected with hypothesis of any sort," and one therefore in which "the phenomena of all parts of the globe must be viewed in the abstract as possessing an equal importance" (p. 32). These claims about the empirical nature of geomagnetic science may have reflected Sabine's personal preference as well as his political ambitions; he was happiest as a collector of data, and as an organizer of observers. But he did find it necessary to give some mathematical account of the relation between specific observations and the magnetic vectors to which they were directed.

Sabine likewise devoted a technical appendix to a familiar problem in making magnetic observations at sea, that of the deviation caused by the ship's iron. The problem was acute, and became more so as the construction of ships moved more and more to iron; hence the publication of the *Admiralty Manual for Ascertaining and Applying the Deviation of the Compass Caused by the Iron in a Ship,* which went through seven editions by 1901.[64]

---

[63]  Lefroy and Richardson, *Magnetical and Meteorological Observations at Lake Athabaska and Fort Simpson, by Captain J. H. Lefroy . . . , and at Fort Confidence, in Great Bear Lake, by Sir John Richardson* (London, 1855).

[64]  The volume was edited by F. J. Evans and Archibald Smith (London, printed for the Hydrographic Office, 1862). Archibald derived his formulas "from the fundamental equations of M. Poisson's theory, in his 'Mémoire sur les Déviations de la Boussole produites par le Fer des Vaisseaux' " [Sabine in the *Manual,* Herschel, ed., p. 49]. See also G. B. Airy, "Accounts of Experiments on Iron-built Ships, instituted for the Purpose of Discovering a Correction for the Deviation of the Compass produced by the Iron of the Ships," *Phil. Trans. 129* (1839) 167–213; C. H. Cotter, "The Early History of Ship Magnetism: the Airy-Scoresby Controversy," *Annals of Science 34* (1977) 589–99; and G. W. Barber and A. S. Arrott, "History and Magnetics of Compass Adjusting," unpublished paper, Dept. of

*Meteorology.* Astronomy and geomagnetism, the two branches of the sciences on which navigation principally depended, were the only ones in which sailors were expected to have some mathematical proficiency. Other sciences were less mathematically informed and more essentially descriptive. Some descriptive sciences were clearly peripheral to naval enterprise, and others were central. Meteorology was prominent among the latter; knowledge of winds was crucial to sailing, and knowledge of temperatures would help in predicting the onset of the formation of sea ice. In the *Manual,* Herschel himself undertook the presentation of instructions for meteorological observations. The basic requirement was a

definite, systematic process, known as the *"keeping a meteorological register,"* which consists in noting at stated hours of every day the readings of all the meteorological instruments at command, as well as all such facts or indications of wind and weather as are susceptible of being definitely described and estimated without instrumental aid. Occasional observations apply to occasional and remarkable phenomena, and are by no means to be neglected; but *it is to the regular meteorological register, steadily and perseveringly kept throughout the whole of every voyage, that we must look for the development of the great laws of this science.* (p. 281)

Observations were to be made at least four times daily, at 3 and 9 a.m., and at 3 and 9 p.m., but "in voyages of discovery, where scientific observation is a prominent feature, the register ought to be enlarged, so as to take in every *odd* hour of the twenty-four" (p. 284). Here was a rhythm of observation ideally suited to sea voyages. Arctic weather was rightly thought, by Banks and others, to be somehow related to more southerly climes like Britain's; and even those arctic expeditions that were deficient in all other branches of science generally brought back thorough meteorological logs. In spite of this assiduity, the development of the "great laws" that Herschel anticipated was not soon forthcoming, but given the quintessentially Humboldtian character of meteorology,[65] its conformity to early Victorian norms of Baconian science, and its ready involvement of the disciplined amateur, theoretical frustration scarcely mattered.

What did matter, apart from the disciplined cooperation of observers, was a good set of instruments, and the knowledge of how to use them. The

Physics, Simon Fraser University. Scoresby's magnetic work was not just in the North: he swung a ship in Port Phillip Bay, Melbourne, Australia, in 1856 (personal communication from Prof. R. W. Home).

[65] S. F. Cannon, *Science and Culture: The Early Victorian Period* (New York, 1978) chap. 3, and, more briefly, T. H. Levere, "Elements in the Structure of Victorian Science, or Cannon Revisited," in J. D. North and J. J. Roche, eds., *The Light of Nature: Essays in the History and Philosophy of Science presented to A. C. Crombie* (Dordrecht, 1985) pp. 446–7.

instruments included a good barometer, appropriately suspended, with an attached thermometer; a delicate and precise reference thermometer, against which to check other thermometers, among them a self-registering thermometer (e. g., Six's), and a thermometer for solar radiation, having its bulb blackened with Indian ink; hygrometers, of which the best and sturdiest type used two thermometers, one with a dry bulb, the other being wet; a rain gauge; an anemometer, Lind's[66] being the preferred instrument on shipboard; and actinometers, for occasional use to measure solar radiation.

Among the occasional phenomena appropriate for observation and recording were auroral phenomena, and *"halos, parhelia, mock suns,* and other luminous phenomena of the kind" (pp. 328–30). For aurora, detailed descriptions were to be supported by an exact record of time, and of place determined in relation to a known star.

The northern lights had a theoretical interest that Herschel chose not to go into here. The connection between sunspots and geomagnetism had been noticed by the French physicist Arago. Various arctic explorers had since documented a correlation between auroral displays and geomagnetic disturbances, and there had also been inquiries into the possible meteorological significance of auroral displays. Russian and French collaboration showed that magnetic storms preceded such displays in places separated by more than 47° of longitude. The possible interconnections of atmospheric electricity, solar activity, geomagnetism, aurora, and climate were wonderfully attractive to von Humboldt and others seeking unification among diverse natural phenomena. The observations of naval officers on arctic service could contribute to one of the most complex and attractive areas of nineteenth-century science.[67]

Parhelia and other such phenomena were to be "noted, delineated with care if complicated, and their dimensions measured with a sextant, or otherwise, by bringing the limb [i. e., edge] of the sun or moon (noting which limb) in contact with the two edges of the phenomenon in succession. Their colours also and their order should be described" (p. 329).

Beaufort's concern with atmospheric tides[68] was maintained in the *Manual*. A separate section (pp. 337–56) was devoted to these and other regular "atmospheric waves and barometric curves." The author, William Radcliff

---

[66] "The first satisfactory syphon wind gauge, or manometer, is that of James Lind (1736–1812) invented by him in 1775" (Turner, *Nineteenth-Century Scientific Instruments*, p. 244, where a fuller description of the apparatus may be found).

[67] Cawood, "Terrestrial Magnetism," pp. 569–70.     [68] See chap. 2.

Birt, had been an assistant to Herschel at the Kew Observatory,[69] "record-ing and reducing barometric data." His work came shortly before the in-troduction of official weather reports in 1853, an exercise in which naval reports were important.[70]

Arctic data were not going to be adequate for any prediction for a very long time, but keeping a meteorological record was a favorite occupa-tion of arctic navigators and explorers, from Parry's observations of para-selenae and parhelia, to Back's painstaking and detailed accounts of northern lights. Such work contributed to the realization of Herschel's long-term goals.[71]

### SCIENCES OF THE SEA

Hydrography in the *Admiralty Manual* came immediately after geomag-netism, an indication of its high importance in naval science. It was pre-sented by Beechey, who had been elected F.R.S. the year before sailing to the Arctic as commander of H.M.S. *Blossom* in 1825. He was to become vice-president of the Royal Society in 1854, president of the Royal Geographical Society from 1855 until his death in 1856, and Royal Academician in 1854.[72] His command of an exploring expedition with a subsidiary scien-tific mandate, his technical expertise as a representational artist, the rec-ognition of surveying as part of the scientific enterprise, and the status of geography as one of the sciences, are all brought together in his career.

The hydrographer[73] of the navy was in charge of cartography. Hydrog-raphy was the art and science of making maps for navigation, including whatever would best serve to guide subsequent mariners on their passage: information about tides, winds, and currents, the location of distinctive land forms, the presence or absence of major beds of vegetation, numbers and direction of flight of flocks of birds, ocean temperatures, and any and

---

[69]  The Kew Observatory carried out magnetic, meteorological, and to a lesser extent astro-nomical observations.

[70]  Knight, "Introduction," *Manual* p. xix.

[71]  For a valuable compilation, see *Contributions to our Knowledge of the Meteorology of the Arctic Regions*, vol. I (Her Majesty's Stationery Office, 1885), listing meteorological ob-servations of thirty-six expeditions between 1819 and 1858.

[72]  Bershad, "The Drawings and Watercolours by . . . Beechey," *Arctic* 33 (1980) 116–67. His earlier naval duties had included making drawings on Parry's first expedition.

[73]  Ritchie, *The Admiralty Chart;* A. Day, *The Admiralty Hydrographic Service, 1795–1919* (London, 1967).

every other potentially useful observation. The discipline in its best and full-est sense incorporated a great variety of subtle observation and interpreta-tion, and deserved the name of science.[74]

The study of winds and currents, and their effects upon vessels, "is one of the most useful inquiries a seaman can make; and as both (wind and cur-rent) perform an important part in the economy of nature, an additional interest attaches to a correct knowledge of them" (p. 53). To detect a cur-rent required careful attention to compass navigation, the determination by sextant and chronometer of the ship's position, the use of a log to calculate the ship's speed through the water, and the subsequent comparison of the ship's observed position with its position calculated by dead reckoning.

Once the current's direction and speed had been noted,

it is very desirable to connect the temperature of the surface of the sea, for it has been by such observations that we have been able to trace, with a certainty amount-ing almost to proof, the continuous course of the same body of water for thousands of miles over the troubled surface of the ocean, and that other curious and important facts in physical hydrography have been ascertained. (p. 54)

One current that needed and over the ensuing decades received careful de-lineation, was the broad flow from Bering Strait across the Arctic Ocean and down Davis Strait and into the North Atlantic; ice drift as well as more navigable currents became increasingly important to arctic navigators as the century wore on. Meanwhile, Beechey instructed observers to record air temperature at the same time as water surface temperature; hydrography and meteorology thus became interdependent branches of science.

Where masses of seaweed were noticed, deep-sea soundings should be attempted, since such soundings were the best and safest way to locate shoals (pp. 55–6). Not only the surface temperature of the water, but also the temperatures of submerged strata should be recorded, using self-registering thermometers such as Six's[75] as "it is only by numerous well-recorded observations of this nature that we shall ever be able satisfactorily to define the limits of the various zones of moving water which sweep over the face of the globe, mingling the waters of the Polar Seas with those of the equatorial regions . . . " (p. 57).

[74]   M. J. Dunbar, "The History of Oceanographic Research in the Waters of the Canadian Arctic Islands," in Zaslow, ed., *A Century of Canada's Arctic Islands* (Ottawa, 1981) pp. 141–52 gives a broad survey here.

[75]   James Six, "An Account of an Improved Thermometer," *Phil. Trans.* 72 (1783) 72–81; *On the Construction and Use of a Thermometer* . . . (Maidstone, 1794); J. Austin and A. Mc-Connell, "James Six, FRS: Two Hundred Years of the Six Self-Registering Thermometer," *Notes and Records of the Royal Society of London* 35 (1980) 49–65.

Approaching a coast, everything that facilitated navigation should be re-corded and described "as graphically as possible." Here geography and hy-drography became so nearly allied "that it is scarcely possible to avoid encroaching upon the province of the sister branch" (p. 65). Beechey's sketches made while Parry sailed along Lancaster Sound had been models of their kind; landmarks were clearly delineated, as were bearings, while the portrayal of cliffs and promontories was precise not only in outline, but stratigraphically, so that these navigational records could also serve as geo-logical field notes. All the natural sciences could be brought to bear upon the hydrographer's task. In harbor, for example, not only the contours of the sea bed and the appearance of the land, but also precise location were required. With a transit instrument and a good achromatic telescope, "the longitude by occultations, moon-culminating stars, and eclipses of Jupiter's satellites, will form a valuable addition to that by observations of lunar dis-tances with the sextant" (p. 71). In harbor, when a survey was to be made, a tide pole was necessary so that soundings could be reduced to the low-water standard. Arctic expeditions that spent some time in harbor – gen-erally those that passed one or more winters in the ice – almost always set up a tide pole or tide gauge, and had the ungrateful task of regularly chop-ping it free of ice in order to take readings.

Such nuisances were a reminder, if one were needed, that the Northwest Passage was not readily navigable. In 1848, when the Hydrographic De-partment of the Admiralty made a return to Parliament, it gave no promi-nence to the Arctic. Almost at the end of the return came the observation that

The United States are carrying on an elaborate survey of their own coasts; and to northward of them a part of the Bay of Fundy has been done by ourselves, as well as all the shores of Nova Scotia, Canada [i. e., Upper and Lower Canada, corre-sponding to Ontario and Quebec), and Newfoundland; and when these surveys are finished, we shall only want to complete the eastern coast of America, those of La-brador and Hudson Bay, which, being in our possession, ought to appear in our charts with some degree of truth.[76]

Ideally, coastal surveys would include tides, which were clearly part of hydrography. In the *Manual,* tides received the dignity of a separate section. One reason for this was the advocacy of William Whewell,[77] a friend of

[76]  *Manual,* pp. 106–7, from Appendix "Coasts and Islands of which our Hydrographical Knowledge is imperfect. Abstract from a Return made to the House of Commons 10th February 1848, from the Hydrographic Department of the Admiralty."

[77]  The most recent study is M. Fisch and S. Schaffer, eds., *William Whewell: A Composite Portrait* (Oxford, 1991), in which Whewell's studies of tides are discussed on pp. 13–16,

Herschel, statesman and philosopher of science, spokesman for a liberal education embracing the sciences, former professor of mineralogy at Cambridge, and recently elected professor of moral philosophy there. He would shortly become Master of Trinity college, Cambridge, and was a leading member of the clerisy of science who directed the fortunes of the British Association for the Advancement of Science.[78] Within that organization, Whewell had especially concerned himself with tides; and now he added to Beechey's general remarks detailed directions for tide observations (pp. 108–30). He preferred a float carried on an upright tube to the simpler tide pole, but this was scarcely practicable in arctic waters, where ice would have rendered the float useless. His principal concern was to obtain an observational base for the reference of tides to the motions of the moon. He gave instructions for frequent and regular observations of tide height. These observations were to be plotted on a graph against time, fleshed out by interpolated values (a procedure absolutely vetoed in the *Manual*'s instructions for astronomical and meteorological observations), and translated into smooth curves. He viewed tides as resulting from the progress of very wide waves, which could be recorded by observations at a series of places in the region under consideration, but such observations were seldom practicable in the Arctic. The development of a theoretical basis for the prediction of tides would have been valuable for the navy, but it was about as much within the reach of contemporary science as the prediction of the variation in geomagnetic vectors. The inclusion of this section in the *Manual* says much about Whewell's influence in matters scientific; it also testifies to the Admiralty's willingness to contribute to the advancement of science, a matter for which peacetime freed men and ships, and without which naval unemployment would have been even greater.

## GEOGRAPHY

Hydrography overlapped with geography, which, in the mid-nineteenth century, was prominent among the descriptive sciences.[79] There was a powerful interaction between the Royal Society and the Royal Geographical Society, which was to become more prominent in the second half of the century than

---

87, 96–8: Whewell "amassed a body of data which placed tidal studies on a new footing" (p. 13); between 1833 and 1850, he wrote "some fifteen papers on the subject of 'tidology' (a name, like so many, of his own creation)" (p. 96).

[78]  J. Morrell and A. Thackray, *Gentlemen of Science.*

[79]  D. R. Stoddart, "The RGS and the 'New Geography': Changing Aims and Changing Roles in Nineteenth-Century Science," *Geographical Journal* 146 (1980) 190–202.

in the first, testifying to the scientific status of geographical inquiry. Advocacy for scientific exploration in general, and for polar exploration in particular, came from both learned societies, often in close cooperation. Geography first emerged as a powerful section in the British Association for the Advancement of Science in 1860, when it was linked to ethnology; Murchison was its first president. The section devoted to geography in the *Manual of Scientific Enquiry* was written by William Hamilton, a geographer and politician. He was for many years secretary of the Geological Society, and twice its president. He was also thrice president of the Royal Geographical Society; he had worked in the Foreign Office, which had its own uses for geography, and was soon to serve as a member of Parliament.

Hamilton began by pointing out the recent parallel advances in geographical science and in its estimation by the public. The drawing of accurate maps, showing physical features and political boundaries, was

but the commencement of our science. The most perfect maps are but the skeleton or groundwork of geography, taken in the higher or more extended sense in which it should be cultivated. Its application to the progress and development of civilization, and to the knowledge of the animal and vegetable productions of the earth, of the distribution of the different races of the human family, and the various combinations which have arisen from their repeated intercourse, are subjects of the highest consequence.... [W]e should aim at ... the improvement of man's moral culture by a more extended knowledge of the productions of different climes, and by bringing before him, on a large tabular scale, the moral and physical conditions of his race. (pp. 131–2)

Physical geography shaded into hydrography and geology, and had consequences observable in meteorology. Any and every observation was of potential interest; and Hamilton recommended, in addition to the systematic description of regions, the recording of natural curiosities, such as grottoes and natural bridges.

Political geography embraced "all those facts which are the immediate consequences of the operations of man" (p. 135), where possible given statistical form, and thus overlapping both with the science of statistics,[80] and

[80]  T. M. Porter, *The Rise of Statistical Thinking 1820–1900* (Princeton, 1986). For other sources, and for a good brief account, see Ian Hacking, "Probability and Determinism, 1650–1900," in *Companion to the History of Modern Science*, R. C. Olby, G. N. Cantor, J. R. R. Christie, and M. J. S. Hodge, eds. (London and New York, 1990) pp. 690–701. The *Manual's* final section is on statistics, and begins with the statement: "The population of any place or country must be considered as the groundwork of all statistical inquiry concerning it. We cannot form a correct judgment concerning any community until we shall have become acquainted with the number of human beings of which it is composed, nor until we shall have ascertained many points that ascertain their condition, not only as they exist at the time of inquiry, but comparatively also at former times" (p. 480). There

with ethnology. Hamilton divided this branch of geography into the following categories:

1. Population: different races of inhabitants
2. Language: words and vocabularies
3. Government: ceremonies and forms
4. Buildings: towns, villages, houses
5. Agriculture: implements of labor and peculiarities of soil
6. Trade and commerce: roads, and other means of communication

Not all these categories applied to the native peoples of the Arctic, but
naval explorers recorded observations on as many of them as seemed appropriate, sometimes with disciplined elaboration (e. g., John Ross's tabulation of Inuktitut vocabulary). In regions unsettled by Europeans, political
geography was essentially early ethnology; in settled regions, it would have
been closer to espionage.[81] Where the results were simply uncritical lists of
data, they were more reminiscent of unsytematic collections of nature
notes. Systematic natural history deserved, and received, recognition as an
important branch of the sciences.

### NATURAL HISTORY, INCLUDING GEOLOGY

The nineteenth century was the heyday of natural history, above all in Britain. Every class of society was encouraged to participate, and what had
been the preserve of the country clergyman, the rare professional naturalist,
or the collector of curiosities, grew very much into a social phenomenon,
pursued avidly at home and in the farthest reaches of the empire. The notebook became an adjunct to the gun, or even a substitute for it, among collectors, and in no other branch of the sciences was the role of the amateur
of such great importance. The way in which William Hooker built up a network of plant collectors in North America was a particularly successful ex-
emplar of a widespread phenomenon.[82] Little specialized equipment was

was a statistical section in the BAAS from 1833. This section in the *Manual* was by G. R.
Porter, Head of the Statistical Dept. of the Board of Trade from 1834, the year in which
he was a founder of the Statistical Society; Porter was elected F.R.S. in 1838, a comment
on the status of statistics. The instructions in the *Manual* are directed to visitors to civilized countries, and the data that Porter wanted reduced to statistical form were compounded by geographical lore and espionage.

[81] Knight, *Admiralty Manual*, pp. xvii–xviii. For an illustration of this double role, see Jefferson's public and secret instructions to Lewis and Clark for their western voyage of
exploration.

[82] For an antipodean example, see A. M. Lucas, "Baron von Mueller; Protégé turned

needed; mathematical skills were redundant; and museums and institutions proliferated as centers for the collection of specimens and observations, and for their integration into the fabric of the sciences. Natural history had no immediate utility for the navy, but naval officers were not immune from its attractions, as the range of arctic collections and collectors shows.[83]

*Geology.* Geology, both in its economic aspect and as a part of natural history, had been prominent among the sciences that were of interest to arctic explorers and navigators, from Frobisher's fool's gold, to Richardson's systematic observations on Franklin's second overland expedition. In the early nineteenth-century Royal Naval voyages, Jameson's dominating role had ensured that geological reports were couched within the framework of Wernerian language, with precipitation from environing seas the primary cause of the formation of land. Against this had been Huttonian theory, based on continuity of past and present processes, and with the agency of fire and heat given far more of a role than in Werner's neptunist approach. More recently, the principal geological debate had been between catastrophism, a doctrine embracing discontinuity and based upon evidence of extinction in the fossil record, and some form of uniformitarianism. British geologists had pretended to get on with their science without too much theoretical debate, hence the severely factual tone of some of the papers presented to the Geological Society, masking vigorous debate behind the scenes. Then came the publication of Charles Lyell's *Principles of Geology.*[84] Here was a work at once codifying the achievements of past practitioners of geology, and offering a model and method for the future. The book signaled in Britain the broad dominance of uniformitarianism, which now embraced paleontological and stratigraphic evidence.[85]

The new geology was presented in the *Manual* by Charles Darwin, who, after the publication of his *On The Origin of Species* in 1859, became the

Patron," in R. W. Home, ed., *Australian Science in the Making* (Cambridge, 1988) pp. 133–52.

[83]   For background and context, see D. A. Allen's admirable *The Naturalist in Britain: A Social History* (London, 1976), and L. Barber, *The Heyday of Natural History* (New York, 1980).

[84]   Charles Lyell, *Principles of Geology,* 3 vols. (London, 1830–33; reprinted Chicago, 1990, 1991).

[85]   R. Porter, *The Making of Geology: Earth Science in Britain 1660–1815* (Cambridge, 1977); R. Laudan, *From Mineralogy to Geology: The Foundations of a Science, 1650–1830* (Chicago, 1987); James A. Secord, *Controversy in Victorian Geology: The Cambrian-Silurian Dispute* (Princeton, 1986).

world's best known naturalist.[86] But when the *Manual* was published, Darwin had recently returned from the *Beagle* voyage, and his reputation in zoology lay in the future.[87] His first impressive zoological publication was to be *A Monograph of the Sub-class Cirripedia with Figures of all the Species* (2 vols., London, 1851–4), a technical work establishing his credentials, but far from making the impact of his writings on evolution. In the late 1840s he was known as the author of a narrative of the *Beagle* voyage, and of a fanatically Lyellian monograph on the formation of coral reefs,[88] and had a reputation as a promising young geologist. As he had written in 1836 to his old teacher John Henslow, Professor of Botany at Cambridge, he was "much more inclined for geology, than the other branches of Natural History."[89] From Rio de Janeiro he wrote to Henslow that "Geology and the invertebrate animals will be my chief object of pursuit through the whole voyage."[90] To another correspondent, a few years previously, he had written that "Geology carries the day; it is like the pleasure of gambling, speculating on first arriving what the rocks may be."[91] He was thinking seriously of geology as a profession, and in his notes on marriage, where he was contemplating a Cambridge professorship in either geology or zoology as appropriate employment for a married man, he noted that "I could not systematize zoologically so well." But if geology was still his first choice, *he* was Herschel's second choice as author of the section on geology, just as he had been second choice for the *Beagle* voyage; one of Darwin's teachers, Adam Sedgwick, Professor of Geology at the University of Cambridge, was the intended author.[92]

[86]   The best recent biography of Darwin is P. J. Bowler, *Charles Darwin: The Man and his Influence* (Oxford and Cambridge, Mass., 1990).

[87]   Not only that, but his notebooks, which worried away at the problem of species, were at first secret, and then, in the 1840s, were revealed in confidence to only a small circle of intimates: he opened his first notebook on the transmutation of species in 1837, made the first "reasonably comprehensive outline" of the theory of natural selection in 1842, and by the mid-1840s had shared his thoughts with at least half a dozen scientists, including Lyell and Joseph Hooker (Bowler, *Charles Darwin*, pp. 68, 95). He had by then edited *The Zoology of the Voyage of HMS Beagle*, 5 parts (London, 1838–43).

[88]   C. R. Darwin, *Journal of Researches into the Geology and Natural History of the Various Countries Visited by H.M.S. Beagle* (London, 1839); *The Structure and Distribution of Coral Reefs* (London, 1842). Darwin's other early geological writings included *Geological Observations on the Volcanic Islands Visited During the Voyage of H.M.S. Beagle* (London, 1844) and *Geological Observations on South America* (London, 1846).

[89]   Darwin to J. S. Henslow [30–31 Oct. 1836], in F. Burkhardt and S. Smith, eds., *The Correspondence of Charles Darwin*, vol. 1 (Cambridge, 1985) p. 514.

[90]   Ibid., p. 237; cf. p. 436: "Since leaving Valparaiso, during this cruize [*sic*] I have done little excepting in Geology."

[91]   Darwin to W. D. Fox, May 1832, Burkhardt and Smith, *Darwin Correspondence*, p. 232.

[92]   Knight, *Admiralty Manual*, p. xviii.

Darwin, in geology as in the study of living nature, was clear about the role of informing questions. Now that Lyell's *Principles of Geology* had given retrospective form to geological observations and method, there was no excuse for the illusion that mere collecting constituted science:

> it may be doubted whether the mere collection of fragments of rock without some detailed observations on the district whence they are brought, is worthy of the time consumed and the carriage of the specimens. The simple statement that one part of a coast consists of granite, and another of sandstone or clay-slate, can hardly be considered of any service to geology; and the labour thus thrown away might have been more profitably spent, and thus saved the collector much ultimate disappointment. (p. 167)

Geology required an understanding of taxonomy and the theories supporting it; to practice geology required discipline and precise observation. But here was a branch of the sciences more accessible than most:

> In order to make observations of value, some reading and much careful thought are necessary; but perhaps no science requires so little preparatory study as geology, and none so readily yields, especially in foreign countries, new and striking points of interest. Some of the highest problems in geology wait on the observer in distant regions for explanation; such as, whether the successive formations, as judged by the character of their fossil remains, correspond in distant parts of the world to those of Europe and North America, or whether some of them may not correspond to blank epochs of the north, when sedimentary beds either were not there accumulated, or have been subsequently destroyed. Again, whether the lowest formation everywhere is the same with that in which living beings are first present in the countries best known to geologists. (p. 168)

Darwin was advocating a search for explanations, whereas other authors of the *Manual* wrote of collecting facts. He was also questioning the widely accepted notion that mineral and stratigraphic classifications derived from British and European models could be applied around the world. But even where that notion might be wrong, it was methodologically convenient, enabling the observer to make sense of unfamiliar terrain:[93] the order of strata in one place could serve as the type for sequences elsewhere.

Darwin strikingly included North America as among the geologically known regions. It is true that, by mid-century, the Geological Survey of Canada had been under weigh for almost a decade, and so had several state surveys in the United States.[94] But Canada was still restricted to two prov-

---

[93]  R. A. Stafford, "The Long Arm of London: Sir Roderick Murchison and Imperial Science in Australia," in Home, ed., *Australian Science in the Making*, pp. 69–101 at 91.

[94]  M. Zaslow, *Reading the Rocks* (Ottawa, 1974); Zeller, *Inventing Canada*, part I. The question of what was the oldest form of life, and in what strata it occurred, remained a lively one in Canada thanks to William Dawson; see C. F. O'Brien, "*Eozoön Canadense:*

inces along the Saint Lawrence and the Great Lakes, and knowledge of the geology of the Canadian Arctic was limited to what Richardson and a very few others had reported. Arctic geology was to prove complex and elusive until well into this century.[95]

Darwin discussed the geologist's equipment, beginning with a basic library:[96] Lyell's *Principles,* the first volume of which Darwin had been given by Henslow for the *Beagle* voyage, and also Lyell's *Manual of Geology;* a treatise on mineralogy; W. H. De la Beche's *Researches in Theoretical Geology,* "particularly desirable from discussing many of the questions which ought especially to engage the attention of a sea voyager" (pp. 169–70); and, in polar and temperate regions, Agassiz's work on glaciers.[97] Then there were tools: heavy and light geological hammers; chisels and a pickax for fossils; a pocket lens with three glasses, "to be incessantly used"; a clinometer, for measuring the inclination of strata; a blowpipe for mineralogical analysis; and a mountain barometer.

The first task, on landing on a new coast, was to develop a stratigraphic section. The sketches made for navigation by Beechey and other junior officers assigned to this duty were often geologically valuable in precisely this respect. Paleontology and stratigraphy had to be incorporated with mineralogy and the identification of formations. The collection of fossils, and the careful identification of the strata in which they were found, would help to date and arrange strata, and facilitate their comparison with strata elsewhere. Once the rocks and their stratigraphic succession had been identified in one part of the region being explored, then making traverses parallel to the outcrops of strata, noting how those strata were inclined to the horizontal, and combining these data with a knowledge of topography, would enable one to extend one's knowledge of the geological structure of a region.[98] Specimens collected and labeled should be correlated with

---

The Dawn Animal of Canada," *Isis* 61 (1970) 206–23.

[95]  R. L. Christie and J. W. Kerr, "Geological Exploration of the Canadian Arctic Islands," in Zaslow, ed., *A Century of Canada's Arctic Islands* (Ottawa, 1981) pp. 187–202 includes a brief account of exploration 1819–85.

[96]  Darwin's library aboard the *Beagle* is listed in appendix IV, Burkhardt and Smith, *Darwin Correspondence,* vol. 1 pp. 553–66.

[97]  J. L. R. Agassiz, *Etudes sur les Glaciers* (1840).

[98]  Cf. J. A. Secord, *Controversy in Victorian Geology: The Cambrian-Silurian Dispute* (Princeton, 1986) pp. 4, 24–7. Other studies of geology as a field science in nineteenth-century Britain include Secord, "The Geological Survey of Great Britain as a Research School, 1839–1855," *History of Science* 24 (1986) 223–75; M. J. S. Rudwick, *The Great Devonian Controversy: The Shaping of Scientific Knowledge Among Gentlemanly Specialists* (Chicago, 1985), and D. R. Oldroyd, *The Highlands Controversy: Constructing Geological Knowledge through Fieldwork in Nineteenth-Century Britain* (Chicago, 1990).

sectional diagrams, and "an infinity" of interesting observations should be accumulated. Ideally, the totality of observations should be summarized in a geological map.

Darwin's recipe for acquiring the necessary power of observation was "to acquire the habit of always seeking an explanation of every geological point met with; for one mental query leads on to another, and this will at the same time give interest to his researches, and will lead him to compare what is before his eyes, with all that he has read of or seen" (p. 174). The recipe is one that he certainly adopted. A letter from him to Richardson in 1837 is one long exuberant succession of questions, mostly directed to gaining historical and geographical understanding in zoology; questions were intended to lead to explanations.[99]

Darwin devoted considerable space to considering the action of ice, a field in which arctic navigators had exceptional opportunities for observation. How did the results produced by coastal ice and by true icebergs differ?

A polar shore, known from upraised organic remains to have been lately elevated, would be eminently instructive. Do great icebergs force up the mud and gravel at the bottom of the sea in ridges like the moraines of glaciers? Can shells, or other marine animals, live in a shallow sea, often plowed up and rendered turbid by the stranding of icebergs? The dredge alone could answer this. The means to distinguish the effects of ancient floating ice from those produced by ancient glaciers is, at present, a great desideratum in geology. (p. 190)

The arctic geologist would find himself preoccupied by such questions, together with those concerning the elevation and depression of land, and corroborating evidence from marine remains, for example seashells. Recent researches by the zoologist and oceanographer Edward Forbes had led to a close correlation of species of shellfish with particular depths, so the presence of particular shells at or above sea level would suggest the extent of the elevation of the soil or rocks containing them above the level at which the shellfish were originally living.[100] Questions of the elevation of land in geologically recent time were repeatedly to concentrate the minds of arctic naturalists, for example in their study of the shores of Ellesmere Island in the

---

[99]  Darwin to Richardson, 25 July 1837, SPRI MS 1503/16/1, not yet published in *The Correspondence of Charles Darwin*.

[100]  E. J. Mills, "Edward Forbes, John Gwyn Jeffreys, and British Dredging before the *Challenger* Expedition," *Journal of the Society for the Bibliography of Natural History* 8 (1978) 507–36. The sequel is told in Mills, *Biological Oceanography: An Early History 1870–1960* (Ithaca, N.Y., 1989).

second half of the century.[101] And, Darwin being Darwin, and being indebted to Lyell, it is not surprising to find him explaining that, as

geology includes the history of the organic inhabitants, as well as of the inorganic
materials, of the world, facts on distribution come under its scope. Earth has been
observed on icebergs in the open ocean; portions of such earth ought to be collected, washed with fresh water, filtered, gently dried, wrapped up in brown paper,
and sent home by the first opportunity to be tried, with due precautions, whether
any seeds still alive are included in it. (pp. 192–3)

Biogeography and geology interpenetrated one another in Darwin's work
on the Galapagos and more generally on the origin of species; and once his
theory had given authority to the historical and dynamic study of biogeography, naturalists looked to the Arctic to resolve such questions as those
linking glaciation to the distribution of species – witness for example the
work of Joseph Dalton Hooker.[102]

*Mineralogy.* Darwin had stressed how accessible to the nonspecialist the
science of geology was; and he had included mineralogy among its ancillary
disciplines. Sir Henry De la Beche wrote the *Manual's* section on mineralogy. He had conducted the Geological Survey from its inception in Britain
in 1832, and since 1840 had been in charge of the Ordnance Survey, an operation whose name proclaims the role of the military in surveying. Elected
to the Royal Society in 1819, he had recently been president of the Geological society, and had played a major role in recommending geologists for
colonial surveys, including William Logan for the Geological Survey of
Canada.

In general, minerals may be distinguished by their chemical composition
and their crystalline form. Simple correlation is complicated by dimor-
phism, in which a single chemical species can occur in two crystalline
forms, and by isomorphism, in which chemically distinct species occur in
closely similar crystalline forms, and can even form mixed crystals. Physical
properties such as luster, transparency, and hardness also help to distinguish minerals.

De la Beche was concerned to help mariners identify the most common
and geologically most significant minerals using a combination of chemical

101  See, e. g., the work of H. W. Feilden on the British Arctic Expedition of 1875–6, discussed in chap. 7 and in Levere, "Henry Wemyss Feilden, Naturalist on H. M. S. *Alert*
1875–76," *Polar Record* 24 (1988) 307–12, and "Henry Wemyss Feilden (1838–1921)
and the Geology of the Nares Strait Region, with a Note on Per Schei (1875–1905),"
*Earth Sciences History, 10* (1991) 213–18.
102  See notes 122–6, and W. B. Turrill, *Pioneer Plant Geography: The Phytogeographical Researches of Sir Joseph Dalton Hooker* (The Hague, 1953).

and crystallographic evidence. The most useful chemical tool was the blowpipe, by means of which the jet of an oxidizing flame could be blown on to a mineral specimen, either held in platinum forceps or upon charcoal: the resulting phenomena gave clues about chemical composition. Strontium, for example, gave a red tint to the flame. The angles between crystal faces were to be read with an instrument called a goniometer. De la Beche was concerned not only with minerals constituting the main strata of a formation, but also with veins that ran through them, and which often contained economically significant ores. There were also rocks of no clear crystalline form: in what was increasingly the age of steam, coal was especially important among these, and could extend the range and scope of a naval voyage.[103]

*Zoology.* Mineralogy and geology were part of natural history. But among naval officers, especially surgeons, the sciences of living nature were more widely cultivated. John Richardson was the preeminent naval arctic naturalist of the nineteenth century. At the meeting of the British Association for the Advancement of Science held at Bristol in 1836, he had presented a comprehensive report on North American Zoology.[104] He was at once modest and precise in delimiting his report, and in distinguishing his labors from the taxonomic initiatives of others:[105]

As it leaves untouched the principles of systematic arrangement, structure, physiology, and in fact the fundamental doctrines of the science, the only subjects coming properly within its scope appear to be, an *enumeration of the animals inhabiting North America;* the *peculiarities of the fauna which they constitute* when contrasted with those of the other zoological provinces into which the earth may be divided; and *the geographical range of groups or individual species,* with the circumstances which tend to influence its extent, such as *the configuration of the land, climate, vegetation, &c. (Report,* p. 121)

Richardson was tackling more than a descriptive list of species; by seeking to characterize features special to North America, and by relating geographical distribution to environmental factors, he was touching on the

103  A good example is the coal seam found near H.M.S. *Discovery's* winter quarters during the British Arctic Expedition 1875–6: see chap. 7.

104  Richardson, *Report of the Sixth Meeting of British Association for the Advancement of Science 5* (1837, for 1836) 121–224.

105  See Swainson's account of Macleay's system in vol. II of the *Fauna Boreali-Americana,* and his *A Treatise on the Geography and Classification of Animals* (London, 1835). See also D. M. Knight, "William Swainson: Naturalist, Author and Illustrator," *Archives of Natural History* 13 (1986) 275–90.

nascent discipline of biogeography,[106] as well as on questions that would later be incorporated into ecology. The notion of biological provinces, in which sets of species were related both to one another and to their geographical region, had emerged toward the end of the eighteenth century. The concept had been developed by J. R. Forster, a naturalist with Captain Cook on the *Resolution*.[107] By the time the *Admiralty Manual* was put together, botanical and zoological studies alike addressed the question of geographical distribution. Numerous authors, from Willdenow on, had noted the importance of temperature in characterizing a region, and had shown that mountain tops in southerly latitudes might have similar climates and similar floras and faunas to more northerly regions; in particular, the similarities between alpine and arctic floras were well noted. The phenomena of distribution raised questions that could be answered in terms of either pattern or process,[108] and also raised questions about species and variation. It is little wonder that Darwin, in his geological instructions, drew attention to those questions about distribution that were to be so crucial for his account of *On The Origin of Species*.[109]

Such questions were not, however, urgent for Richardson. He was a field naturalist, not a theoretician, and his first and main concern, here as in the *Fauna Boreali Americana,* was the accurate identification and description of species. Even at this basic level, a necessary prelude to more sophisticated treatments, "great uncertainty still exists as to many species, the original descriptions being so obscure that they do not enable us to recognize the animals" (*Report,* p. 122).

In his report, Richardson considered the North American region as extending to the Tropic of Cancer in Mexico, where he followed William Swainson in seeing the meeting of North and South American faunas. He identified the Rocky Mountains as the most significant physiographic feature and barrier to animal migration in the region, which he characterized with vivid economy:

---

[106]   Janet Browne, *The Secular Ark: Studies in the History of Biogeography* (New Haven and London, 1983).

[107]   M. E. Hoare, ed., *The* Resolution *Journal of Johann Reinhold Forster 1772–1775,* 4 vols., Hakluyt Society 2nd Series vols. 152–5 (London, 1982): see chapter 1.

[108]   Browne, *The Secular Ark,* p. 107.

[109]   Also important in this context was the work of Edward Forbes, notably his paper "On the Connexion Between the Distribution of the Existing Fauna and Flora of the British Isles, and the Geological Changes Which Have Affected Their Area, Especially During the Epoch of the Northern Drift," *Memoirs of the Geological Survey of Great Britain 1* (1846) 336–432. Forbes noted the presence of arctic plants on British and European mountains, and explained this in terms derived from Lyell's *Principles of Geology;* in 1859, so did Darwin; see Browne, *The Secular Ark,* pp. 115–31.

We thus perceive that to the eastward of the Rocky Mountains there is an immense longitudinal valley extending from the Arctic sea to the Gulf of Mexico, crossed by no dividing ridges of note, but forming three separate water-sheds; the southerly one having, in addition to a general easterly declination to the Mississippi, also a descent from the 49th parallel towards the outlet of the latter river in the Gulf of Mexico; the northerly one having an inclination towards the Arctic sea, commencing between the 53rd and 54th degrees of latitude, and the central one, which is necessarily the most elevated, having merely an easterly descent towards Hudson's Bay. This configuration of the land evidently gives great facilities for the range of herbivorous quadrupeds from north to south, and is the line of route pursued by many migratory birds; and while the Mackenzie furnishes a channel by which the anadromous fish of the Arctic sea can penetrate 10 degrees of latitude to the southward, the Mississippi offers a route by which those of the Gulf of Mexico can ascend far to the north. (*Report*, p. 125)

The relation between physical geography and the distribution and migration of animals was wonderfully illustrated by the caribou. The most northerly land was treeless, and hence was named the "barren grounds"; south of the tree line, running roughly W.N.W. from the 60th parallel in Hudson's Bay to Great Bear Lake on the 65th parallel, were the woods and lakes of the Canadian Shield, much of which was affected in its climate and hence in its potential for supporting animal life by the great inland sea of Hudson's Bay. On the Pacific side of the Rockies, Alaska and the Aleutian Islands appeared "similar in geological and zoological characters, as far as has been ascertained, to the eastern barren grounds" (*Report*, p. 127). The qualification was a necessary admission of ignorance: "The mountain system of Russian America is unknown. The peninsula of Alaska and the Aleutian Isles, extending towards Asia, separate from the Pacific the sea of Kamtschatka, which nourishes several fish of very peculiar forms and some singular cetaceous animals" (*Report*, p. 128).

Climate, soil, and permafrost were all important to the naturalist. Richardson presented a digest of temperatures, noting that in the high Arctic, long cold winters were succeeded by twenty-four hour daylight, which made possible an accelerated burst of surface thaw and growth. East of the Rockies, the arctic fauna was pretty well known, almost entirely as a consequence of British expeditions, with Richardson's work, although he was too modest to say it, being the most extensive and broadly reliable of all. The cooperation of the Hudson's Bay Company had also been of major importance. But Richardson was concerned with lacunas even more than with achievement, for he wanted to direct naturalists to those areas most needing attention. The land west of the Rockies, from Icy Cape southward, was still "*terra incognita* to zoologists" (*Report*, p. 134). Very little had been published on the fauna of Russian America since Steller's work.[110]

After these introductory remarks, Richardson embarked on a brisk fauna, following the order of Cuvier's *Règne Animal*. For a goodly number of northern species, he refers to his own *Fauna Boreali Americana*, and to the natural history appendixes of other arctic explorers. The pygmy shrew, for example, comes from the *Fauna*, and as Richardson remarks, is found within the Arctic Circle to the tree line. Polar and grizzly bears had been observed by numerous naval expeditions. Richardson wanted to unscramble identifications, and to be clear about the relations of North American species with European ones. Weasels provided both opportunity and challenge here:

The American weasels and martins have greatly perplexed naturalists, and their synonyms are involved in much confusion; yet we can pretty confidently assert that five species only are known in the fur-countries from latitude 50°N. to the Arctic sea. . . . It is very probable that both the ermine and stoat of America are different from those of the old continents . . . [as shown by, e. g.] the much smaller skull of the American animal. . . . The "fisher," or "wejack" (*mustela canadensis*), is found up to the 60th parallel. Its synonymy is embroiled in confusion, which is attempted to be unravelled in the *Fauna Boreali-Americana*. . . . (*Report*, pp. 143–4)

He went on, through carnivores and herbivores, to North American deer, "still very imperfectly known." He rightly identified caribou with European reindeer, "the most northern ruminating animal, being an inhabitant of Spitzbergen, Greenland, and the remotest arctic islands of America" (*Report*, p. 160). In summary, he noted that the orders of carnivores, rodents, ruminants, and cetaceans were all represented in "the most northern known lands or coasts" (*Report*, p. 163).

Moving on to ornithology, he pointed out that birds, "having always been objects of interest to collectors and artists, are better known than the other animal productions of North America." New ornithological observations, and the description of hitherto unknown species of birds, enriched the narratives of successive Royal Naval expeditions, including Richardson's own *Fauna*, which could be considered as a zoological appendix to Franklin's narrative. As for Russian America, and even California, the ornithological appendix to Beechey's narrative was still the only scientific list (*Report*, pp. 164–5).[111]

---

[110]   G. W. Steller, *Journal of a Voyage with Bering, 1741–42* (Stanford, 1988). L. Hess, *Georg Wilhelm Steller, the Pioneer of Alaska Natural History* (Cambridge, Mass., 1936).

[111]   N. A. Vigors in J. Richardson et al., *The Zoology of Captain Beechey's Voyage: Compiled from the Collections and Notes Made by Captain Beechey, the Officers and Naturalists of the Expedition . . .* (London, 1839).

Concerned here, as in the case of the mammals, to compare the fauna of Europe and North America, Richardson stressed that not only was there close resemblance or even identity of the generic forms, but that one-third of the species were common to the two avifaunas, albeit unevenly divided among the various orders.

Richardson pursued the comparison between the European and North American avifaunas, explored the distribution of different orders and groups of birds, and took broad cognizance of migration patterns. He noted that many birds, especially the "soft billed waders" took different routes in spring and fall migration, because of varying surface conditions (*Report*, p. 187).

He considered the influence of climate and behavior on distribution. The brevity of arctic summers,

taken in conjunction with the time necessary to complete the process of incubation, the growth of plumage, and, in the case of the *anatidae* [swans, geese, and ducks], the moulting of the parent birds, serves to limit the northern range of the feathered tribes. The waders, which seldom make a nest, and the water-birds, which lay their eggs among their own down, and obtain their food on the sea or open lakes when the land is covered with snow, breed farthest north. The ptarmigans, which breed in very high latitudes, and moult during the season of reproduction, migrate only for a short distance, and by easy flights; and their food moreover being the buds or tips of willows and dwarf birch, can be obtained amidst the snow. (*Report*, p. 187)

Snow buntings and lapland longspurs find food on their arrival, before any significant melting of the land's snow cover, thanks to the sudden and severe onset of the previous fall's frost, which preserves seeds and shoots. As the snow melts, new growth is accompanied by the hatching of myriads of insects. Richardson recognized the roles of seasonally abundant food and flexible diets in enabling so many birds to enter the arctic fauna (p. 188).

Most kinds of food were only seasonally abundant in the Arctic. With increasing cold, supplies of food generally diminished, and it was this, rather than the cold itself, which led to the scarcity of species of birds residing in the Arctic through the winter. Thus, for example, when Parry wintered at Melville Island, the only bird seen "was a white one, supposed to the [snowy owl], or it may have been a wandering *falco islandicus* [gyr falcon], both these birds preying on small quadrupeds." This single observation, and its imprecision, points to the paucity of winter data; few naturalists, and no expert zoologists, had spent a winter observing in the high Arctic. Still, observations were accumulating, Richardson's prominent among them. "In the pools of water which remained open all the year in the arctic seas, the [black guillemot and Brunnich's murre], are to be found at the coldest periods, the [auks, murres, and puffins], consequently, are the

most northerly winterers." Many such winterers went far south then. Regular winter residents among the land-birds in the extreme North were per-
egrine and gyr falcons, snowy owls and, slightly further south, hawk owls,
and the raven, all carnivorous, and the ptarmigan (*Report*, pp. 189–90).

Richardson was at his best in discussing the birds; first-hand field observations came to the aid of wide reading, and enabled him to relate behavior to migration and distribution, pinned on to a taxonomic skeleton.

When it came to fish, Richardson was more frustrated. The third volume
of his *Fauna Boreali Americana* was devoted to fish, and contained "a
considerable proportion of the species which inhabit the fresh waters of
the fur countries; it is, however, very deficient in marine fish, and even in
the fresh water ones of New Caledonia [Nova Scotia] and [Upper and
Lower] Canada, owing to the author's attempts to procure specimens
from these countries having failed" (*Report*, p. 203). Richardson's own
collections had been among the most significant in the Arctic. The trouble
was that there had been no other recent major study of North American
fish. Previous works, for example Pennant's, hit a major problem: "the
Linnean genera are so ill adapted for the reception of many of the
forms peculiar to the New World, and the specific descriptions of the old
writers are so brief and indeterminate, that the labours of these naturalists
are often altogether unavailable to modern cultivators of science. Cuvier
and Valenciennes[112] had begun to embrace "all the determinable species
noticed by preceding naturalists," but it remained uncompleted, the "death
of its great projector having retarded its progress" (*Report*, pp. 202–3).
But as far as it had gone, it was for Richardson "the only trustworthy guide
for general ichthyology," and he handled very briefly families untreated
therein (*Report*, p. 212). His interest in fish embraced comparative anatomy and taxonomy. As he wrote to Joseph Dalton Hooker, who had been
with James Clark Ross on the antarctic voyage of 1839–43, comparative
anatomy was the surest route to a professional reputation in the scientific
world. Among fish,

The structure of many interesting species & even tribes is very imperfectly known
and much instruction is to be gathered from a study of the whole class. The fish
seem to have been the earliest created of the vertebrata and they are the simplest, but
without a thorough knowledge of them it is impossible to understand the reasoning
of modern anatomists on the vertebral elements and the analogies of the various
parts of the skeleton.[113]

[112] G. Cuvier and A. Valenciennes, *Histoire Naturelle des Poissons,* 22 vols + 3 vols. plates
(Paris, 1828–49).
[113] Richardson to J. D. Hooker, 2 August 184[7], Royal Botanic Gardens, Kew, J. D. Hooker
correspondence 1839–1845(!) no. 224.

Anatomy was the key to taxonomy. In looking at the classification of fishes, Richardson found that northern families of fish had a greater proportion of genera common to the New and Old worlds than more southerly families. This was unsurprising, since "the conditions of the waters as well as of the land and atmosphere of the arctic regions of the two hemispheres is more alike than in the more temperate parallels" (*Report,* p. 209). At least zoologically, it made sense to think of unifying aspects of the circumpolar north.[114]

Richardson went through the fishes in taxonomic order. It is worth noting that among the *salmonoideae,* one genus was named for Hearne, one for Ross, and one for Hood, whereas "the genus or sub-genus" *Stenodus Mackenzii,* which "differs from the other *salmones* in the teeth, was first named in the Appendix to Captain Back's narrative of his journey to the mouth of the Thleweechoh," now the Back River (*Report,* pp. 214, 216).

Richardson's report was a major contribution to North American zoology, and did good service for the Arctic. Making a list of species was, as he said, an essential step in a continent where so many remained undescribed. The constant comparisons between European and North American faunas not only indicated "the variations of animal life in different localities, and in different circumstances, under the same parallels of latitude, but also, though more obscurely and merely by analogy, the tribes of animals of which new species will be most probably hereafter detected in North America" (*Report,* p. 223). Richardson's feel for these analogies was part of his imaginative grasp of the unity of natural history. His questions and insights are not those that come with merely academic training; rather, they show something of Darwin's inquiring spirit, although they lack Darwin's passion for explanation. Richardson was a field observer, and he accepted Cuvier's remark that zoology was and would for years remain a science

of observation only, and not of calculation; and no general principles hitherto established will enable us to say what are the aboriginal inhabitants of any quarter of the world. It seemed therefore hopeless to attempt to elicit the laws of the distribution of animal life from results yielded by a fauna so very imperfectly investigated as that of North America. (*Report,* p. 223)

Richardson was, however, thinking about those laws, and offering his data in a way that might come to illuminate them. He had also, in summarizing the zoology of North America, provided an admirable factual summary of the current state of knowledge. When Herschel commissioned

---

[114]  T. Armstrong, G. Rogers, and G. Rowley, *The Circumpolar North* (London, 1978); B. Stonehouse, *Polar Ecology* (Glasgow and London, 1989); B. Sage, *The Arctic and its Wildlife* (New York and Oxford, 1986).

an essay on zoology for the *Admiralty Manual,* Richardson's report to the British Association complemented that essay.

Zoology in the *Manual* was presented by Richard Owen, a major figure in the Royal Microscopical Society[115] and a formidable comparative anatomist who "had made his reputation by reconstructing the moa from a single bone brought from New Zealand."[116] Owen was a long-standing opponent of transmutation, who would thus become one of Darwin's most dogged adversaries;[117] after 1856, his institutional role as superintendent of the Natural History Department of the British Museum helped to make him a formidable opponent. His advice in the *Manual* came under four headings: first and most copiously, how to collect and preserve animals; second, how to transport living animals for the Zoological Society of London;[118] third, how to use the microscope on board ship (drawing on the experience of Charles Darwin, and of Thomas Henry Huxley, who at that date was an assistant-surgeon in the Royal Navy); and fourth, general directions to be observed during a voyage.

Only remarks under the first heading, the collection and preservation of animals, had special arctic relevance. Owen began with marine invertebrate organisms. First came algae, sponges, corallines, and corals, some to be preserved in dried form, others in fluids. Then came "infusorial animalcules. . . . The important relations of these minutest forms of animal life to great questions in geology, to the alteration of coast-lines, and to the phenomena of oceanic luminosity, make it indispensable to include them in directions for collecting facts in natural history" (*Manual,* pp. 363–4). Acalephs, comprising jellyfish and other floating marine gelatinous animals, were to be preserved in fluid, but needed to be drawn in color immediately on capture, in order to preserve their "brilliant but evanescent hues" (p. 365). John Ross had complained of the lack of a skilled naturalist-illustrator, and of the resulting loss of knowledge about the collections made on his expeditions. Similarly, Richardson had written to Joseph

---

[115] G. L'E. Turner, *God Bless the Microscope!: A History of the Royal Microscopical Society over 150 Years* (Oxford, 1989).

[116] Knight, *Admiralty Manual,* p. xix.

[117] A. Desmond, "Richard Owen's Reaction to Transmutation in the 1830s," *British Journal for the History of Science 18* (1985) 25–50. The nature and context of Owen's opposition to Darwin are well brought out in Desmond's *Archetypes and Ancestors: Palaeontology in Victorian London 1850–1875* (London, 1982).

[118] The Zoological Society of London ran the London Zoo; see W. Blunt, *The Ark in the Park: The Zoo in Nineteenth-Century London* (London, 1976), but they did far more. See P. C. Mitchell, *Centenary History of the Zoological Society of London* (London, 1929), and J. Bastin, "The First Prospectus of the Zoological Society of London: New Light on the Society's Origins," *Journal Soc. Bibliography Nat. Hist. 5* (1970) 369–88.

Dalton Hooker: "Your skill with the pencil qualifies you for understanding that branch and most of the tender acalepha can be examined with advantage only with the pencil in the hand as the anatomical characters can scarcely be preserved by any means in our power."[119] Among the remaining invertebrates were deep-sea shellfish, for which the dredge was indispensable. Owen gave practical hints for the use of the dredge – how much rope to use, where and how to weight the line, and how to sieve the contents. "Besides shells, numbers of crabs, star-fishes, sea-urchins, worms, corals, zoophytes, algae, &c., are procured by the dredge" (p. 379).

Fishes, which were Richardson's forte,[120] constituted the majority of marine vertebrates. They should be preserved entire in spirits, but, as with acalephs, their colors should be accurately recorded before death. Similarly, "neither shape nor colour can be preserved in the dried skins of whales, porpoises, &c., nor can they be ascertained from skins alone, without the aid of drawings taken from the specimens in a fresh state" (p. 396). Naturalists had to be artists as well as anatomists and taxidermists, or else they needed appropriately skilled assistants. Midshipmen, for example Hood and Back under Franklin, were often assigned to make sketches of natural history specimens, in addition to their sketches for the hydrographer's office, and naval surgeons, skilled in dissection, often doubled as naturalists.

The human race received special attention from Owen, as constituting the most important branch of natural history. The chief points to which the "philosophic and zoological voyager" should direct his attention ranged from comparative anatomy and the discrimination of races, as practiced by Petrus Camper and Blumenbach at the end of the eighteenth century,[121] to questions that are as much appropriate to the emergent disciplines of anthropology and ethnology as they are to natural history. For example,

How are widows treated?
Describe the kind and materials of dress; and any practice of tattooing or otherwise modifying the person for the sake of ornament or distinction.
What is the received idea respecting a future state?

[119]   Richardson to J. D. Hooker, 2 August 184[7], Royal Botanic Gardens, Kew, J. D. Hooker correspondence 1839–1845(!) no. 224.

[120]   His numerous publications on the fishes are listed in the bibliography in R. E. Johnson, *Sir John Richardson, Arctic Explorer, Natural Historian, Naval Surgeon* (London, 1976) pp. 145–57.

[121]   J. F. Blumenbach, *De Generis Humani Varietate Nativa* (Göttingen, 1775); *Oeuvres de P. Camper, qui ont pour Objet l'Histoire Naturelle, la Physiologie et l'Anatomie Comparée*, trans. H. J. Jansch, 3 vols. (Paris, 1883); Camper, *Dissertation Physique sur les Différences Réeles, qui Présentent les Traits du Visage chez les Hommes de Différens Pays, et de Différens Ages* (Utrecht, 1791), trans. from the Dutch.

Note down any illustrative particulars of the government, policy, religion, superstitions, or sciences of the people; their mode of noting or dividing time; their mode of carrying on war, and favourite weapons. (p. 398)

Clearly natural history had a wider reference then than it does today.

*Botany.* Botany shared with ornithology the prize for popularity among amateur naturalists in the nineteenth century.[122] William Hooker wrote about it in the *Manual,* stressing that botanical science required study at home as well as in the field. "For this reason it is highly desirable that persons visiting foreign countries should not only obtain information on the spot respecting the plants and their uses and properties, but that they should transmit to this country ample collections of *well-dried specimens,* with the rarer *fruits* and *seeds,* and all sorts of interesting vegetable *products*" (p. 416). He gave instructions for transporting living plants for cultivation, for preserving plants for the herbarium, and for collecting vegetable products for a "Museum of Economic Botany" (p. 421).

He stressed the importance of developing floras for different parts of the world. Coupling botany and geography was important.[123] Plant geography had been recognized as a distinct field of inquiry by Karl Willdenow, whose *Principles of Botany* (Edinburgh, 1805) formed part of the education of the next generation of British botanists. Among Willdenow's successors were Robert Brown,[124] and Alexander von Humboldt, for whom biogeography was both a passion and a focus for a myriad of studies.[125] Humboldt had written of his botanizing in South America that: "I flattered myself that our investigation might add some new species to those already known, both in the animal and vegetable kingdoms: but . . . the discovery of an unknown genus seemed to me far less interesting than an observation on the geographical relations of the vegetable world, [and] on the migration of the social plants. . . . "[126]

Hooker was interested in new species as well as in the distribution of known ones. He pointed out that even relatively well-known parts of the

[122]  D. E. Allen, *The Naturalist in Britain: A Social History* (London, 1976) pp. 106–16; *The Botanists: A History of the Botanical Society of the British Isles Through a Hundred and Fifty Years* (Winchester, 1986).
[123]  See chapter 5 for more on Hooker and phytogeography.
[124]  See chapter 2 and Mabberley, *Jupiter Botanicus* (Braunschweig and London, 1985).
[125]  Von Humboldt's best-known essay on phytogeography is his *Essai sur la Géographie des Plantes: Accompagné d'un Tableau Physique des Régions Equinoxiales* (Paris, 1805).
[126]  Alexander von Humboldt, *Personal Narrative,* trans. and ed. T. Ross (London, 1907) vol. I pp. ix–xi, quoted in Zeller, "The Spirit of Bacon," *Scientia Canadensis* 13 (1989) 85; M. Nicholson, "Alexander von Humboldt, Humboldtian Science and the Origin of the Study of Vegetation," *History of Science* 25 (1987) 167–94.

world generally had inadequately recorded flora, but also listed regions "unknown alike to the botanist and the geographer" (p. 424).

It is striking, however, that in William Hooker's listing of botanical and geographical lore, the Canadian Arctic, and indeed all of North America, did not figure as unknown. Nor did Hooker list any arctic plant as economically significant and needing further study. He, and Richardson, Drummond, Douglas, and other collectors, had done their work well.

### HUMAN SCIENCES

*Ethnology.* We have seen that zoology in Owen's hands embraced ethnology as part of natural history. It would be equally reasonable, in the mid-nineteenth century, to regard humankind as the culmination of the ascent of life,[127] and to place ethnology and anthropology at the summit of the life sciences. James Cowles Prichard, who wrote the section of the *Manual* on ethnology, was one of the nineteenth-century founders of the science.[128] He had learned medicine in Bristol and London, and received his M.D. from Edinburgh, which was still the premier British academic institution in medicine. He was interested in philology, because the filiation of languages shed light on the relations of human populations. He had a general interest in the natural history of man, his book of that title appearing in the year of his death, 1848. He emphasized the unity of humankind, while exploring the relations between different races, and speculated that Adam and Eve may have been black. His son saw his essay for the *Manual* through the press, and Herschel supplied an appendix on reducing native foreign languages to writing.

Prichard's definition of the scope of ethnology was all embracing:

Under that term is comprised all that relates to human beings, whether regarded as individuals or as members of families or as members of communities. The former head includes the physical history of man; that is, an account of the peculiarities of his bodily form and constitution, as they are displayed in different tribes, and under

---

127　As in S. T. Coleridge, *Hints Towards the Formation of a More Comprehensive Theory of Life* (London, 1848).

128　J. C. Prichard was the foremost British anthropologist and ethnologist of his day. See, e. g., his *Researches into the Physical History of Man* (London, 1813; reprinted Chicago and London, 1973, ed. and intro. G. W. Stocking) and *The Natural History of Man*, 2 vols. (London, 1855). The secondary literature here is dominated by Stocking: see his article on Prichard in the *Dictionary of Scientific Biography*, and his *Victorian Anthropology* (New York and London, 1987). In the mid-nineteenth century there was no clear and consistent distinction between anthropology and ethnology.

different circumstances of climate, local situation, clothing, nutrition, and under the various conditions which are supposed to occasion diversities of organic development. The same expression may also, in a wide sense, comprehend all observations tending to illustrate psychology, or the history of the intellectual and moral faculties, the sentiments, feelings, acquired habits, and natural propensities. To the second division of this general subject, viz., to the history of man as a social being, must be referred all observations as to the progress of men in arts and civilization in different countries, their laws and customs, institutions – civil and religious, their acquirements and traditions, literature, poetry, music, agriculture, trade and commerce, navigation; and, which of all things affords the most important aids in all researches as to the origin and affinities of different tribes or races, their languages and dialects. (p. 439)

Given this range, everything was grist for the ethnologist's mill, and most of the observations that Prichard recommended had no special arctic reference. For example, in recording physical characteristics, not only measurements and verbal descriptions, but also colored portraits were highly desirable. John Ross and George Back were among the arctic explorers who had labored with different degrees of skill in making such portraits of Indians and the Inuit. Other kinds of evidence, for example bringing home a collection of skulls, were less likely to meet with native cooperation.

Prichard was a remarkably objective observer, unusually free from Eurocentric prejudice. In looking at "the hunting state" of society, he observed that it was not always a primitive condition: "even . . . the most destitute of the Esquimaux and other American tribes, display as much ingenuity in following their respective pursuits as nations of much more refined and artificial habits of life" (pp. 444–5).

His emphasis upon language was clear in his definition of the scope of ethnology, and the concluding section of his essay, "Language, Poetry, Literature," was emphatic:

As no other means have contributed so much to the increase of ethnology, and to the ascertaining of the connexions and relationship of different nations, as a comparison of languages, great care should be taken in every newly discovered country, and among tribes whose history is not perfectly known, to collect the most correct information as to the language of the people. (p. 452)

Prichard's questions, his rigor, and his mix of history, geography, philology, comparative anatomy, and other disciplines in the service of ethnology, would serve as a model for arctic explorers throughout the century. Franz Boas[129] is usually cited as the main founding figure of early twentieth-

---

[129] G. W. Stocking, ed., *The Shaping of American Anthropology, 1883–1911: A Franz Boas Reader* (New York, 1974).

century anthropology. In the British tradition, convergent with the German one through a common reading of continental sources, Prichard could equally serve as a prior model. His care for the study of language is at any rate conformable with that of later arctic anthropologists, Diamond Jenness chief among them.[130]

*Medicine and medical statistics.* Last of all among the human sciences were medicine and medical statistics. They were presented in the *Manual* by Alexander Bryson, who had studied at the Universities of Edinburgh and Glasgow, was an assistant surgeon in the navy in 1827, became F.R.S. in 1854, and in the following year was appointed inspector general of Hospitals and Fleets. His main concerns were tropical diseases and fevers, and even where he mentions the correlation between diet and climate, he does so referring to the tropics, rather than to the polar regions, although this issue had been discussed by John Ross following his second arctic voyage. There is no mention here of scurvy and its avoidance,[131] hypothermia, or of what is to be learned from native diets. In Bryson's contribution, more than in any other except Porter's on statistics,[132] the Arctic, and the medical lessons learned by its explorers, simply do not exist.

## CONCLUSION

The experience of arctic explorers in the first half of the nineteenth century had contributed to many sciences, and the advice that leading scientists offered to naval and other arctic explorers through the *Admiralty Manual* drew on that experience in every major discipline except statistics. The Royal Navy was not a representative body in nineteenth-century science, but through Herschel's contributors to the *Manual,* and through particular advice given by men of science in the Royal Society, the Geological Society, the Royal Geographical Society, the Linnean Society, the Dublin Observatory, the Royal Artillery Institution in Woolwich, and other bodies, the directions given to naval officers came increasingly to correspond to the norms of the sciences. This was especially true of those sciences that could use trained and disciplined amateur help, including the inventory sciences.

---

[130]  See chap. 10, and H. B. Collins and W. E. Taylor, "Diamond Jenness (1886–1969)," *Arctic* 23 (1970) 71–81.

[131]  For a good general treatment, see K. J. Carpenter, *The History of Scurvy and Vitamin C* (Cambridge, 1986).

[132]  See note 80.

The technical expertise deployed in London, Dublin, and Woolwich also bore on sciences requiring some mathematical competence, including astronomy and geomagnetism; and because these sciences contributed to navigation, the navy was very ready to provide its officers with some specialized training. Hence arose that distinctive amalgam characteristic of naval science in mid-century. Its application was not restricted to the Arctic, for the navy was active in every ocean. But the increasingly obvious lack of economic and even of military benefits to be obtained from arctic exploration contributed to the prominence of scientific instructions in the Admiralty's orders to its arctic commanders. Since institutions have a momentum as well as an inertia of their own, and since scientists, like Barrow and the hydrographer's office, had a stake in continuing arctic exploration, the sciences were able to benefit. The *Admiralty Manual* gave definition and standing to a set of scientific practices that had evolved in the decades following John Ross's first arctic voyage. Henceforth arctic commanders would sail with the *Manual* and the scientific mandate it embodied.

# 5

## The Navy and the Hudson's Bay Company 1837–1859: John Franklin and the Search Expeditions

By the mid-1830s, the Northwest Passage, although not fully mapped, was less urgent a goal than it had been in the twenties. The Arctic was of continuing scientific interest, but greater attention accrued to the impending antarctic voyage of James Clark Ross.[1] The Russian question in North America, although not finally settled, no longer seemed to be urgent. If in the late thirties Great Britain had any interest in problems in the northern hinterland of its American territories, they concerned boundaries with the United States. The *Morning Post* of 7 March 1838 observed that it was "necessary that some practical means should be found of settling a point which contains the germ of a serious quarrel, and which would at the present moment probably lead to a rupture, if the commercial interest of both nations was not more powerful than their political pride."

Well to the fore on the British side was the commercial interest of the Hudson's Bay Company. Their concern was with profit and trade, especially the fur trade, and they viewed the land very much as their own fiefdom. They had looked on in some exasperation while they provided relief for earlier overland naval explorations. Not only did the company regard the navy's methods as inefficient, but the navy's goals in science and exploration were marginal to the company's concerns.[2]

It is true that the company was not a total stranger to the sciences, nor was it altogether unhelpful in scientific pursuits. Earlier on, it had for a while seemed that the sciences and economic enterprise might go comfortably hand in hand. After all, the Hudson's Bay Company and the Royal

---

[1] J. C. Ross, *A Voyage of Discovery and Research in the Southern and Antarctic Regions during the Years 1839–43*, 2 vols. (London 1847; reprinted Newton Abbot, 1969) vol. I pp. xxii–xlvi.

[2] See chap. 3.

Society of London had both received their charters from Charles II, neither made any early distinction between science and practical life, and they had frequent occasions for cooperation.[3] But by the time the Napoleonic Wars had ended, and the Royal Navy began its peacetime attack on the Arctic, the company was sufficiently distant from the Royal Society to see the navy more as a body of explorers intruding in the company's territory, than as an agent of scientific investigation. Apart from the continued collection of specimens and the sometimes systematic recording of meteorological data, the company was outside the mainstream of science. Even geographical exploration was incidental to fur trading.

## THE REIGN OF GEORGE SIMPSON

Profits, monopoly, control, trade, and empire – the Hudson's Bay Company's empire, to which the British Empire was sometimes seen as auxiliary or accessory – these were the guiding lights of a company that had triumphed over its rivals, absorbing the North West Company in 1821, and reaching far into the lands south and east of the Columbia River. Presiding over this empire from 1820 was George Simpson, who, as the company's governor-in-chief in North America, had so impressed Lefroy. He was tough, flamboyant, and relentless. "A bastard by birth and by persuasion, George Simpson dominated the HBC during four crucial decades. . . . "[4] Simpson wanted power for and through the company, and that included social and cultural standing. The sciences were part of the civilized culture of the nineteenth century, and Simpson saw this circumstance as one conforming to the company's original charter, which noted encouragingly its "Expedicion for Hudsons Bay . . . for the discovery of a new Passage into the South Sea and for the finding some Trade for Furrs Mineralls and other considerable Commodityes." In such an enterprise, the sciences, exploration, and business could be full partners working for private profit and the public good. The company certainly had a mandate for northern exploration before the Royal Navy engaged in arctic science and exploration.

[3] S. Zeller, "The Spirit of Bacon," *Scientia Canadensis* (1969) 81; R. Stearns, "The Royal Society and the Company," *The Beaver* (1945) 8–13; R. H. G. Levenson Gower, "HBC and the Royal Society," *The Beaver* (1934) 29–31, 66.

[4] P. C. Newman, *Caesars of the Wilderness: Company of Adventurers* vol. II (Markham, Ontario, 1987) p. 219; J. S. Galbraith, *The Hudson's Bay Company as an Imperial Factor, 1821–1869* (Toronto, 1957) and *The Little Emperor: Governor Simpson of the Hudson's Bay Company* (Toronto, 1976); F. Merk, ed., *Fur Trade and Empire: George Simpson's Journal . . . 1824–25* (Cambridge, Mass., 1968).

One way in which Simpson manifested his and the company's support for the sciences was through the Natural History Society of Montreal. Simpson "represented the Company's willingness to help the society to collect specimens from within its territories, and he permitted the distribution to company posts" of questionnaires "for a preliminary inventory. Valuable written reports were duly received, but specimens intended for the society were instead ordered home by the London Committee."[5]

More effective was Simpson's engagement to support the Royal Society and Sabine's magnetic program, which enabled Lefroy to travel where he pleased in the Canadian North. Indeed, Simpson's personal encouragement led Lefroy to extend his magnetic survey considerably beyond the bounds of his orders from Sabine; this would be a subsequent source of conflict and frustration.[6]

Following Lefroy's survey, Simpson's orders were extended to anyone who "might be sent to the country for botanical or other scientific pursuits." Besides rendering such aid, company officers were also instructed to make regular meteorological observations, as well as observations on the aurora.[7]

*Peter Dease and Thomas Simpson 1837–1839.* Geographical exploration and the final resolution of the Northwest Passage formed part and parcel of Simpson's interpretation of the company's mandate. No doubt he was encouraged toward that interpretation by the impending expiration of the Company's license to trade beyond its original territorial limits.[8] In 1837 he sent out Peter Dease and Thomas Simpson: Dease had been with Franklin on his second and successful expedition; Thomas Simpson was the governor's young cousin. Their task was to set out from Fort Chipewyan, and to map those stretches of the northern coast, which still had not been explored by Franklin, between Turnagain Point and Fury and Hecla Strait in the east, and between Return Reef and Point Barrow in the west. Simpson pushed westward and eastward harder and further than Dease, and came very near to realizing his ambitions and the company's wishes. Dease and more significantly Simpson had indeed added to the map; geographical science had gained. So had other sciences. Along the way they collected geological specimens, noted erratic boulders, and described the principal stratigraphic

[5] Zeller, "The Spirit of Bacon," 88; McGill University Library, Natural History Society of Montreal (NHSM), "First Report of the Indian Committee," 26 May 1828; NHSM, *Annual Report* (Montreal, 1830) pp. 3–4.
[6] Royal Artillery Institution, Woolwich, Lefroy–Sabine correspondence (bound MS volume).
[7] Zeller, "The Spirit of Bacon," 92–3.
[8] Berton, *The Arctic Grail* (Toronto, 1988) p. 131.

features. They had been provided with astronomical and surveying instruments, and made regular observations to determine latitude and longitude. They also made fairly regular observations of magnetic variation, and carried out measurements of magnetic dip between Cape Barrow and Cape Herschel.[9] Dease bowed out from further exploration, and Simpson proposed to renew the exploration of the remaining stretch between Queen Maud Gulf and Hudson Bay. George Simpson was delighted:

> our Arctic travellers . . . did glorious work during the past summer. Dease has got tired of it but my relative & namesake is quite a Glutton in the way of discovery and is alone to conduct the expedition from the mouth of Great Fish River to the straits of Fury & Hecla. He is fond of scientific pursuits and as our barometrical thermometers were disposed of so unfortunately through . . . bad packing and our worse roads I beg to suggest that you give . . . the necessary instructions to prepare and pack up 2 or 3 more to be sent to Simpson by one of our Bay ships. . . . [10]

Thomas Simpson never learned of his powerful cousin's support. Impatient for authorization, which he could not know was on its way to him, he decided to argue his case in London; but on the way, he died of gunshot wounds, attributed then to suicide, but just as easily attributable to murder. Simpson was hot-tempered, and could easily have engaged in a violent and fatal brawl.[11] The company's commitment to exploration was put momentarily on hold.

*John Rae 1846–1847.* Thomas Simpson had died believing that Boothia was an island, in which case that unexplored stretch of arctic coast would be only about one hundred miles long; but Boothia was a peninsula, and to go round it meant that the gap was in fact nearer seven hundred miles long. But still, *most* of the coast was known, and the final resolution of the passage was surely within grasp. George Simpson wanted the company's grasp to be the one that closed around it. He looked for somebody tough, somebody experienced in the North, a good company man, someone sufficiently educated for the more recondite demands of modern exploration, a skilled hunter, and one young enough to be vigorous in the field. He found his man in Dr. John Rae, an Orcadian who had studied medicine at Edinburgh, and

9  *The Times*, 18 April 1840, quoting Dease and Simpson to HBC, Fort Simpson, 16 Oct. 1839; Thomas Simpson, *Narrative of the Discoveries on the North Coast of America: Effected by the Officers of the Hudson's Bay Company during the Years 1836–1839* (London, 1843): Dease to John Richardson, 3 Dec. 1840, and Journals: HBC instructions 2 July 1836, entries 30 July 1838 and 3 Aug. 1838.
10  G. Simpson to Capt. Washington, RGS, 25 April 1840, RGS HBC letters.
11  Berton, *The Arctic Grail*, pp. 136–7.

obtained the licentiate of the Royal College of Surgeons of Edinburgh in 1833. He had forthwith set out for Hudson Bay as surgeon on the company ship *The Prince of Wales*, and he spent the next decade as surgeon at Moose Factory.[12]

Simpson began discussing needful preparation with Rae. "As regards the management of the people and endurance of toil, either in walking, boating or starving, I think you are better adapted for this work than most of the gentn. with whom I am acquainted." As for technical qualifications, "[w]ith a little practice in taking observations, which might very soon be acquired, I think you would be quite equal to the scientific part of the duty."[13] In July, Simpson ordered Rae to Red River, where, over the winter, and under the guidance of George Taylor, he would "study . . . Astronomy, filling up any spare time you may have in making yourself conversant with Geology, Botany & such sciences as you may have an opportunity of giving attention to."[14] Rae accordingly walked from the Hudson Bay Lowlands to Red River, where, as he reported to Simpson, he first found Mr. Taylor sick, then wandering in his mind, and finally dead, so that

he was incapable of affording me any assistance in my *studies*. At first my intention was to endeavour to gain a knowledge of Astronomy by my own unaided exertions, but a little reflection caused me to give up the idea of doing so, as altho' I might have become sufficiently scientific, yet there could not be that dependence placed on my observations, necessary in a manner of so much importance, without their being tested by a person of experience.[15]

At least a rudimentary knowledge of astronomy was necessary for the use of a sextant, or, better still, of a transit instrument, essential for the accurate determination of place. So, carrying a sextant and a few books, he set out on snowshoes to Sault Sainte Marie, and then went on to Toronto, there to learn what he could from Lefroy at the magnetic observatory. Rae reported to Simpson: "My progress in the study of astronomy has been slow, however I hope to acquire sufficient scientific knowledge for the purpose." Lefroy was altogether helpful, "and his labors have not been entirely thrown away, although they have not proved so successful as they would have done, if he had had a less stupid head to work upon."[16] Back at the

---

[12]   R. L. Richards, *Dr. John Rae* (Whitby, 1985).
[13]   Quoted in Richards, *Rae*, p. 29, Simpson to Rae, 11 May 1844.
[14]   Quoted in Richards, *Rae*, p. 17.
[15]   Rae to [Simpson], Mechipicoton, 25 Feb. 1845, PAC HBC microfilm HBC 4M22; published in The Hudson's Bay Record Society; XVI: *Rae's Arctic Correspondence 1844–55: John Rae's Correspondence with the Hudson's Bay Company on Arctic Exploration 1844–1855*, E. E. Rich and A. M. Johnston, eds. (London, 1953) pp. 4–7.
[16]   Rae to Simpson, Toronto, 9 July 1845 (published in *Rae's Arctic Correspondence*, pp. 10–11) and Buffalo, 20 July 1845 (published ibid., pp. 11–12), PAC HBC microfilm 4M22.

Sault later that summer, Rae had learned enough to realize that the company's dip circle was a very inferior instrument, and that it lacked a needle for determining magnetic force. He asked Lefroy to have one made, and reassured Simpson, who was always parsimonious for the company, that it would cost only a dollar or two. The azimuth compass, a basic navigational and surveying tool, also needed repair, since it would not traverse at all.[17] Rae's vigorous travels, and the lengthy shipping route of instruments from England, combined to test those instruments, sometimes to destruction. That autumn, during Rae's trek to York Factory, the chronometer, essential for accurate determination of longitude, became irregular, but became steady again at the factory. And the use of instruments in general was always a challenge in winter; gloves were an impediment to manipulating the fine screws, and bare skin froze to the metal. Nevertheless, Rae commented, with practice the sextant could be managed at $-40°$.[18]

That winter, Rae trained some of the factory hands as assistants in taking observations, and in skinning birds and other specimens. He wrote privately to Simpson:

My time is principally occupied in taking observations, studying Natural History, Geology, Botany & c. With one branch of Natural History (Ornithology) and to me the most useful one, I am slightly acquainted, having read a little on the subject, but learnt more by observation during many shooting excursions. If I can find time, I purpose employing myself during a part of the month of May, in making a collection of the birds to be found here, as I think there are some which have not yet been described among the birds migrating to Hudsons Bay. Of Geology and Mineralogy I know little or nothing – tis true, I can tell a boulder stone from a brick-bat and a bank of alluvial deposits from a stratum of lime when I see them – and could form a pretty good guess of quartz, slate, granite and some other of the more common minerals, I might also venture a few remarks on the Geological features of a new country, if they were not very complicated, but this is all.[19] I intend making a collection of plants whenever I have time and opportunity, but shall not attempt to name or class them; a much more intimate acquaintance with Botany than I possess would be requisite to do so correctly.[20]

Rae was as well prepared in science as most of his naval predecessors. He was a competent surveyor and navigator, familiar with the use of meteorological and magnetic instruments, an efficient collector of natural history

---

[17] Ibid., Rae to Simpson, Sault Ste. Marie, 29 and 30 July 1845 (published pp. 13–16).

[18] Ibid., Rae to Simpson, 29 Nov. 1845 (published pp. 17–20).

[19] Cf. John Rae to Henry Wemyss Feilden, 19 April 1878, Royal Geographical Society archives, referring to his explorations in the 1840s, "[h]aving no knowledge of geology and pretending to none. . . . my hobby was in a very mild way, natural history."

[20] Rae to Simpson, 24 Feb. 1846, PAC HBC microfilm 4M22 (published in *Rae's Arctic Correspondence*, pp. 22–4).

specimens, although no taxonomist, and something more than a collector in ornithology.[21] Simpson sent him instructions[22] in June, and they were as ambitious in geography, and as encyclopedic and unrealistic as any naval orders had been. Rae was to be the company's answer to a succession of naval expeditions. He, on behalf of the company, was "to complete the geography of the Northern Shore of America, by surveying the only section of the same that has not yet been traced"; Dease and Thomas Simpson had almost secured what the world expected of the company, the achievement of a final settlement of the problem of the Northwest Passage, and Rae's job was to satisfy the world on this score.

Or rather, that was Rae's principal task. But there was also the subsidiary task of beating the navy in fulfilling their and the Royal Society's program. Simpson went on to tell Rae, with wonderful unreality, that

you will devote as much of your attention as possible to various subordinate and incidental duties. You will do your utmost, consistent with the success of your main object, to attend to Botany and Geology; to Zoology in all its departments; to the temperature both of the air and of the water; to the conditions of the atmosphere and the state of the ice; to winds and currents; to the soundness as well with respect to bottom as with respect to depth; to the magnetic dip and the variation of the compass; to the aurora borealis and the refraction of light. You will also, to the best of your opportunities, observe the ethnographical peculiarities of the Esquimaux of the country; and, in the event of your wintering within the Arctic Circle, you will be careful to notice any characteristic features or influences of the long night of the high latitudes in question. These peculiarities, and such others as may suggest themselves to you on the spot, you will record fully and precisely in a journal to be kept, as far as practicable, from day to day, collecting, at the same time, any new, curious or interesting specimens in illustration of any of the foregoing heads.[23]

[21]  Rae was a good field ornithologist, but the labeling of specimens was not his forte. P. A. Taverner noted of nineteenth-century Arctic ornithology in general, and of Rae in particular: "The principal errors to look out for are exact localities of specimens and observations and the identification, especially of those made by sight, of closely resembling species. Of the former, the results of the Rae Expedition to the Arctic Seas in 1846–1847 are an example. The specimens resulting from this expedition found their way into the British Museum and are included in the British Museum Catalogue of Birds as all from Repulse Bay whereas under present knowledge it is evident that many of these must have been taken much farther south along the Hudson Bay coast even to York Factory enroute [*sic*] to or from the more northern point. The uncertainty of locality in this case is particularly unfortunate as considerable interesting material was collected but without exact locality data and originating over some twenty degrees of longitude from within the tree limits to the arctic barren-grounds, much of its value as distributional evidence is lost." The typescript of Taverner's comments on "The Birds of the Arctic Islands" is in the Department of Vertebrate Zoology, National Museum of Natural Science, Ottawa; I am grateful to Prof. Jack Cranmer-Byng for telling me of it.

[22]  Simpson to Rae, 15 June 1846, PAC HBC microfilm 4M22.

[23]  Ibid.

Rae was properly amused: "You appear to think," he rejoined, "that I have got a head stuffed with all sorts of knowledge. . . . The head is big enough certainly outside, but whether there is a large quantity of bone in it or not I have not yet tested." Still, Rae attempted to follow up Simpson's instructions.[24] Geographically, his expedition was successful, although not entirely so; he left unexplored only the west coast of Boothia Peninsula, to which he returned in the next decade. In the sciences, he kept a magnetic and a meteorological journal, and collected natural history specimens; around York Factory, the northern phalarope was so common that "I could have shot twenty in half an hour." His party included two native interpreters, and with their help he had extensive and friendly dealings with the Inuit, describing their appearance, tattooing, and clothing. Rae wintered at Repulse Bay, where he established two observatories built of snow, as Lefroy had advised, with a pillar of ice in each, one for the dip circle, the other for a horizontally suspended needle "to try the effects of the aurora upon it."[25] His over-wintering was different from all previous European ventures, for he achieved it with a small party, living off the land by his and his interpreter's skill as a hunter. He learned from the Inuit, applied that learning to his own experience, and returned with all his men healthy. Their return was welcomed, but it may be doubted whether anything short of total success in the conquest of the passage would have provided what Simpson wanted: the speedy compensation "of the Hudson's Bay Company for its repeated sacrifices and its protracted anxieties."[26] But he should at least have felt some relief as a man of business, for the cost of the expedition was only "the comparatively trifling expense of £1,100 or £1,200 st[erlin]g."[27]

Simpson was satisfied enough to send Rae north again, in the aftermath of Franklin's last and lost expedition. Rae's eventual findings, although definitive, were unwelcome to British ears. Simpson supported Rae, who was to face a stormy reception, not the least for championing the company and himself against the rival claims of naval officers. We shall rejoin Rae in the course of Franklin searches. But first, we need to consider what has become known as *the* Franklin expedition.

[24] Ibid., Rae to Simpson, 5 and 8 July 1846 (published in *Rae's Arctic Correspondence*, pp. 31–5).
[25] John Rae, *Narrative of an Expedition to the Shores of the Arctic Sea in 1846 and 1847* (London, 1850) pp. 12, 90, 76.
[26] Simpson in Rae, *Narrative*, p. 17.
[27] Rae to Simpson, 20 Sept. 1847, PAC HBC microfilm 4M22 (published in *Rae's Arctic Correspondence*, pp. 43–4).

## JOHN FRANKLIN'S LAST EXPEDITION 1845–[28]

George Back's arctic expedition of 1836 had been grueling, harrowing, and almost tragic. The trauma lingered in memory for all who had been associated with it. Rae had found difficulty in finding volunteers for his expedition a decade later, in part because company men who had helped guide Back were persuaded "that the whole party, if not starved for want of food, would run the risk of being frozen to death for want of fuel."[29] The Lords of the Admiralty had something of the same notion of the risks, until James Ross returned successfully from the Antarctic in 1843.

Beechey promptly requested the renewal of arctic exploration. He wanted Ross's antarctic vessels *Erebus* and *Terror,* already strengthened against the ice, to be fitted with screw propellors, and sent to the North Pole. Others among the arctic crusaders argued for the passage first, and then the Pole. John Barrow, on the verge of retirement but still and always urging renewal and revitalization of the search for the passage, conferred with Sabine and with Beaufort, and submitted to the First Lord of the Admiralty a "Proposal for an attempt to complete the discovery of a North-West Passage." Barrow played a familiar tune on a familiar instrument. He stressed England's naval glories since Tudor times, the unwisdom of yielding the palm of discovery to other nations active in the North, the hydrographic gains for Britain, the virtues of such a peacetime use of the navy, and above all the importance of such a voyage for the completion of the magnetic survey of the globe.[30] The Admiralty submitted the plan to the Royal Society, where, unsurprisingly, with its arctic officers and magnetic enthusiasts, approval was prompt. James Clark Ross, on the Royal Society's Council, was surely pleased.[31]

When Ross sailed for the Antarctic in 1839, he had done so largely in response to the recommendation of the British Association for the Advancement of Science in 1838, that the deficiency of southerly magnetic observations be made good.[32] The association had at the same time underscored

[28]  The best general account remains Richard J. Cyriax, *Sir John Franklin's Last Arctic Expedition: A Chapter in the History of the Royal Navy* (London, 1939); otherwise unacknowledged sources in this section are from this book.

[29]  Rae, *Narrative,* p. 3.

[30]  Barrow to Haddington, 27 Dec. 1844, quoted in Cyriax, *Franklin's Last Expedition,* pp. 18–20; PRO Adm. 7.187.

[31]  Members of the council of the RSL at this date included James Clark Ross, Roderick Impey Murchison, William Buckland, and Richard Owen: the full membership of council is given in *Proc. Royal Soc. London 1843–1850 5* (1851) 533–4.

[32]  J. C. Ross, *Voyage,* pp. v et seq.

the importance of other centers for magnetic observation, including Canada. They had in mind both the permanent observatory at Toronto, and the prospect of further observations in the region of the north magnetic pole. The Royal Society fully endorsed the British Association's recommendations, and the Admiralty was persuaded. Now the Royal Society again emphasized that, besides geographical knowledge, the main advantage would derive from magnetic observations. Time was of the essence, for the international cooperative program in geomagnetism was drawing to a close. "[I]f the expedition were deferred beyond the present season, the important advantages now derivable from the co-operation of the observers with those who are at present carrying out a uniform system of magnetic observations would be lost."[33] Franklin, recently returned from Tasmania, Parry, and James Clark Ross were consulted, and all were pleased to encourage another voyage. Parry stressed the high prospects of success in finding the passage. Ross added to such optimism the prospect of economic gains from expanded whale fisheries. Recently discovered fishing grounds on the west coast of Baffin's Bay, Lancaster Sound, and Prince Regent Inlet were already producing more than half a million pounds sterling each year, more than the entire cost of arctic exploration to date. The navy, always conscious of cost, was unlikely to be indifferent to such an argument. Franklin, agreeing that they now knew where to look, either between Cape Walker or Banks Land, or down the Wellington Channel, underlined the incalculable advantages of steam, and the confident prospect of geographical and magnetic gains.[34]

Peel, the prime minister, gave approval. But who should lead the expedition? The Admiralty first approached James Clark Ross, but he refused to go. Barrow had his own nominee, James Fitzjames, who was rejected on the grounds of youth; he was then in his early thirties. Then Franklin threw his hat into the ring. The end of his term in Tasmania had been unhappy. Now he saw an opportunity to complete the task to which he had already devoted two expeditions. He was experienced, a good commander, well liked, familiar with arctic navigation and with a range of scientific observation. If Fitzjames was too young in his thirties, was Franklin too old at fifty-nine? The First Lord conferred with Parry, who is reputed to have said that he was the fittest man he knew for the job, "and if you don't let him go, the man

[33] Cyriax, *Franklin's Last Expedition*, p. 23; PRO Adm. 7.187, Marquis of Northampton to Admiralty, 16 Jan. 1845.

[34] J. C. Ross to Admiralty, 24 Jan. 1845; W. E. Parry to Admiralty, 10 January 1845; J. Franklin to Admiralty, 24 Jan. 1845; PRO Adm. 7.187.

will die of disappointment."[35] Franklin was given command of the expedition, and was appointed to H.M.S. *Erebus;* second in command of the expedition, and captain of H.M.S. *Terror,* was Francis Crozier, who had commanded the same ship on its antarctic voyage; and second in command on *Erebus,* third in the expedition, was Barrow's candidate, Fitzjames.

The proposed scientific program emphasized geomagnetism, but, as Fitzjames remarked, the expedition "was to observe 'everything from a flea to a whale', a task in which each officer was encouraged to take one branch of science under his immediate care . . . "[36] Franklin's instructions[37] were indeed comprehensive. The main goal of the expedition was the completion of the Northwest Passage. The Admiralty referred to the "great variety of valuable instruments" supplied to the expedition:

among these, are instruments of the latest improvements for making a series of observations on terrestrial magnetism, which are at this time peculiarly desirable, and strongly recommended by the President and Council of the Royal Society, that the important advantage be derived from observations taken in the North Polar Sea, in co-operation with the observers who are at present carrying on an uniform system at the magnetic observatories established by England in her distant territories, and, through her influence, in other parts of the world; and the more desirable is this co-operation in the present year, when these splendid establishments, which do so much honour to the nations who have cheerfully erected them at great expense, are to cease. The only magnetical observations that have been obtained very partially in the Arctic Regions, are now a quarter of a century old, and it is known that the phenomena are subject to considerable secular changes. It is also stated by Colonel Sabine, that the instruments and methods of observation have been so greatly improved, that the earlier observations are not to be named in point of precision with those which would now be made; and he concludes by observing, that the passage through the Polar Sea would afford the most important service that now remains to be performed towards the completion of the magnetic survey of the globe.[38]

Here was a remarkable degree of enthusiasm for magnetic work. Certainly the previous quarter century had seen major advances, including the development of an international network of magnetic observatories, coordinated through term-days of shared patterns of intense observation; the development of standardized and key instruments, deriving from Gauss's work at Göttingen, and later modified in Britain at Kew and in Dublin;

---

[35]  Cyriax, *Franklin's Last Expedition*, p. 27.

[36]  "Journal" 6 June 1845, in *Nautical Magazine* (March 1852) 159–60, quoted in H. N. Wallace, *The Navy: the Company, and Richard King*, p. 99.

[37]  Great Britain, Admiralty, Arctic Blue Books, "COPIES of INSTRUCTIONS to Captain Sir *John Franklin,* in reference to the ARCTIC EXPEDITION of 1845; and to the Officers who have been appointed to command EXPEDITIONS in search of Sir *John Franklin*" (Arctic Bibliography 45216); Cyriax, *Franklin's Last Expedition*, chapter 4.

[38]  Admiralty, "Instructions," p. 5.

the relatively systematic training in magnetic observation provided through the Royal Artillery at Woolwich; and the less systematic training sometimes provided by Lloyd in Dublin. Also important was the ancillary development of instruments for use at sea, notably Fox's dip circle, and of devices and methods for eliminating the effects of a ship's iron on the magnetic compass.[39]

The Admiralty bought the argument in its entirety, and asked Sabine, now a lieutenant colonel, to instruct Commander Fitzjames in the magnetic mysteries. Fitzjames would then be in charge of magnetic work during the expedition. Several other officers on the expedition had already received training at Woolwich, and therefore a full program of magnetic observations could be carried out on both ships. If the ships were forced to spend a winter in the North, their portable observatories would be set up and equipped with instruments similar to those employed in the fixed colonial magnetic and meteorological observatories. Observations would then be carried out exactly according to the system used in those observatories.[40]

There were also experiments on atmospheric refraction, and work in hydrography and oceanography, including depth and bottom soundings, and measurement of tides and currents. Made sanguine about polar work by Ross's antarctic voyage, and imbued with renewed optimism about the Northwest Passage, of which so little, it seemed, remained to be explored, the Admiralty was uncommonly well inclined to science:

And you are to understand that although the effecting of a passage from the Atlantic to the Pacific is the main object of this expedition, yet, that the ascertaining the true geographical position of the different points of land near which you may pass, so far as can be effected without detention of the ships in their progress westwards, as well as such other observations as you may have the opportunities of making in natural history, geography, & c. in parts of the globe, either wholly unknown or little visited, must prove most valuable and interesting to the science of our country; and we therefore desire you to give your unremitting attention, and to call that of all the officers under your command to these points, as being objects of high interest and importance.[41]

*Erebus* and *Terror* sailed from London on 19 May 1845. Their transport left them in July, carrying papers home, including a letter from Franklin to the Admiralty (dated 12 July) indicating that the magnetic instruments had

---

[39] See chap. 4. See also A. McConnell, *Geophysics and Geomagnetism: Catalogue of the Science Museum Collection* (London, 1986); R. P. Multhauf and G. Good, *A Brief History of Geomagnetism and A Catalog of the Collections of the National Museum of American History* (Washington, D.C., 1987).

[40] Admiralty, "Instructions," p. 6.          [41] Ibid.

been used and found most satisfactory. Later that month, two whalers saw his ships heading for Lancaster Sound.

## THE FRANKLIN SEARCH EXPEDITIONS FROM 1848[42]

Franklin's ships had been strengthened against the ice, and had been well tried in the Antarctic. They were equipped with auxiliary steam engines and screws and although they were expected to complete the passage in a season, they were provisioned so that one or even two winters in the ice would cause no concern. Parry and Barrow wrote an essay on the practicability of reaching the North Pole, and the gains that would thereby accrue to science.[43] John Ross also proposed an expedition to the pole; the Russians and Swedes had lately measured an arc of the meridian, and he felt Britain should follow suit, preferably at Spitsbergen. Ross offered to lead an expedition, perhaps using Swedish sledges and horses – shades of Captain Scott before his time. Beaufort was discouraging: "I confess that the only temptation to me in the whole scheme is the measurement of an arc of the meridian . . . – and that would require so many instruments to be prepared – so many arrangements to be made here before you start – and so many observers to accompany you that I could not in conscience [ . . . support] your request."[44]

In November 1846, Ross raised the issue of a search for Franklin in 1847, unless he returned or was heard of before then, in which case the Spitsbergen project should be revived.[45] But there seemed as yet no need to worry about Franklin; Ross's concerns were not echoed at the Admiralty, which at that stage was more concerned about bills from the Hudson's Bay

---

[42]   There is an extensive literature on "the" Franklin expedition and searches. See Maurice Hodgson, "The Literature of the Franklin Search," and Alan Cooke, "A Bibliographical Introduction to Sir John Franklin's Expeditions and the Franklin Search," in P. D. Sutherland, ed., *The Franklin Era in Canadian Arctic History 1845–1859* (Ottawa, 1985) pp. 1–11 and 12–20 respectively. The entire volume provides a good multidisciplinary account, drawing on history, bibliography, archeology, forensic medicine, glaciology, meteorology, and other disciplines.

[43]   *Edinburgh New Philosophical Journal* 40 (1846) 294–301.

[44]   F. Beaufort to John Ross, 21 May 1846, McCord Museum MS 2833. John Ross to ? (marked copy to Sir Chas. Adam), 27 Nov. 1846, SPRI MS 486/5/1.

[45]   Ross letter 27 Nov. 1846, and letter to Marquess of Northampton [President of the Royal Society of London], 19 Nov. 1846, Museum of the History of Science, Oxford, MS Buxton 2 no. 22.

Company and the geomagnetic program of the Royal Society.[46] In January, February, and March 1847, John Ross peppered the Admiralty with warnings. Franklin would soon be facing his third winter in the Arctic, and running low on supplies. A relief party would need to find Franklin that summer to save him from such a winter. Ross, nearing seventy, offered to lead such a party. But the Admiralty's chief advisors, Barrow, Richardson, James Clark Ross, and Sabine – the group that was to be known as the Arctic Council[47] – still saw no reason for alarm.

By the fall of 1847, when neither Franklin nor news of him had come, John Ross's concern for his friend had spread. The Admiralty decided on a search by four ships and an overland party in 1848. James Ross would take two ships among the arctic islands from the east; two other ships would enter from Bering Strait in the west; John Richardson would lead an expedition down the Mackenzie to the coast; and, with optimism far exceeding experience, the Admiralty assumed that the naval expeditions by sea and land would link up, and that Franklin's fate would soon be known. There were in the end more than forty search expeditions; weaving through the arctic archipelago in space and time, they would establish Britain's claim to the islands, and subsequent Canadian sovereignty.[48]

*Richardson and Rae 1847–1849.* John Richardson was Franklin's old confidant, correspondent, and fellow explorer. Like Franklin, he was then sixty years old. As he reported to Lady Franklin, he had heard of John Rae's recent successful exploration of the bottom of Prince Regent Inlet, and went in pursuit of him when they were both in London. He liked what he saw.[49] A few days later, he wrote to Rae, inviting him to join his proposed expedition as second in command. There had been numerous volunteers from the navy, the army, and civilian occupations, but Richardson had waited until now in hopes of hearing news of Franklin. He made it clear to Rae that he had no authority from the Admiralty, but was merely sounding him out. The object was "to descend the Mackenzie next summer with 4 boats and 20 men, to communicate with the Esquimaux and examine the coast and islands between that river and the Coppermine, seeking for tidings of the Discovery ships and being guided by what we may learn."

[46] F. Beaufort, Hydrographical Diary, Huntington Library MS FB 42, 12 June and 12 July 1847.
[47] "The Arctic Council of 1851," *Polar Record* 6 (1952) 385–9.
[48] L. H. Neatby, *The Search for Franklin* (London, 1970).
[49] Richardson to Lady Franklin, Haslar Hospital [Portsmouth], 8 Nov. 1847, SPRI Richardson–Voss collection MS 1503/35/44.

He planned to winter on Great Bear Lake, and then make another summer excursion to the arctic coast. "You ought also to be made aware," he warned Rae, "that the Admiralty require officers employed by them to place all collections of objects of natural history etc. at their disposal."[50] Rae replied from Orkney that he was willing to go under Richardson's command, the Admiralty agreed, and the Hudson's Bay Company granted Rae the necessary leave.[51]

Finding Franklin came first, but Richardson was also a naturalist who would seize his opportunities. He wanted a good microscope, in a case convenient for traveling, and wrote to Paris, ordering one with Chevalier's excellent lenses, and magnifying from 40 to 250 diameters.[52] Through the oceanographer Forbes in Edinburgh he requested thermometers, also from Paris, and discussed the best barometers to take. Fox Talbot, pioneer photographer, supplied Richardson with the apparatus for his art.[53] William Hooker was glad to learn that Rae was going with Richardson, since Rae's plants from his own expedition "are with me, & very fair as to amount of species & exceedingly good as to preservation." Hooker also reminded Richardson that he had paper for mounting and preserving plants waiting for him at Kew Gardens.[54] Sabine supplied magnetic instruments, and the Admiralty provided sextants and other instruments essential for navigation.[55] For a light and fast-moving search expedition, they went well equipped for science.

Richardson and Rae's journey began according to plan. They reached the mouth of the Mackenzie in August 1848, searched the coast to the Coppermine River, and wintered at Fort Confidence, which John Bell of the Hudson's Bay Company had been building for them on the shores of Great Bear Lake. This was one expedition where, prepared for scientific work as they were, science took a back seat, especially for Richardson. "I find," Rae reported to Simpson on the journey to the coast,

50    Richardson to Rae, 12 Nov. 1847, PAC HBC microfilm Addenda M2, E 15/5.
51    R. L. Richards, *Dr. John Rae*, p. 59; Rae to Richardson, Stromness, 12 Dec. 1847, SPRI MS 1503/35/56.
52    For microscopes in this period, see G. L'E. Turner, *The Great Age of the Microscope: The Collection of the Royal Microscopical Society* (Bristol, 1989).
53    Richardson to Oberhäuser, 14 Dec. 1847; James D. Forbes to Richardson, 22 Dec. 1847; H. F. Talbot to Richardson, 25 Dec. 1847: SPRI Richardson–Voss collection MSS 1503/35/57, 58, [59].
54    Hooker to Richardson, 5 Jan. 1848, SPRI Richardson–Voss collection MS 1503/38/4.
55    Richardson to [Sabine], 25 Oct. 1848, McCord Museum, Richardson MSS, folder 4; Rae to Simpson, 5 July 1848, PAC microfilm HBC 4M22/E.15/5.

Figure 22. "Barrow Instrument," unifilar vibration apparatus and variation compass used by John Richardson. Science Museum London 1876-790, neg. no. 59/79.

that Sir John has almost entirely forgot the mode of taking observations so that the greater part of that duty devolves on me – not but that by a very little practice he would become very possibly the better observer of the two, but his time is taken up with other duties and he is so anxious to get forward that he cannot fix his attention sufficiently to study the subject. The sextants supplied by Government are neither of them any good, and are not at all well suited for taking the most accurate observations – such as Lunar Distances.[56]

Richardson was indeed impatient to be moving and searching, and magnetic observations, which occupied much of their time, especially Rae's at Fort Confidence, were almost entirely neglected until they settled there for the winter. Richardson wrote to Sabine in October, confirming his expectation that the magnetic apparatus had not been used at all that summer:

in fact the rapidity of our march to . . . save time for the sea voyage, precluded me from carrying much luggage, and the instruments being necessarily left to come up in a heavily laden boat did not reach this place until a fortnight after our return from the sea. Almost all the instruments both astronomical and magnetical were much shook by the usage the packages received on the American railways and in their steam boats, which did them more damage than the portage work which is rough enough. We found on unpacking them that most of the screws [were] loose and the glass case & the frame of the dipping needle broken to pieces. The wood work of this nicely finished instrument was much too delicate for such a journey and not at all calculated for the dry climate of America, where wooden frames of every kind made in England, shrink and loosen at the joints. Fortunately my companion Mr. Rae has like the other residents in this country acquired the power of repairing such accidents & by the help of some glass intended for the windows of our residence, the frame has been made good and the instrument seems to work well. I hope that we may get a proper number of determinations by the Unifilar & Dipping needle between this time and the end of April at this place and also keep a record of the movements of the needle in the Declinometer. . . .

Since there is no record of the repaired instruments being restandardized, the results must have been of doubtful value. Richardson confirmed that he was happy to leave the bulk of geophysical and astronomical work to his companion:

Observations with such instruments being a novelty to me, I do not make them quickly, but Mr. Rae who is a more practiced observer is more *au fait* and between us I hope we shall procure a satisfactory series. . . . Every clear night since our arrival here, now five weeks ago, we have seen the Aurora, but since the Declinometer was set up, we have noticed the needle to be affected in two or three instances only.[57]

It was reasonable to use the winter for science, and the summer for traveling. There was no way that magnetic term-days could be combined with

[56]   Rae to Simpson, Portage la Loche, 5 July 1848, PAC microfilm HBC 4M22/E.15/5.
[57]   Richardson to [Sabine], 25 Oct. 1848. For the magnetic instruments etc., see chap. 4.

vigorous travel, since they involved the constant observation of one instrument after another at intervals of two and a half minutes throughout the twenty-four hours. There were besides meteorological and astronomical programs of observation to maintain. But all this was no substitute for travel. Richardson, Rae, and their crews had covered the arctic coast with exemplary efficiency, but had found no trace of Franklin and his ships. Richardson had made natural history observations, which later served as the basis for a fine essay on physical geography, and another on the geographical distribution of plants.[58] But his main object was unrealized. Too old for another season of heroic traveling, frustrated and physically stressed, Richardson determined to leave in the spring, while Rae would renew the search. The renewal was initially abortive. Rae reached the coast, but ice stopped him from crossing to Wollaston Peninsula, and he returned to Fort Simpson. He would resume the search in 1850 under the auspices of the Hudson's Bay Company, but in his later travels science was even more firmly subordinated to travel than it had been in his expedition with Richardson.

Simpson's earlier encyclopedic directives for scientific observation yielded to the imperatives of searches that became more urgent and more despairing as the months and years passed. Franklin's discovery ships might well have carried their crews alive through three winters, but thereafter most search parties recognized that they were seeking evidence of Franklin's fate, not mounting his rescue.

*Other Royal Naval searches.* Geography was ultimately the main beneficiary of the Franklin search expeditions. By their end, a route through the archipelago was known, and the coastlines of most of the islands had been mapped, except in the extreme North and Northwest. This part of the story has been often and well told.[59] The international magnetic crusade had peaked, and although the Royal Society saw continued observations as desirable, there was no longer the urgency that had equipped Franklin and his crews, and sent them to vanish in the ice. As the bills for more and more searches mounted, the Admiralty rediscovered parsimony. Science was not a priority in these years. Yet some naval expeditions to the Arctic still made significant contributions to natural science.

---

[58] J. Richardson, *Arctic Searching Expedition: A Journal of a Boat-Voyage through Rupert's Land and the Arctic Sea, in Search of the Discovery Ships under Command of Sir John Franklin: With an Appendix on the Physical Geography of North America*, 2 vols. (London, 1851).

[59] E. g., Neatby, *The Search for Franklin* (Edmonton, 1970); Berton, *The Arctic Grail.*

Henry Kellett had sailed in 1845 on H.M.S. *Herald* for surveying in the Pacific Ocean. In 1848, 1849, and 1850 *Herald* headed for Bering Strait to help in the search for Franklin, finding that Franklin had not gone west of the Mackenzie. *Herald,* equipped as a survey vessel, also produced a goodly scientific haul. Botanical collections were made whenever possible; a journal entry of 21 August 1849 records a landing twenty-five miles south of Cape Lisburne, where the naturalist, Berthold Seeman,[60] had a good harvest of plants. Seeman and Kellett's officers made various collections in natural history, including paleontology. Richardson subsequently discussed fossil remains of large animals found in the Russian and American Arctic by Kellett's expedition, by Rae and the officers of the Hudson's Bay Company, and by Beechey on *Blossom* following earlier findings by Kotzebue.[61] The dredge was used on *Herald,* where circumstances allowed. Magnetic observations were made where possible, in less than perfect conditions.[62] The bulk of British exploration in the Arctic had been from the east, so that observations from Kellett's voyage, even when fragmentary, were especially welcome.

Meanwhile, James Clark Ross had sought Herschel's advice about what scientific work might be done "for our occupation and usefulness" should his ships pass a winter in the Arctic.[63] That summer, his ships entered the archipelago via Lancaster Sound, then wintered on Somerset Island where ice held him until the following August. They made spring sledging journeys across Barrow Strait and Prince Regent Inlet; the Strait was blocked by ice, and ice carried the ships back to Baffin Bay, whence they returned to England, and to public if not professional acclaim.[64]

Scientifically Ross's voyage had achieved little, and as for finding Franklin, he had achieved nothing. He had persisted in the old naval style, using relatively large ships and correspondingly large crews. John Ross, an advocate of smaller ships and smaller parties, and the only one whose ships had successfully passed four consecutive winters in the ice, had constantly bad-

[60]  B. Seeman, *Narrative of the Voyage of H.M.S. Herald during the Years 1845–51, under the Command of Captain Henry Kellett, R. N., C. B., being a Circumnavigation of the Globe, and Three Cruises to the Arctic Regions in Search of Sir John Franklin,* 2 vols. (London, 1853).

[61]  J. Richardson, "Zoology of the voyage of H.M.S. Herald," in his *Vertebrals, Including Fossil Mammals* (London, 1854).

[62]  Kellett, in unidentified newspaper cutting, PRO Adm. 7.189.

[63]  Ross to Herschel, 23 April 1848, RSL MS HS 14.430.

[64]  See, e. g., Robert Burford, *Description of Summer and Winter Views of the Polar Regions as seen during the Expedition of Capt. James Clark Ross, Kt, F.R.S., in 1848–9, now exhibiting at the Panorama, Leicester Square: Painted by the Proprietor, Robert Burford, from Drawings Taken by Lieut. Browne, of H.M.S. "Enterprise"* (London, 1850).

gered the Admiralty about the best way to mount a rescue expedition. On 28 April 1849 the hydrographer's journal noted "Reports all day on John Ross and other plagues"; a related entry for 8 June notes simply "Lady Franklin!"[65] John Ross was as exasperated as he was exasperating, and not less so for being more in the right of it than the Admiralty. He exclaimed that "The return and complete failure of my nephew Sir James Ross, owing, I believe, to the absurd size of his ships, has left the position and fate of poor Franklin still undecided." He advocated cooperation with the Hudson's Bay Company, something uncongenial to both the Admiralty and the company, the former appalled at the latter's lack of military discipline and respect for hierarchy, the latter dismayed at the unfitness of sailors for heavy work on land, let alone for living off the land. But John Ross, no favorite with the Admiralty, wanted them to work with "the intrepid Dr. Rae" to find Franklin, and "at any rate in surveying the remaining 140 miles and thereby determine the existence of a N W passage."[66]

Rae was sent out again by the company in 1850–1. Although he managed to reach Victoria Island on this occasion, he came no nearer to discovering news of Franklin. More surprising was the company's contribution of £500 toward John Ross's proposed expedition, which had for its object "*solely* to search for and relieve Sir John Franklin and his brave companions." If altruistic, the company's action was admirable; if partly calculating, it was good public relations. On 14 February 1851, the Hudson's Bay Company informed the editor of the *Times* of their contribution, and indicated willingness to receive subscriptions for Ross. Everything hinged on public subscription, which was soon raised.[67] Franklin's fate absorbed the nation, and the hydrographer had to prepare charts of the polar sea showing the tracks of the search expeditions for the information of Queen Victoria, the perfect embodiment of her people's concerns and aspirations.

Shortly before Ross's departure in May 1850, Jane Franklin, who previously had little time for Ross, now wrote to urge him on; no doubt he would make interesting geographical discoveries, "tho' I am sure no discoveries you can possibly make will be any inducement to you compared with the hope of saving your poor countryman."[68] Ross shared her priorities. Before he left, he sought a farewell interview with his estranged wife: his

65 F. Beaufort, hydrographical diary, Huntington Library MS FB 44. November and December saw Beaufort working hard on polar charts and polar affairs.

66 PAC microfilm HBC 4M23.E.15/7, Ross to A. Barclay, 7 Nov. 1849.

67 Ibid., correspondence, subscriptions, and miscellaneous 1849–51, concerning Sir John Ross's third arctic expedition.

68 Jane Franklin to John Ross, 21 May 1850, Museum of the History of Science, Oxford, MS Buxton 2 no. 36.

domestic life had been as rough as his professional career, he was seventy-three years old, and he was not confident that he would return.[69] Then he sailed, in close company with H.M.S. *Resolute,* which, with H.M.S. *Assistance,* H.M.S. *Intrepid,* and H.M.S. *Pioneer,* constituted one of the major naval search expeditions of those years.

The naval expedition discovered Franklin's winter quarters on Beechey Island, and Ross helped to search the island. At the same time, William Penny sailed with two ships, H.M.S. *Lady Franklin* and H.M.S. *Sophia;* Penny entered the archipelago via Lancaster Sound, joined Horatio Thomas Austin's ships[70] at Beechey Island, helped to explore it, and found graves of Franklin's men. He wintered on Cornwallis Island, next to Ross's expedition.

Penny's expedition was remarkable because he was a whaler, indeed the leader of the Davis Strait whalers, who had been in the Arctic for thirty seasons, and had commanded a whaling ship for sixteen of them. Jane Franklin had maneuvered the Admiralty into having him as expedition commander, in spite of the view that even if he succeeded, the results "would prove inimical to the strict rules of government service."[71] There was friction between the whaler and the naval officers, and afterward some of this got back to London. Penny and Austin had clashed over the question of going up Wellington Channel, with Penny wanting it and Austin refusing it. "I put," said Austin, "some close questions to Capt. Penny – but not by any means so close as I should have put to Capt. Ommanney[72] *or anyone of his rank.*" *The Athenaeum* exploded:

[T]he royal navy captain scorned to take information or advice from the captain of a "mercantile" Expedition, though sailing like himself under Admiralty orders, and engaged, with him, at great national cost, on a common work of humanity. Sir John Franklin and his gallant companions might lie and rot . . . – and the yearnings of a generous country after its long lost sons be spurned and disregarded – rather than the former commander of a whaler should show the way to rescue.[73]

[69]  Ross to [John Jones], April 1850, Museum of the History of Science, Oxford, MS Buxton 2 no. 34.
[70]  Austin had been a lieutenant on Parry's third expedition in 1824–5, and was now captain of H.M.S. *Resolute,* having sailed in 1850 as the commander of four ships, the largest Franklin search expedition.
[71]  P. C. Sutherland, *Journal of a Voyage in Baffin's Bay and Barrow Straits, in the Years 1850–1851, Performed by H.M. Ships "Lady Franklin" and "Sophia," under the Command of Mr. William Penny, in Search of the Missing Crews of H.M. Ships Erebus and Terror,* 2 vols. (London, 1852), vol. i p. xx.
[72]  Commanding H.M.S. *Assistance.*
[73]  20 Dec. 1851. For the context of Penny's involvement in controversy, see C. A. Holland,

There were, however, plenty who had a high regard for Penny's judgment and experience, and among them was Peter Cormack Sutherland, surgeon of the *Sophia*. Sutherland was in the tradition of naval surgeon-naturalists, and contributed to making natural history collections and observations ranging from ethnology to geology.[74] Thanks to him, Penny's expedition managed a good deal of such scientific work. Geology was best served by Sutherland; his discoveries included Silurian invertebrate fossils in Baffin Bay. Murchison, Siluria's discoverer and champion, was now president of the Royal Geographical Society, involved in the plans for almost every arctic expedition, including the Franklin searches. Impressed by Sutherland's discoveries, he later helped him to secure the post of geological surveyor of Natal.[75]

Besides the geological material, which was written up by J. W. Salter, assistant naturalist of the Geological Survey of Great Britain, there were important findings in zoology, including the discovery of four new species of brittle stars, described by the ever-active Edward Forbes. Botanical specimens were described by William Hooker, and algae by Sutherland's old friend and teacher Dr. Dickie, professor of natural history at Queen's College, Belfast. The standard meteorological log was published too, with Sutherland's added emphasis:

The subject of climate (or I should rather say meteorology) – equally interesting to the physical geographer, the geologist, and the naturalist, in whose domain it works marvellous changes; as it is important to the merchant, . . . and to the physician, who can neither understand nor counteract its subtle and frequently pernicious influence upon the human race – has hitherto been treated with too much indifference by travellers generally to enable those deeply interested in it to arrive at correct inductions.[76]

*Sophia* returned, and Sutherland promptly shipped out again, this time on *Isabella* under Edward Augustus Inglefield. Again there were botanical, geological, and meteorological reports; and "to those persons who understand the value of such research, the examination of the sea-water will suggest ideas with reference to the whale fisheries that may be worth more than passing consideration – especially now that our fisheries are so rapidly

"The Arctic Committee of 1851; a Background Study," *Polar Record* 20 (1980) 3–19, 105–118.

[74] Sutherland, "On the Esquimaux," *Journal of the Ethnological Society of London* 4 (1856) 193–214; "On the Geological and Glacial Phenomena of the Coasts of Davis Strait and Baffin Bay," *Quarterly Journal of the Geological Society of London* 9 (1853) 296–312.

[75] Stafford, *Scientist of Empire* (Cambridge, 1989).

[76] Sutherland, *Journal of a Voyage*, vol. i pp. xxxviii–xxxix.

declining."[77] This was to be a spur to further northern voyages, especially on the part of the Americans, who for a while came to dominate the route between Greenland and Ellesmere Island. But at least in the early 1850s, the northern searching enterprise remained predominantly British.

Many explorers and expeditions added incrementally to arctic science. Francis Leopold McClintock ranks high among them. Born in 1819, he entered the navy in 1831, had served on the South American station, in the Pacific, and in 1848 had been to the Arctic under James Ross. Tough, active and energetic, he became the unquestioned champion of European arctic sledge travelers, and collected valuable geological specimens during his journeys. As Darwin had pointed out, this required a good grasp at least of the principles of geology. Since rocks and fossils are heavy and impede efficient travel, McClintock's collections bespeak a determined commitment to geological science.

In 1849, he sailed under James Ross, wintered on Somerset Island, and undertook a series of sledge journeys that he described as brutally demanding. Nevertheless, he managed to collect what he dismissed as a few "waistcoat pocket" specimens, which he later presented to the Royal Dublin Society's Museums. He wintered at Port Leopold, where, eleven hundred feet above sea level, he found many natural casts of a fossil gastropod, hitherto unknown and named after him as *Loxonema M'Clintocki*. There were also corals, and other fossils "of a decidedly Upper Silurian type." Murchison must have been well pleased. McClintock collected numerous specimens of minerals and fossils, and did a good job of geological mapping on Somerset Island.[78]

---

[77]   E. A. Inglefield, *A Summer Search for Sir John Franklin, with a Peep into the Polar Basin: With Short Notices by Professor Dickie, on the Botany, and by Dr. Sutherland, on the Meteorology and Geology; and a New Chart of the Arctic Sea* (London, 1853). The Franklin searches were to extend whaling grounds northward; in 1853, Penny went whaling in the *Lady Franklin*, and wintered in Cumberland Sound, the first time that a whaler had deliberately chosen to winter in that region. This successful venture established a new pattern for the whalers. John Barrow's correspondence (British Library Add. MS 35, 306 f. 136ᵛ) contains an untraced newspaper cutting headed: "THE NEW DAVIS STRAITS' FISHERY." "The successful voyage of the Lady Franklin, promises to introduce a new era in the history of our Davis Straits' whale fishery. Captain Penny has shown how the whales of the Arctic regions may be followed up, and the treasures of the frozen ocean turned to good account. And this discovery is no less opportune than it is important. The oleaginous products of Russia are not legitimately at command, and the stoppage of our usual supplies of oils and other fatty substances, has come on us at a time when science and art, in their endless variety of applications, have made them more than ever necessary for manufacturing and commercial purposes. . . . "

[78]   F. L. McClintock and Samuel Haughton, "Reminiscences of Arctic Ice-Travel in Search of Sir John Franklin and his Companions, by Captain F. L. M'Clintock. With Geological

By 1851, he had enough experience to tackle longer sledging journeys, including one of 900 miles in 80 days, taking in Byam Martin Island and Melville Island. His observations were interpreted by the geologizing clergyman Samuel Haughton, Professor of Geology in the University of Dublin, and President of the Geological Society of Dublin. Haughton thought it likely that "the sandstones, limestone, and coal of Byam Martin Island, and the corresponding rocks of Melville Island, Baring Island, and Bathurst Island, are low down in the Carboniferous System, and that there is in these northern coal-fields no sub-division into red sandstone, limestone, and coal-measures, such as prevail in the west of Europe."[79]

The summer of 1852 saw McClintock again at Melville Island on his third expedition, in command of H.M. screw steamer *Intrepid*. This time he estimated his longest sledging trip at 1400 miles. "It is now," he wrote, "a comparatively easy matter to start with six or eight men, and a sledge laden with six or seven weeks' provisions, and to travel some 600 miles across desert wastes and frozen seas, from which no sustenance can be obtained. There is *now* no known position, however remote, that a well-equipped crew could not effect their escape from by their own unaided efforts."[80] The long journeys were made by man-hauling sledges, in a tradition that culminated in Robert Falcon Scott's last expedition.[81] As McClintock wrote to Barrow, he believed that "dogs are most useful for short & rapid journeys, but the resources of the country cannot be depended upon for their maintenance";[82] he never considered using the dogs for food, either for their fellows or for the men. These journeys, like their predecessors, yielded a haul of observations and fossils; among them, at Wilkie Point on Prince Patrick's Land, were lias fossils including a new species of ammonite named after McClintock.[83]

Lacking the drama of the sledging exploits, but still valuable for geology, was the expedition of Robert John Le Mesurier McClure, a reserved disciplinarian who had been educated for the army at Sandhurst, entered the navy at the ripe old age of seventeen, and had twice sailed to the Arctic,

Notes and Illustrations, by Rev. Samuel Haughton," *Journal of the Royal Dublin Society* 1 (1856–7; published in 1858) 183–250. This includes geological notes on the illustrations of 20 fossil invertebrate species collected during the journeys.

[79] Ibid., 199.     [80] Ibid., 237–8.

[81] C. S. Mackinnon, "The British Man-Hauled Sledging Tradition," in P. D. Sutherland, ed., *The Franklin Era*, pp. 129–40.

[82] McClintock to Barrow, R. Naval College, Portsmouth, 9 Feb. 1855, British Library Add. MS 35, 306 ff. 299$^r$–302$^v$ at f. 300$^v$.

[83] McClintock and Haughton, "Reminiscences," 238.

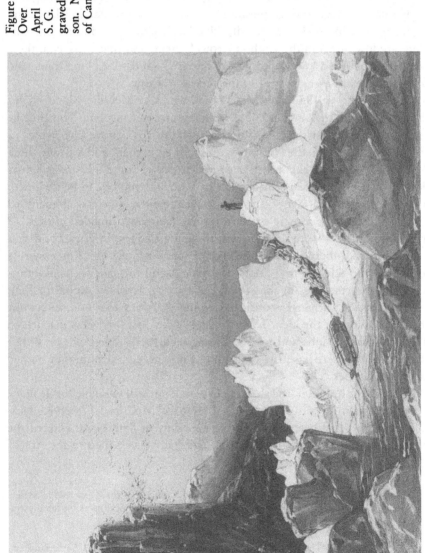

Figure 23. "Sledging Over Hummocky Ice, April 1853," by Lieut. S. G. Cresswell, engraved by William Simpson. National Archives of Canada C-041013.

under Back in 1836, and under James Ross in 1848. In 1850 he set out in command of H.M.S. *Investigator,* spent three winters in the ice, and was rescued with his crew and returned to England in 1854. McClure's narrative of the voyage includes natural history notes, and an appendix by Murchison on wood fragments (evidence of ocean currents and changing climates) and rocks, many of them from his beloved Silurian system.[84]

Other sciences were favored by other explorers. Rochfort Maguire, commander of H.M.S. *Plover* since 1852, and stationed near Bering Strait in case Franklin should head for the Pacific, sailed to Icy Cape in Alaska, and thence made a boat journey to Point Barrow. In 1852, 1853, and 1854 he and his officers carried out a program of magnetic observations at Point Barrow, which was important for providing a more sustained record than had hitherto been available from the western Arctic.[85] The Royal Artillery at Woolwich had equipped Maguire with two dip circles and a declinometer, which were set up in an observatory of ice lined with seal skins. Observations were made from November 1852 to June 1853, and from October 1853 to the following June, when the *Plover* set out for England. The results showed clearly that at Point Barrow, as at the Toronto observatory and elsewhere, there were periodic variations. Disturbances to east and west, however, appeared to follow different laws. Similar observations from other stations in the Arctic, coupled with measurements of the intensity of the magnetic field, might go far toward explaining such changes and laws. Sabine, who wrote up the results of Maguire's expedition, observed:

The instrumental means and the processes of observation are sufficiently simple. Instruments similar to those which have been so usefully, and so honourably to themselves, employed by Captain Maguire and his officers, have been sent with nearly all the expeditions which, in the last twelve years, have wintered within the Arctic circle; but the maintenance of a routine of hourly observation during several months of compulsory detention, in the absence of authoritative direction or professional encouragement, requires perhaps a greater amount of private zeal and devotion than can be expected, except in such exceptional cases as [this one]. In one of these expeditions in particular (the only one that unhappily has not returned in safety), the well-known zeal of its commander SIR JOHN FRANKLIN in the cause of science, and

[84] R. J. Le M. McClure, *The Discovery of the North-West Passage by H.M.S. 'Investigator', Captain R. M'Clure, 1850, 1851, 1852, 1853, 1854,* Sherard Osborn, ed., 2nd. ed. (London, 1857).

[85] J. Bockstoce, ed., *The Journal of Rochfort Maguire 1852–1854: Two Years at Point Barrow, Alaska, aboard H.M.S. Plover in the search for Sir John Franklin,* Hakluyt Society 2nd. series vols. 169–70, 2 vols. (London, 1988): a biographical notice of Maguire is on p. 52. The principal scientific information in the journal is about the Inuit, and there is also an appendix, "Dr. John Simpson's essay on the Eskimos of northwestern Alaska," vol. 2 pp. 501–50.

the anxiety of his officers to cooperate with him in every useful and honourable work, gave reason for hopes of the highest promise.

Sabine was confident that at least in their first winter, Franklin and his crews had erected observatories and carried out the planned magnetic observations. If only the records of their observations could be recovered![86]
Some naval officers had an enthusiastic but less disciplined view of science. Edward Belcher, who had been a lieutenant and assistant surveyor in H.M.S. *Blossom* under Beechey in 1825–8, was now Sir Edward, captain of H.M.S. *Assistance* in 1852–4, and in general command of a fleet of five ships. His expedition furnished a goodly haul in natural history.[87] On his way north in search of Franklin, he wrote to John Barrow from Disco, on the west coast of Greenland:

Indeed I have many interesting experiments to make, in which I think I shall be ably seconded by my surgeon [& o]thers, here, that our "idle time" will be profitably engaged. Amongst the vicious propensities, I [am] thinking of capturing Bears, wolves, or foxes, either by detonating silver,[88] of Gun cotton; – but I am not yet quite sure that the temperature of the animals will be sufficiently high to explode the Gun Cotton. We have on board a man who was employed making "Jones' Promethane" and by the use of sulphuric acid, confined in glass globules, [one] can easily cause the explosion of Detonating Silver.[89]

Whether the scientific activity was eagerly inventive, as here, or highly disciplined, as in Maguire's case, it was subordinated to the now vanishing hope of finding Franklin, or at least of discovering his fate. Scientists on these expeditions were a luxury, and it was only a happy accident if an officer or surgeon was also a skilled naturalist or geophysicist. The keenest observers were generally untrained in all but those sciences that bore di-

---

[86]   E. Sabine, "On Hourly Observations of the Magnetic Declination made by Captain ROCHFORT MAGUIRE, R. N., and the Officers of H.M.S. 'Plover,' in 1852, 1853 and 1854, at Point Barrow, on the Shores of the Polar Sea," *Phil. Trans. Roy. Soc. London 147* (1857) 497–532 at 506–7.

[87]   Edward Belcher, *The Last of the Arctic Voyages: being a Narrative of the Expedition in H.M.S. Assistance, under the Command of Captain Sir Edward Belcher, C. B., in Search of Sir John Franklin, during the Years 1852–53–54: With Notes on the Natural History, by Sir John Richardson, Professor Owen, Thomas Bell, J. W. Salter, and Lovell Reeve*, 2 vols. (London, 1855).

[88]   Detonating or fulminating silver, discovered by C. L. Berthollet, was produced by pouring liquid ammonia over silver oxide; the resulting compound detonates on heating, or when mixed with a strong acid.

[89]   Sir Edward Belcher to John Barrow, Lively, Disco, 9 June 1852, British Library Add MS 35, 307 f. 91ᵛ.

rectly on navigation. Typical was Sherard Osborn, the lieutenant commanding *Pioneer* in its search.[90] He wrote to Barrow in 1854:

I fear you will say my observation lies all in one line – Geography and Search – and that there are many other interesting points to which my attention might be profitably turned. None perhaps more generally appreciable than that of the Natural History of the North. Unhappily my *forte* does not lie therein, and those horrid Naturalists who seem to delight in making every science a fortified space of hard names, which it requires all one's time to besiege and master, renders men less thick-skinned than a Rhinoceros somewhat diffident of approaching the subject.

If one did report one's observations with less than scientific rigor, one was subject to ridicule. If, on the other hand,

we say we have seen nothing! we are likely . . . to be smote hip & thigh by our zealous Scotch naturalist rising to his legs and assuring a respectable society – "That though he'd ha no doubt Capt. such a one would do to Bombard Algiers, [he'd] no send him to look for animal life in the Arctic Regions.[91]

And yet, although untrained and not technically proficient in natural history, Osborn could and did record his observations vividly:

I wish you could see the Bear in his glory amongst the Ice of these Regions. Never was a creature more beautifully adapted to the life he leads. He seems like a true sailor, half flesh half fish – and (as he rolls along) when walking at his ease you are struck with the look of latent energy and power evinced in every action. . . . Amongst Packed Ice . . . their activity is wonderful . . . their powerful paws, and hind legs enabling them to keep springing from piece to piece, scaling one fragment and sliding down another, with the activity of a huge *monkey* rather than that of a cat. . . .

Their sense of smell must be very acute. . . . I have seen them in Baffin's Bay, running down the scent from the ships, exactly as a hound would do – and the snorting noise made by the brutes as they approach, indicates how much they were trusting to the nose. . . . In Wellington Chan.[l] I more than once saw, in the distance, a Bear going at the hard swinging pace peculiar to the brute, the head of the animal up, and occasionally it would stop, raise its long ungainly neck as if to inhale a fresh whiff of the distant seal, and then again, it would resume its course as straight to its point as an arrow.[92]

90  P. Berton, *The Arctic Grail*, p. 203 describes him as "the most prolific and also the most enthusiastic of the literary explorers"; the enthusiasm is manifest in Osborn's *Stray Leaves from an Arctic Journal; or, Eighteen Months in the Polar Regions, in Search of Sir John Franklin's Expedition, in the Years 1850–51* (London, 1852).
91  Osborn to Barrow, March 1854, British Library Add MS 35, 309 ff. 29[r]–34[v.]
92  Ibid.

It was not just living nature that Osborn recorded. His journal contains a fine record of geological, meteorological, and zoological observations.[93]

In the midst of some notes on animal migration, in which he conjectured that several species, including the lemming, musk-ox, and ptarmigan sometimes wintered in the high North, Osborn made it clear that the purpose of arctic exploration in general, and of Franklin's expedition in particular, was the increase of natural knowledge. "I look upon Franklin and his men," he wrote, "as a sacrifice, made in the great cause of general knowledge, and many a greater has been made in far worse cause!"[94] Osborn was of course right; magnetic and geographical science between them had provided the principal motives for Franklin's final expedition.

He was also right in assuming in 1854 that Franklin and his men were by now long dead. The Admiralty, having previously reached the same conclusion privately, now made that conclusion public. On 20 January 1854, the *London Gazette* published a notice from the Lords of the Admiralty that if there was no news of Franklin and his men by the end of March, their names would be removed from the navy list. Jane Franklin protested; no presumption of their deaths was justified until their fate was known. She begged that the searches should continue, for the sake of the men, and for the sake of science.[95] If they were now given up, perhaps it would be the wonder of a future generation that not only were the men deserted, but that

any discoveries of great scientific interest and importance should have been abandoned by the Government at the conspicuous moment when it had at its disposal a fleet of invulnerable ships, fit, and fit alone, for Arctic service, and still afloat in Arctic seas, and a host of trained and brave explorers, better disciplined for their work than ever, a combination such as was never seen before, and may never be seen again.[96]

The Admiralty needed no reminding of the size of the search fleet; it had begun to reckon up the totals of ships, officers and men, and money devoted to arctic exploration since 1818, and look especially closely at the cost of arctic expeditions since the loss of Franklin. Even a generous and philan-

---

93   E. g., Capt. Austin's Expedition, Proceedings of Lieut. Sherard Osborn, Steam Tender Pioneer, PRO Adm. 7 193, 21 April 14 and 19 May 1851. Osborn, "Remarks upon the Amount of Light Experienced in High Northern Latitudes during the Absence of the Sun," *Journal of the Royal Geographical Society* 28 (1858) 371–5; *Stray Leaves*, passim.

94   Osborn to Barrow, no. 7, *Pioneer* [1854], British Library Add. MS 35, 309 ff. 35ᵛ–39ʳ at 36ᵛ.

95   E. F. Roots, "Anniversaries of Arctic Investigation: Some Background and Consequences," *Trans. Roy. Soc. Canada* (1982) 373–90 at 383.

96   Jane Franklin to the Lords Commissioners of the Admiralty, 24 February 1854, in Arctic Blue Books, Arctic Bibliography no. 45242. Lady Franklin had also protested immediately on 20 Jan. 1854.

thropic institution, which by and large the Admiralty was not, would have quailed at the cost. By 1857, when the navy had put a definitive stop to its searches, the figures were impressive, and constituted an overwhelming argument in the minds of the Lords of the Admiralty to cry halt. Between 1818 and 1844, they had sent out twelve naval expeditions involving sixteen ships (counting a ship once for each expedition in which it sailed),[97] with complements ranging from twenty-eight to sixty-two. There had also been four overland expeditions, one of them mounted by the Hudson's Bay Company. Since Franklin's departure, there had been another fifteen Royal Naval expeditions, apart from overland expeditions, and not counting private expeditions or, more worryingly, American expeditions, the chief of which we shall consider below. These numbers were large, as were the numbers of ships that had been left in the Arctic. But the largest number was the estimated cost of arctic expeditions since Franklin's loss, a staggering £610,520.[98] Any further searches for Franklin, and, for a good while, any arctic exploration at all, would have to look elsewhere than to the Admiralty for its support.

*Elisha Kent Kane.* Jane Franklin had merely stated the case when she referred to a fleet of well-equipped ships. The early 1850s saw more ships and sledges crisscrossing the high Arctic than had ever been seen there before, or would be seen again for more than half a century. The Royal Navy had sent out nine ships in 1850, and more were to follow. But besides the British presence west of Greenland and in the arctic archipelago, there was for the first time an American naval presence.

Jane Franklin had not only lobbied in England, she had also sought American aid in searching for her husband, and Henry Grinnell, a financier and shipping magnate from New York, had responded handsomely. He bought two vessels, and presented them to his government to be used by the United States Navy. Thus originated the United States Franklin Search Expedition, otherwise known as the first Grinnell Expedition. They entered Lancaster Sound, helped to search Beechey Island, came up with Penny's expedition at Griffith Island, were caught in the ice, and drifted until they were released in Davis Strait. The expedition had no dramatic achievements to record, but it introduced Elisha Kent Kane, a sickly but determined young gentleman from Philadelphia, to the Arctic. He was a surgeon on that expedition,[99] and was to become the most famous arctic explorer of his day.

[97] Thus *Griper* counted as three, and *Isabella* as only one.
[98] PRO Adm. 1 5676, 1 5685.
[99] E. K. Kane, *The United States Grinnell Expedition in Search of Sir John Franklin: A Personal Narrative* (London, 1854).

Grinnell sponsored a second expedition, this time jointly with other sponsors; Kane was its leader.

True, this was to be another Franklin search expedition. Kane would head north along the west coast of Greenland through Smith Sound, beyond the farthest north that the British had so far achieved, and would search where no search party had been before. But this would put him well on his way to the pole, and on the way to settle the old question of whether or not there was an open polar sea – a dream that the experience of half a century of exploration in the most difficult ice conditions had failed to extinguish. Kane talked of Franklin, but his eye was also on the prize of geographical[100] and other scientific discovery. A newspaper of the day proclaimed: "This is not merely a voyage of search. The vast advantages which will accrue to science from this expedition cannot be too highly estimated."[101] Grinnell and Barrow corresponded about the proper equipment for Kane's venture. As Grinnell observed, "with regard to stowage, as there is plenty of money appropriated by the Government I do not see why every thing requisite should not be put on bd. the Exp., to warrant the best success."[102]

Kane set off as an untried commander whose qualities of leadership were to be taxed and found wanting; the only constant in his make-up was an oversized ego, but that carried him far. His orders came from the secretary of the navy, his complement included a German astronomer, August Sonntag, and a surgeon, Isaac Israel Hayes, and he was equipped for search and science. On the northern journey, he stopped along the Greenland coast to pick up dogs for sledging, and an Inuk hunter and interpreter.

They carried out deep-sea soundings in Baffin Bay, recording a depth of 1,900 fathoms;[103] they noted the remains of an ancient Inuit settlement and burial on the Littleton Islands, so named by Captain Inglefield of the Royal Navy, which was located further north than they had expected to find traces of human habitation.[104] On 7 August 1853 they passed the most northerly latitude achieved by Inglefield.[105] By 29 August, at 78°43' N, they were frozen in for the winter. Besides navigating in awkward ice, they had botanized and geologized along the way. A week before the ship was halted,

100  Naturally he had stressed this in a paper read before the Geographical Society before his departure; see Kane, *Arctic Explorations in Search of Sir John Franklin* (London, 1892) p. 12.
101  Newspaper cutting, source untraced, in PRO MS Adm. 7 194 f. 252.
102  Grinnell to Barrow, n.d., British Library Add. MS 35, 306 ff. 305–6ᵛ.
103  Kane, *Arctic Explorations*, p. 14.        104  Ibid., pp. 31–2.
105  Kane to Secretary Dobbin [? Nov. 1855], SPRI MS 246/430/3.

Kane noted that they had collected 22 species of flowering plants from the shores of the bay; and "it was not without surprise and interest that I recognized among its thoroughly Arctic types many plants which had before been considered as indigenous only to more southern zones."[106]

Sledging from the ship, they found limestone terraces, in one of which they found the skeleton of a musk-ox, of which the bones seemed well on their way to mineralization. Kane thought of fossils in alluvial deposits in Siberia, and, nearer to their temporary home, in the cliffs at Escholtz Bay in Alaska.[107]

Sledging failed to reveal any accessible site better than the ship's fortuitous refuge for winter, which was fast approaching. Sea swallows or arctic terns had been abundant at the end of August, but ten days later they were all gone. So too were the young burgomasters or glaucous gulls that had lingered after them. Only snow buntings remained, the last migrants to fly south. It was time to set about the necessary preparations, for survival, for recreation, and for science.

Work on the astronomical observatory, where the transit instrument and theodolite were set on pedestals of stone cemented by ice, was started on 12 September. Granite walls made this the most substantial arctic observatory to date.[108]

Great care was taken by Mr. Sontag ... in determining our Geographical position. The results for the determination of longitude, as based upon moon culmination, are in every respect satisfactory: they are corroborated by occultations of planets and the late Solar Eclipse of May 1855. An occultation of Saturn simultaneously observed by Mr. Sontag and myself at temperatures of [−]60° and [−]53° differed but 2 seconds. This is the lowest temperature at which such an observation has ever been taken. The position of our Observatory may be stated as in
Lat. 78° 37′ 0″
Long. 70 40′ 6″

Then there was a magnetic observatory, stone built, with wooden floor and roof, and a copper fire grate – high comfort compared with the Royal Navy's comparable observatories. Still, Kane complained that the cold made it almost impossible to maintain a regular program of observations. They had a good unifilar magnetometer for measuring declination, similar to the instruments that Kew and Dublin had furnished for the Royal Navy,

---

[106] Kane, *Arctic Explorations*, p. 48.
[107] Ibid., pp. 55–6. See the account above of John Richardson's discussion of "Vertebrals, including Fossil Mammals," in the *Zoology of the Voyage of H.M.S. Herald* (1854) pp. 1 et seq., "Observations of the Fossil Bone Deposit in Eschscholtz Bay."
[108] Kane, *Arctic Explorations*, pp. 64, 69.

on loan from Professor Bache, who since 1844 had been head of the United States Coast Survey. For measurements of dip and intensity, they used a Barrow's circle from the Smithsonian Institution rather than the Fox instrument that Sabine had predictably recommended.[109] American preferences in instrumentation were the ones that counted here, and the two major mid-century centers of American geomagnetic research contributed directly.[110] There were "weekly determinations of variation of declination, extending through the twenty-four hours, besides observations of intensity, deflection, inclination, and total force," as well as the concentrated observations of term-days.[111] The brunt of the program of observations was carried out by Sonntag and Hayes, with Kane assisting only toward the end of winter. But when he did his share, it seemed to him that the arctic magneticist's life was a close approximation to frozen hell.

Imagine it a term-day, a magnetic term-day.
The observer, if he were only at home, would be the 'observed of all observers.' He is clad in a pair of seal-skin pants, a dog-skin cap, a reindeer jumper, and walrus boots. He sits upon a box that once held a transit instrument. A stove, glowing with at least a bucketful of anthracite, represents pictorially a heating apparatus and reduces the thermometer as near as may be to 10° below zero. One hand holds a chronometer, and is left bare to warm it; the other luxuriates in a fox-skin mitten. The right hand and the left take it 'watch and watch about.' As one burns with cold, the chronometer shifts to the other, and the mitten takes its place.
   Perched on a pedestal of frozen gravel is a magnetometer; stretching out from it, a telescope; and, bending down to this, an abject human eye. Every six minutes said eye takes cognizance of a finely-divided arc, and notes the result in a cold memorandum book. This process continues for twenty-four hours, two sets of eyes taking it by turns; and, when twenty-four hours are over, a term-day is over too.
   We have such frolics every week. . . . 'A grateful country' will of course appreciate the value of these labours, and, as it cons over hereafter the four hundred and eighty results which go to make up our record for each week, we never think of asking, 'Cui bono all this?'[112]

Kane obviously asked himself what the point was, time and again. Their meteorological observatory rubbed in the lesson by recording temperatures as low as −70° F. His officers and crew also wondered why they were there; indeed, a substantial party at one stage left for the South, rejecting Kane's leadership, to which, when beaten by cold and hunger, they returned with

[109]   Sabine to Beaufort, 24 Feb. 1853, concerning Kane's expedition, PRO MS Adm. 7 195.
[110]   For a description of such instruments, and for an introduction to such American research, see R. P. Multhauf and G. Good, *A Brief History of Geomagnetism and a Catalog of the Collections of the National Museum of American History* (Washington, D.C., 1987).
[111]   Kane, *Arctic Explorations*, p. 89.          [112]   Ibid., p. 98.

help from the Inuit. None of them would have survived two winters without the Inuit,[113] who supplied the Americans with meat from their settlement at Etak, some seventy miles to the south – the most northerly human settlement in the world. It was not just cold and hunger, nor the scurvy that afflicted them through the winters, but also the loss of the majority of their dogs to a convulsive illness. Kane, whose men dressed in furs like the Inuit, not in wool, canvas, and flannel like the Royal Navy, learned to eat raw meat, and although never fully competent in driving the dogs, they nonetheless emphatically preferred them over manpower for journeys across the ice. They were faster learners in arctic sledging than the Royal Navy, although not than the Hudson's Bay Company. Kane purchased additional dogs from the Inuit in April to make up some of the losses, and noted that their value for travel could hardly be over-estimated. "The earlier journeys of March April and May proved incomparably more arduous and exposing than those performed with dogs, while their results were entirely disproportionate to the labors they cost us. It was invariably the case that the entire party on its return from the field passed at once upon the sick list."[114]

Kane's reports of sledging journeys along the Greenland coast were full of accurate accounts of the geological strata he encountered. In 1854 he mapped the coast, naming the land on the western shore of the great basin that now bears his name after his patron, Grinnell. From their furthest position north (81°22' N), he looked down as he thought upon open water, stretching to the horizon, frequented by hordes of pelagic birds, and with a pronounced swell – perhaps there was a polynya, but he jumped from his immediate observations, erroneous or not, to the idea of an open polar sea. He suggested that the Gulf Stream, traced already in one of its branches to the coast of Novaya Zemlya, might there be deflected and flow into the polar regions.

The moment of false discovery of an open sea was for Kane the high point of the expedition. The following winter, which saw the breakdown of leadership, a mutiny peacefully but not happily resolved, near starvation, and scurvy, brought them all to a state of misery, and would, as we have seen, have killed them all, but for the Inuit.

In 1855 they abandoned their ship, and made their way on foot and by boat south along the Greenland coast to Godhavn. From there, they were taken home as heroes. Kane's report glossed over problems – he claimed that scurvy, for example, had not attacked them after all – and reinforced

---

[113]  Ross's crews had similarly depended upon Inuit help. See chap. 2, note 166 above.
[114]  Kane, SPRI MS 248/430/3.

the image of success.[115] The scientific results were published in extenso by the Smithsonian Institution, in a series of volumes containing tidal, astronomical, meteorological, and magnetic data.[116] Science made up for almost everything.[117]

But what of Franklin? His fate, discovered meanwhile elsewhere in the Arctic, virtually disappeared from American minds in the excitement of Kane's return.

THE DISCOVERY OF FRANKLIN'S FATE:
JOHN RAE 1853–1854 AND FRANCIS
LEOPOLD MCCLINTOCK 1857–1859

While Kane and his men were enduring their expedition, John Rae was once again in the Arctic. He had set out from York Factory in June 1853, to complete the survey of the coast of North America, along the west coast of Boothia Peninsula. He wintered at Repulse Bay, and in the following spring heard Inuit reports of a party of white men walking toward Back River in 1850, and of numerous corpses found along the river later in the season. He purchased some of Franklin's relics from the Inuit, was back at Repulse Bay on 26 May, and returned to York Factory on 31 August. He wrote promptly to Archibald Barclay, secretary of the Hudson's Bay Company in London:

I arrived here yesterday, with my party all in good health. . . . Information has been obtained, and articles purchased from the natives, which prove beyond a doubt, that

[115]   Ibid., and *Proceedings of the Geographical Society 1* (1856) 17–20; Kane, *Arctic Explorations: The Second Grinnell Expedition in Search of Sir John Franklin, 1853, 1854, 1855,* 2 vols. (Philadelphia and London, 1856).

[116]   Kane, *Tidal Observations in the Arctic Seas, Made during the Second Grinnell Expedition in Search of Sir John Franklin* (Smithsonian Contributions to Knowledge, vol. 13, art. 2, Smithsonian Institution Publication no. 130) (Washington, D.C., 1860); *Astronomical Observations* . . . (vol. 13 art. 2, Smithsonian Institution Publication no. 129) (1860); *Meteorological Observations* . . . (vol. 11 art. 5, Smithsonian Institution Publication no. 104) (1859); *Magnetic Observations* . . . *Reduced and Discussed by C. A. Schott* (vol. 10 art. 3, Smithsonian Institution Publication no. 97).

[117]   Kane's reputation was stronger than his leadership. Apart from serving as inspiration to subsequent American expeditions (see chap. 6), he also had a masonic lodge named after him. Robert E. Peary was a member, and he used the lodge's note paper, headed by a circle made of chain links, surrounding a polar scene: a night sky, jagged icebergs in the background, ship's mast and prow encountering ice, and, in the foreground, a bearded figure in a parka with hood, planting a flag in the snow. The flag has stars on a dark ground, and the masonic compass and square.

a portion, if not all, of the survivors of the long lost and unfortunate party under Sir John Franklin, had met with a fate as melancholy and dreadful as it is possible to imagine.[118]

That fate was death from starvation and disease, compounded by lead poisoning,[119] and a final desperate resort to cannibalism. On 21 October 1854, the day after the publication of Rae's findings, *The Times* of London spoke for the nation: "We have had quite enough of great Arctic expeditions; since Sir Edward PARRY'S first voyage in 1819–20, with the single exception of Captain M'CLURE's, they have invariably resulted in disappointment and disaster."

The Admiralty had its answer, justifying its removal of Franklin's name from the navy list. Franklin's fate, added to the cost of arctic explorations, would for more than two decades discourage the Royal Navy from scientific or other exploration in the high Arctic. Jane Franklin urged and begged and lobbied for a continuation of the naval search. She was joined in her suit not only by arctic explorers, but by some of the leading men of science of the day. On 5 June 1856 a letter with thirty-six signatories was sent to Lord Palmerston, the prime minister, urging renewal and completion of the search in the region identified by Rae. Explorers and sailors one expects as signatories: in addition to those who signed, eighteen Royal Naval officers were listed as having been unable to sign because they were absent from London, but as having previously expressed themselves as favorable to the final expedition. The list of scientists is truly impressive: Murchison and Sabine, of course, William and Joseph Hooker, unsurprisingly, but also William Whewell, the physicists Charles Wheatstone and J. P. Gassiot, the Astronomer Royal, George Biddell Airy, Charles Daubeny, the Oxford chemist who also wrote about meteorology and vulcanology, and others of note. The scientific establishment was solidly on Jane Franklin's side. The government, however, like the Admiralty, would not budge.[120]

There were, as we have seen, other channels than official ones to send men north. Philanthropy, science, whaling, and more than half formulated

---

[118]   Rae to Barclay, 1 Sept. 1854, *The Hudson's Bay Record Society XVI: John Rae's Correspondence with the Hudson's Bay Company on Arctic Exploration 1844–1855*, E. E. Rich and A. M. Johnston, eds. (London, 1943) p. 265.

[119]   Owen B. Beattie, *Frozen in Time: The Fate of The Franklin Expedition* (Saskatoon, 1987).

[120]   The memorial from scientists and explorers, and the supporting letter to Palmerston from Jane Franklin (2 Dec. 1856) are printed as appendixes II and I respectively in McClintock, *The Voyage of the 'Fox' in the Arctic Seas: A Narrative of the Discovery of the Fate of Sir John Franklin and his Companions* (Boston, 1860) pp. 318–33 (London 1859; reprinted Edmonton, 1972).

Figure 24. John Rae showing map and arctic Franklin relics after bringing them home. Byrne & Co. Photographers, National Archives of Canada PA-147990.

territorial ambitions had already started the Americans on what was to be a succession of voyages up the Smith Sound route. This indeed became known as the American route, with the land along its western edge known, at least to Americans, as the American shore. British whalers were also

Figure 25. Henry Edwin Landseer, "Man Proposes, God Disposes." Royal Holloway and Bedford New College, University of London.

pushing north along that route. Friendship for Franklin had driven John Ross, who, spurned by the Navy, was well supported by public subscription and the Hudson's Bay Company. Jane Franklin maintained her determination to encourage exploration until all was known, and to correct the wicked story that Rae had brought back. She raised a private subscription for a final effort. Captain McClintock agreed to lead it, as commander of the *Fox*, which Lady Franklin had purchased for the search.

McClintock was a good choice, with his sledging experience gained on three arctic expeditions between 1848 and 1854. He was also keen to accommodate scientific interests, and wrote in May to the secretary of the Royal Society of London, asking "that the Council will afford me such information and instruction as will enable me, in the continually advancing state of physical knowledge, to make the best use of the opportunity afforded by the voyage, for the prosecution of meteorological, magnetical, and other observations."[121]

McClintock also expressed the hope that, because the expedition was to be privately funded, the Royal Society might contribute to the purchase of instruments. The Royal Society contributed £50. Sabine undertook the task of recommending magnetic instruments, and explaining the necessary observations; Joseph Hooker undertook on William Hooker's behalf to furnish McClintock with the necessary means and instructions for collecting and preserving plants. Joseph Hooker was also requested to order two "Wardian Cases" for "bringing home roots of Arctic Plants for the Royal Gardens at Kew," which was both a herbarium and a living collection. The Board of Trade undertook to supply meteorological instruments.[122] The Admiralty granted leave of absence on half pay to officers who joined the *Fox*.[123] Among the complement was the surgeon-naturalist, Dr. David Walker, who also had the responsibility of using the ship's photographic equipment.[124]

By summer, the *Fox* had been refitted as an arctic vessel, which mostly involved strengthening it against the ice, and replacing its light propellor with a heavy iron one, and its boiler with a larger one. All was made ready, and the expedition left Aberdeen in July 1857. Although Jane Franklin had dreams that members of Franklin's expedition might still survive, living off

---

[121]  Royal Society MSS, Council minutes 28 May 1857.
[122]  Royal Society CMB 25.                              [123]  SPRI MS 248/439/1.
[124]  McClintock, *The Voyage of the Fox*, p. 7. P. 82 refers to Walker taking a photograph of the ship in March using the albumen process on glass, rather than the more familiar gelatin process; the temperature was below 0° F. I have not found any of Walker's photographs. Who in the previous decade was the first arctic photographer?

the land with the Inuit, there was little hope of finding survivors; there was, however, the not unrealistic hope that the expedition's papers might have been cached somewhere, and that they, including valuable magnetic data, might indeed be recovered.[125]

The expedition had a frustrating start. Frozen into the ice in Melville Bay on the outward voyage, they spent their first winter drifting southward down Davis Strait. McClintock used his time and his journal in the myriad ways pioneered by Parry, including a good deal of scientific observation and speculation. Geology, anthropology, and archeology came together in what was to become a recurrent association. The *Fox* had the invaluable aid of Johan Carl Christian Petersen, who had previously been a hunter, interpreter, and dog driver on other Franklin searches. McClintock, in his journal for 11 December 1857, noted:

Although it has been abundantly proved by the existence of raised beaches and fossils, that the shores of Smith's Sound have been elevated within a comparatively recent geological period, yet Petersen tells me that there exist numerous ruins of Esquimaux buildings, probably one or two centuries old, all of which are situated upon very low points, only just sufficiently raised above the reach of the sea; such sites, in fact, as would at present be selected by the natives. These ruins show that no perceptible changes ha[ve] taken place in the relative level of sea and land since they were originally constructed. At Petersen's Greenland home, Upernivik, the land has sunk, as is plainly shown by similar ruins over which the tides now flow.[126]

Here was evidence not only of the antiquity of Inuit settlement, but also of the insignificance of the time scale of human history when compared with that of geology. Arctic evidence was reinforcing Lyell's evidence from Sicilian volcanoes, extending the age of the earth even while Darwin was writing his essay *On the Origin of Species*, published in the year that McClintock returned home.[127]

Then there were questions about animals wintering in the high Arctic. Foxes were common, birds had mostly or all gone south for the winter, hares were "comparatively scarce"; so what did the foxes live on for eight months of winter? Petersen made observations about the animals storing provisions in summer in holes and crevices, and recovering them in winter. There was also a pattern of fox tracks following those of polar bears, suggesting (accurately) that foxes scavenged the remains of bear kills on

[125] Murchison, preface p. xii in McClintock, *The Voyage of the Fox*.
[126] McClintock, *The Voyage of the Fox*, pp. 69–70.
[127] M. J. S. Rudwick, "Lyell on Etna, and the Antiquity of the Earth," in C. J. Schneer, ed., *Toward a History of Geology* (Cambridge, Mass., 1969) pp. 288–304; Rudwick, *The Meaning of Fossils: Episodes in the History of Paleontology*, 2nd, ed. (New York, 1976).

the ice.[128] They overlooked the importance of lemmings as winter food on the tundra.[129]

On 18 December, there were long lasting and fine displays of the aurora, and Walker called McClintock "to witness his success [in detecting atmospheric electricity] with the electrometer. The electric current was so very weak that the gold-leaves diverged at regular intervals of four or five seconds. Some hours afterward it was strong enough to keep them diverged."[130]

Once released from the pack, they sailed north again, entered Lancaster Sound, and stopped at Beechey Island, formerly a busy focus of Franklin searches. McClintock notes that there were ten vessels there together in late August 1850. Peel Sound was blocked by ice, and they turned around and headed for Bellot Strait, intending to proceed to King William Island. Again they were stopped by ice at the eastern entrance of the Strait, and prepared to winter there.

By mid-October, they had built a magnetic observatory out of ice blocks, on the ice away from the ship's iron. For more than five months, from October to March, they made hourly observations of variation with their declinometer. They were near the spot where James Ross had found the magnetic pole, and their observations were thus of special interest to Sabine.[131]

In spring, they set out on a sledging expedition, with Walker botanizing as he went, and he and McClintock collecting geological specimens. They also made infrequent measurements of magnetic dip made near the magnetic pole.[132] A sledging party found a record left by one of Franklin's parties while still healthy, on 28 May 1847: "Having wintered in 1846–7 at Beechey Island, . . . after having ascended Wellington Channel to lat. 77°,

[128]  McClintock, *The Voyage of the Fox*, p. 70.

[129]  L. S. Underwood, "Notes on the Arctic Fox (*Alopex lagopus*) in the Prudhoe Bay Area of Alaska," in J. Brown, ed., *Ecological Investigations of the Tundra Biome in the Prudhoe Bay Region, Alaska*, Biological Papers, Special Report no. 2, pp. 144–9 (University of Alaska, Fairbanks, 1975).

[130]  McClintock, *The Voyage of the Fox*, p. 72. The electrometer was a gold leaf electroscope, with two gold leaves hanging from a conductor. When the conductor received a charge, the leaves, being similarly charged, repelled one another, and diverged. The Royal Society was still interested in possible links between the aurora, magnetic disturbances, and atmospheric electricity.

[131]  Sabine, "Results of the Hourly Observations of the Magnetic Declination Made by Sir F. L. M'Clintock, and the Officers of the Yacht 'Fox,' at Port Kennedy, in the Arctic Sea, in the Winter of 1858–59," *Phil. Trans. Roy. Soc. London 153* (1863) 649–63.

[132]  McClintock, *The Voyage of the Fox*, pp. 280–1.

and returned by the west side of Cornwallis Island. /Sir John Franklin commanding the expedition./ All well."[133]

Near the magnetic pole, McClintock's party met a group of Inuit, with whom they bartered for relics of Franklin's expedition, and from whom they gathered information that bore out Rae's principal statements.[134] This recognition, given in McClintock's narrative, makes all the more striking his disparaging reference in his introductory remarks to "the traces accidentally found by Dr. Rae," as well as Murchison's prefatory statement that McClintock had furnished the only authentic intelligence of Franklin. Rae was not to be forgiven for his horrid news about cannibalism, nor was the navy and its associated scientific establishment likely to show generosity to an officer of the Hudson's Bay Company.

In August, the *Fox* left Bellot Strait and sailed and steamed home. They had done useful work in several departments of science.[135] Besides the geomagnetic data, there were essays on the tides and upon geology, published with McClintock's narrative.

Observations upon Halos, &c., with the Polariscope, have been sent to Professor Stokes; a series of earth temperatures, to Dr. Jos. Hooker, of Kew Botanic Gardens, as also the specimens of dried and living plants. Natural history specimens have also been made over to scientific friends of the Expedition, my sole object being, to render our labors subservient to scientific ends, and with the least possible delay.[136]

There was also a systematic meteorological record, of which the data were reduced by the American Charles Schott, who had done similar service for the second Grinnell expedition. The results were published by the Smithsonian Institution, a striking testament to the emergence of that body as a focus and clearing house for northern science.[137] Publication of the results in England was ruled out because of the time and expense involved; one is reminded of Lefroy's complaints about the fate of his magnetic data. Joseph

---

[133] Ibid., quoted and reproduced on p. 256 and facing p. 255.

[134] Ibid., pp. 208–12, March 1859. Apart from the relics that they purchased from the Inuit, McClintock's expedition found numerous relics (ibid., appendix III, pp. 334–40), including a six-inch dip circle.

[135] For a summary, see "Report of Scientific Researches Made during the Late Arctic Voyage of the Yacht 'Fox,' " *Proceedings of the Royal Society of London* 10 (1860) 148–51.

[136] McClintock, *Meteorological Observations in the Arctic Seas: By Sir Francis Leopold McClintock, R. N., Made on Board the Arctic Searching Yacht 'Fox' in Baffin Bay and Prince Regent's Inlet in 1857, 1858, and 1859. Reduced and Discussed . . . by Charles A. Schott* (Smithsonian Contributions to Knowledge, vol. 13 art. 2) (Washington, D.C., 1862) p. xi. Ornithological results are summarized in D. Walker, "Ornithological Notes of the Voyage of the 'Fox' in the Arctic Seas," *Ibis* 2 (1860) 165–8.

[137] McClintock, *Meteorological Observations.* See also chap. 9.

Henry, head of the Smithsonian, wrote the preface. McClintock was also concerned to explore any correlation between ice movements, winds, and currents; in fact he exaggerated the role of the wind: "The long drift of the Terror through Hudson's Straits in 1836–37 appears to me to be another instance of the effect of wind upon the ice, as in this case it does not seem possible that any considerable current could always, that is to say all winter, set out of Hudson's Bay" (*Meteorological Observations*, p. xi).

### THE RECORD OF THE ROCKS

The geological results were written up by Samuel Haughton, who was able to prepare a geological map of a good deal of the archipelago, based largely on collections made by McClintock on his four Franklin search expeditions, but drawing also on other expeditions devoted to search or exploration.[138] Haughton described the archipelago in terms of five principal formations: granitic and granitoid rocks; the Upper Silurian rocks; the Carboniferous rocks; the Lias rocks; and superficial deposits.

Granite and granitoid rocks formed a good part of northern Greenland, as well as the eastern end of Devon Island, the northern shore of the entrance to Lancaster Sound. Silurian rocks in the archipelago overlay the granitoid rocks, "with a remarkable red sandstone, passing into coarse grit, for their base" (*The Voyage of the Fox*, p. 349). Alternating beds of hard limestone and soft shale, just like the Upper Silurian rocks in England and America, produced striking buttresses and castellated formations, the result of unequal weathering of the limestone and shale.[139] Some of the limestones were rich in fossils. McClintock and his officers had been assiduous in collecting them; many of them were named for him, and for other arctic officers.

The succeeding rocks, which were in fact Devonian but misidentified as carboniferous,[140] were of practical interest, because coal seams could provide fuel to extend the range of steamships. Some of the coal beds occurred

---

[138]   Haughton in McClintock, *The Voyage of the Fox*, appendix IV, pp. 341–72, with map facing p. 341; a brief account is also given in Haughton, "On Fossils Brought Home from the Arctic Regions in 1859, by Captain Sir. F. L. M'Clintock," *Royal Dublin Society Journal 3* (1860, pub. 1862) 53–8.

[139]   Beechey had sketched such formations during Parry's first voyage. See the illustrations facing p. 35 of Parry, *Journal of a Voyage*.

[140]   R. L. Christie, "Geology at the Top of the World: Concepts Change as Exploration Advances," *Geos 11* 11–14.

lower in the sequence of strata than did those of Europe, suggesting that they were among the first carboniferous rocks.

Fossiliferous Lias rocks were mainly found in the west of the archipelago, on Prince Patrick Island; one of the fossils McClintock brought back was a fine ammonite, a species that in life must have required, like its modern successors, ice-free water and relatively warm air, since it floated on and near the surface of the sea. Haughton dismissed theories of a warmer global climate, for contemporary theory suggested that if arctic seas were raised to a temperate clime, whether by central heat, by increased solar radiation, or by a different distribution of land and water, then the seas around the equator must have been almost hot enough to boil an egg,[141] a circumstance that rightly appeared risible.

The only speculation that Haughton thought capable of solving the puzzle posed by the presence of fossil ammonites in the Arctic was the hypothesis of a change in the earth's axis of rotation, "the admission of which, as a geological possibility, is mathematically demonstrable" (*The Voyage of the Fox*, p. 367). Such a hypothesis would solve not only the arctic problem, but also the problem of evidence of glacial action in Britain and India at periods "long antecedent to those in which ice transport is commonly supposed to have commenced" (*The Voyage of the Fox*, p. 367). Any theory that brought together previously unconnected phenomena through a unifying explanation has special charms for scientists; and methodologically it had been dignified by William Whewell as the consilience of inductions.[142] Because the motion of the earth's axis would reconcile all the facts known, "it must be regarded as a geological desideratum to determine its amount and direction, and to assign the cause of such a movement. The solution of this problem I regard as quite possible."[143] Continental drift, which we now accept, was too far-fetched to warrant consideration by sober nineteenth-century scientists.

The final and most recent geological formation, scarcely worthy of the name, was that of superficial deposits. Specimens of well-preserved wood, as well as the occurrence of Siberian mammoths entombed in ice and frozen soil on both sides of Bering Strait, suggested that the Ice Age in Europe was very different from that in Asia and America, and that, "while glaciers clothed the sides of Snowdon ... , pine forests flourished in the Parry

---

[141] Contemporary theory was wrong.

[142] See Whewell's account of consilience in his *Philosophy of the Inductive Sciences* (reprinted New York, 1967). A good account of Whewell's philosophy is M. Fisch, *William Whewell, Philosopher of Science* (Oxford and New York, 1991).

[143] Haughton in McClintock, *The Voyage of the Fox*, p. 267.

Islands, and the Siberian elephants wandered on the shores of a sea washed by the waves of an ocean that carried no drifting ice."[144]

The occurrence of partly fossilized shells at various elevations well above sea level pointed to the submersion of much of the archipelago in recent times.

All in all, the mapping of formations through the arctic islands, the historical use of fossils and the sequence of strata, the implications for climate of the presence of particular fossils, and the evidence from partial fossilization of recent major changes in sea level, all combined to give to arctic geology a dramatically new sense of process. Their convergence also underlined the unity of the natural sciences.


DARWIN, JOSEPH HOOKER, AND THE GEOGRAPHY OF
ARCTIC PLANTS


The union of the life sciences and the earth sciences in geology, geography, and paleontology was not a new one. It had received a major methodological impetus from the publication of Lyell's *Principles of Geology* (1830–3). When Darwin sailed on H.M.S. *Beagle* in December 1831, he took with him the first volume of Lyell's work, and applied Lyell's methods beyond Lyell's limits. Lyell had argued for the fixity of species, that had been created and could become extinct; he believed that in no other way could the variation of the earth's flora and fauna be explained. Evolutionary theory, in its Lamarckian form, or in the related form proposed by Charles Darwin's grandfather Erasmus, he flatly rejected. Darwin, however, on his historic voyage, began an inquiry that by 1838 had led him to outline his own theory of the origin of species through natural selection. The voyage made questions of geographical distribution particularly urgent for him, and in 1837 he was writing eagerly to Richardson about the tree limit and related issues. His object, he explained, was to compare the quantity of vegetation in South America with that in the far North.[145]

Darwin became friends with Lyell on returning from South America, and gave him hints about his ideas in the 1840s. It was not until 1856 that Darwin fully revealed his theory to Lyell, who urged him to write a book about it. Darwin began the book that year, and suffered sorely as an au-

[144]  Ibid., p. 370.
[145]  Darwin to Richardson, 25 July 1837, SPRI MS 1503/16/1; not yet published in *The Correspondence of Charles Darwin*, F. Burkhardt and S. Smith, eds. (Cambridge, 1985–).

thor. As he had written about his journal from the voyage of the *Beagle*, "I . . . get on very slowly: – building a pyramid is an insignificant task to writing a book; I had no idea what a hard working wretch an author, even on the humblest scale, [must] be."[146] Two years later, he received a letter from Alfred Russell Wallace, who, in collecting evidence in the Malay archipelago as Darwin had done in the Galapagos, had arrived at essentially the same theory as Darwin. Thanks to Lyell and to Joseph Dalton Hooker, with whom Darwin had also become friendly on his return home from the Antarctic with James Ross, and who became Darwin's able critic and one of his first converts,[147] Darwin's and Wallace's ideas were presented at the same session of the Linnean Society in 1858, and published later that year. Darwin then gave up the larger work on which he had been engaged for the previous two years, and instead devoted his energies to writing what he called an abstract instead. This was published in November 1859 as *On the Origin of Species by Means of Natural Selection*. Successive chapters in that work explored "The geological succession of organic beings," "geographical distribution," and problems of classification.[148]

Darwin's theory thus had its first public statement shortly after the *Fox* sailed and steamed north at Jane Franklin's behest; and the *On The Origin of Species* was published just two months after the *Fox* returned. Just as Darwin had learned method from Lyell, and turned it into new science, so Hooker learned from Darwin.[149] Floras, like faunas, were dynamic, and needed to be examined with an eye to reconciling history with taxonomy.

In 1856 Darwin had asked Hooker to comment on the section of the manuscript of his "big species book" that dealt with the similarities between arctic and alpine floras. Darwin argued that in colder periods, such as the ice ages, arctic plants had migrated south. Then, as the climate once again became warmer, arctic plants in southern regions retreated both northward via the lowlands, and upward, toward mountain tops where an arctic climate survived in relatively southern latitudes. Plants followed their

146 Darwin to Richardson, 25 July 1837.
147 P. J. Bowler, *Charles Darwin: The Man and his Influence* (Oxford and Cambridge, Mass., 1990) pp. 95, 101.
148 The literature on Darwin is enormous. A good starting point is Bowler, *Charles Darwin*. See also D. Ospovat, *The Development of Darwin's Theory: Natural History, Natural Theology and Natural Selection, 1838–1859* (Cambridge, 1981).
149 W. B. Turrill, *Pioneer Plant Geography: The Phytogeographical Researches of Sir Joseph Dalton Hooker* (The Hague, 1953). Janet Browne, *The Secular Ark: Studies in the History of Biogeography* (New Haven and London, 1983) pp. 132 et seq. Hooker was also indebted to Robert Brown and Alexander von Humboldt; indeed, Brown's biographer goes so far as to claim that Hooker "merely extended the trail" that they had blazed. (Mabberley, *Jupiter Botanicus* [Braunschweig and London, 1985] p. 405).

native climates, and so changed their geographic range with climate over time. It seemed very probable that mountain species in Asia and North America were "remnants of the same arctic vegetation that had populated Europe."[150]

Hooker was very taken with Darwin's picture of distribution changing with climate, but argued that land bridges provided the only significant and sure vehicles for the migration of plants, in contrast with Darwin's additional appeal to icebergs as such vehicles. He thus proposed a Darwinian dynamism, with different emphases when it came to the means of migration. In 1860, Hooker presented his views to the Linnean Society in a paper, "Outlines of the Distribution of Arctic Plants,"[151] drawing on the collections made by a full generation of arctic explorers.

Hooker wanted to show how the present distribution of arctic flora in a circumpolar belt of some ten to fourteen degrees of latitude north of the Arctic Circle could be accounted for "by slow changes of climate during and since the glacial period." The belt showed no sudden change except across Baffin Bay, "whose opposite shores present a sudden change from an almost purely European flora on its east coast, to one with a large admixture of American plants on its west." As a whole, the arctic flora was predominantly Scandinavian; arctic Scandinavia contained three-quarters of the entire arctic flora, and Hooker considered that almost all the genera of arctic Asia and America were also Scandinavian, "leaving far too small a percentage of other forms to admit of the Arctic Asiatic and American floras being ranked as anything more than subdivisions, which I shall here call districts, of one general arctic flora."[152] Hooker argued for a single original creation of plants in Europe, followed by their migration to America and Asia. Subsequently, in this dynamic model, the Asian and American species were themselves introduced everywhere except in Europe and Greenland. All this occurred through two causes, the ebb and flow of glaciation, and the rise and fall of land bridges in the oceans.

Hooker assumed that the original distribution of Scandinavian vegetation was more uniform around the pole than it is now. During the Ice Age, that vegetation was everywhere driven southward, and when warmer weather returned, "those species that survived both ascended the moun-

150   Browne, *The Secular Ark*, p. 127.
151   J. D. Hooker, "Outlines of the Distribution of Arctic Plants," *Transactions of the Linnean Society* 23 (1861) 251–348.
152   Ibid., p. 251. "The globe is divided into five principal areas; or rather the species are traced in five directions, as follows: I. ARCTIC DISTRIBUTION.–1. *Arctic European*. . . . 2. *Arctic Asia*. . . . 3. *Arctic Western America*. . . . 4. *Arctic Eastern America*. . . . 5. *Arctic Greenland*..." (ibid., p. 281).

tains of the warmer zones, and also returned northward, accompanied by aborigines of the countries they had invaded during their southern migration" (p. 253). Here was an explanation not only of present arctic distribution, but of the presence of arctic species in more southerly latitudes. The model was close to that formulated first by Edward Forbes and then by Darwin (p. 253). It provided a convenient explanation of the paucity of the Greenland flora:

If it be granted that the polar area was once occupied by the Scandinavian flora, and that the cold of the glacial epoch did drive this vegetation southwards, it is evident that the Greenland individuals, from being confined to a peninsula, would be exposed to very different conditions to those of the great continents. In Greenland many species would, as it were, be driven into the sea, that is, exterminated; and the survivors would be confined to the southern portion of the peninsula, and not being there subjected to competition with other types, there could be no struggle for life among their progeny, and consequently no selection of better adapted varieties. On the return of heat, these survivors would simply travel northwards, unaccompanied by the plants of any other country.[153]

The relative wealth of the continental flora in North America could be explained by the fact that that flora had been able to move southward and then retreat northward overland, entering the while into competition with and enrichment from southerly alpine flora. Darwin's theory was being well used, and in the process gaining added confirmation.

Perhaps the most striking aspect of Hooker's paper is its enthusiastic Humboldtian character, combining different sciences in a broad geographical sweep. Hooker drew on geographical researches and an Admiralty north circumpolar chart, the fruit of a vigorous half-century's exploration; on circumpolar meteorological data, also newly available, and making possible a map showing isotherms besides lands and seas; and on an impressively complete arctic flora. He was able to make the remarkable claim that

With regard to the probable completeness of the flowering plants of the arctic zone, I think it is pretty certain that there are few or no new species to be discovered. The collectors in the numerous voyages undertaken since 1847[154] in search of the Franklin expedition have not added one species to the flora of the Arctic American islands, and but one to that of Arctic Greenland. The Lapponian [Lapland] region is, of course, as well known as any on the globe; but further east, and especially in

[153] Ibid., p. 254.
[154] Hooker had earlier summarized much of that work of collecting in "On Some Collections of Arctic Plants, Chiefly Made by Dr. Lyall, Dr. Anderson, Herr Miertsching, and Mr. Rae, during the Expeditions in Search of Sir John Franklin, under Sir John Richardson, Sir Edward Belcher, and Sir Robert M'Clure," *Journal of the Proceedings of the Linnean Society. Botany 1* (1857) 114–19.

Arctic Siberia, much remains to be done; not perhaps in the discovery of new plants, but in ascertaining the southern limits of various Siberian ones that probably cross the arctic circle. Of Arctic Continental America the same may be said.[155]

The Franklin searches had thus put the seal of completeness on the arctic flora compiled by William Hooker, and based on the collections of Richardson, Drummond, and other arctic naturalists. That flora, even more than the magnetic work that had provided so powerful a motive for ever-extended arctic exploration, contributed to Hooker's early perception of the integrity of circumpolar science. Another two decades were to elapse before that scientific perception became widely shared, and before politicians were ready to underwrite cooperative circumpolar science.[156] Meanwhile, national rivalries were more evident than international cooperation.

[155] Hooker, "Distribution of Arctic Plants," 280.
[156] See chap. 8 on the International Polar Year of 1882–3.

# 6

The Arctic Crusade: National Pride, International
Affairs, and Science

The Royal Navy's Franklin searches since 1851 had been directed by a
group appointed by the Admiralty, known as the "Arctic Council."[1] This
council was made up of ten of the most prominent arctic officers, eight of
whom were Fellows of the Royal Society of London, elected because of their
contributions to geography and other sciences. Back, Parry, James Clark
Ross, Beaufort, Sabine, Richardson, and Beechey, all F.R.S., we have already encountered. John Barrow, F.R.S., second son of recently deceased Sir
John Barrow,[2] had been placed by his father in the Admiralty, where he favored the arctic enterprise that his father had helped to initiate. Sabine, who
had been involved in naval expeditions to the Arctic since 1818, had now
risen to the vice-presidency of the Royal Society, and was also a power in the
British Association for the Advancement of Science. The only members of
the council not established as members of the scientific community were
Captain Edward J. Bird and Captain W. A. Baillie Hamilton. Bird had
thrice served under Parry, and had been to the Antarctic with James Clark
Ross on the most thoroughly prepared polar scientific expedition to date.
He had also commanded a Franklin search vessel on James Clark Ross's
expedition of 1848–9. Hamilton had served in the Mediterranean, then
served as private secretary to his relative Lord Haddington, First Lord of the
Admiralty, when Franklin sailed on his last expedition; when Sir John Barrow died, Hamilton succeeded him as Second Secretary of the Admiralty.
The council as a whole was clearly experienced in and favorable to arctic
exploration, and broadly supportive of its scientific component.

[1]   Stephen Pearce painted a group portrait of the Arctic Council of 1851, the subject of an
      engraving, to which a descriptive key was given by W. R. O'Byrne. See "The Council of
      1851," *Polar Record* 6 (1843) 385–9, and Fig. 26.
[2]   d. 1848.

Figure 26. The Arctic Council, mezzotint from painting by Stephen Pearce. National Archives of Canada C-23538.

It is scarcely surprising that the instructions to the Franklin search expeditions, and their conduct at sea and on the ice, produced a wide range of scientific observation and collection. Nor is it surprising that the Royal Society soon had its own arctic committee, which took over where the Admiralty left off, for example advising McClintock and supplying him with apparatus for his expedition on the *Fox*. When Franklin's fate was discovered, the Admiralty sought increasingly to disassociate itself from further arctic exploration. It was too expensive, in money and in lives, and the potential gains to knowledge were outweighed by the cost.

The sciences, however, constitute an enterprise impossible to complete, and the many inquiries contributing to science in the Arctic would remain as powerful motives for renewed exploration. There were other motives too. Access to Hudson Bay remained crucial for the fur trade of northern Canada. Whaling was pushing ever further into the Arctic. Expansion in the eastern Arctic came as a result of exploration between Greenland and the archipelago, and following Penny's demonstration that wintering in or near the whaling grounds worked.[3] Whaling in the west grew primarily as a result of increasing American activity through Bering Strait and into the Beaufort Sea and beyond.[4]

## SCIENCE AND POLITICS

The American presence was of economic and political import to Britain, particularly by the late 1850s, when as a result of over-fishing, the Davis Strait fishery had year after year proved a failure.[5] There were other potential sources of Anglo–American friction. In several cities in the United States, British consuls had recruited troops for the Crimean War, thereby violating American neutrality. Vigorous diplomatic controversy ensued. There was an added source of tension in the form of growing pressures for southern secession from the Union. If, as it seemed likely, and was indeed soon to transpire, the United States descended into civil war, there were real prospects of British involvement, given British confederate sympathies.[6]

[3] W. Gillies Ross, "Whaling, Inuit, and the Arctic Islands," in M. Zaslow, ed., *A Century of Canada's Arctic Islands 1880–1980* (Ottawa, 1981) pp. 33–50.

[4] For Penny, see chap. 5. For the western Arctic, see John R. Bockstoce, *Whales, Ice, and Men: The History of Whaling in the Western Arctic* (Seattle and London, 1986).

[5] "THE NEW DAVIS STRAITS' FISHERY," *The Times*, cutting (1857/58) in John Barrow letters, BM Add MS 35306 f. 136ᵛ.

[6] *The Times*, 17 Nov. 1859: "A war between England and America would be almost a civil war. . . . "

What that would mean for the whaling fleets was unclear but alarming: arctic activity, for knowledge or commercial gain, had a sensitive political context that dictated caution.

For the next few years, British activity in the Arctic was muted. There was at least one Anglican missionary, the Reverend William West Kirby, who for almost a decade from 1859 was active in the Northwest, establishing the first church on the banks of the Mackenzie.[7] Father Séguin, a Roman Catholic missionary, traveled for a while with Kirby, and spent the winter of 1862–3 at Fort Yukon.[8] There were other missionaries in the North, but not in large numbers, who were important neither for science nor for politics. The Hudson's Bay Company maintained its presence, supplying existing forts and even adding others.[9] Employees of the company continued to collect natural history specimens, sometimes for American or British museums, and they also, somewhat erratically, kept meteorological records.[10] Oceanography had received considerable stimulus in the work of Edward Forbes, whose notions of a depth below which there was no life in the sea were by the 1860s coming increasingly into question.[11] But the nearest the Royal Navy came to the Arctic, or to arctic oceanography, was in McClintock's hydrographic survey made by carrying out deep-sea soundings between the Faeroe Islands and Labrador, as a prelude to laying a North Atlantic telegraph cable.[12] The Admiralty sought guidance from the Royal Society.[13] A cable was not laid along this route, because they could find no suitable landing on the Labrador coast. The soundings, however, were important, conclusively demonstrating the existence of life at great

---

[7]  F. A. Peake, "William West Kirby, Missionary from Alaska to Florida," *Historical Magazine of the Episcopal Church* 34 (1965) 265–76.

[8]  P. E. Breton, *Irish Hermit of the Arctic: The Life of Brother J. Patrick Kearney, O.M.I.* Translated by J. S. Mullaney (Edmonton, 1963).

[9]  J. K. Stager, "Fort Anderson: The First Post for Trade in the Western Arctic," *Geographical Bulletin* 9 (1967) 45–56. For Fort Nelson, established in 1865, see D. M. Powell, "Western Line I (B. C.) District," *Moccasin Telegraph* 30 (1970) 126–8.

[10] See chap. 9.

[11] E. L. Mills, "Edward Forbes, John Gwyn Jeffreys, and British Dredging before the *Challenger* Expedition," *Journal of the Society for the Bibliography of Natural History* 8 (1978) 507–36; M. Deacon, *Scientists and the Sea* (1971) pp. 276–332; Daniel Merriman, "Speculations on Life at the Depths: A XIXth-Century Prelude," *Bulletin de l'Institut Océanographique de Monaco*, special issue no. 2 (1968) 377–85.

[12] F. L. McClintock, "Surveys by H.M.S. Bulldog," *Proc. Royal Geographical Society* 5 (1861) 62–70; *Remarks Illustrative of the Sounding Voyage of H.M.S. Bulldog in 1860* (London, 1861).

[13] Admiralty to the Secretary, Royal Society of London, 12 June 1860, Royal Society MC.6.93.

depths in the ocean,[14] and thereby contributing to a major line of research that was to have its nineteenth-century culmination in the oceanographic and hydrographic work on the *Challenger* expedition in the next decade. There was also a British naval survey of the Labrador coast, seeking new fishing grounds and harbors; and there were whaling voyages.

The voyage of H.M.S. *Bulldog* in 1860 was not only important for marine biology, but made a major contribution to the physical geography of the sea. An important book on the latter subject was written by Matthew Fontaine Maury, published in 1855, and constantly revised and enlarged.[15] Maury was the United States' preeminent oceanographer and hydrographer;[16] by mid-century, that country played a major role in deep-sea science.[17]

The United States was also the major political presence in the American[18] high Arctic through the 1860s. Not only did American naturalists go north in search of specimens for the Smithsonian and other institutions,[19] but a succession of American explorers and scientists pushed further and further north through the archipelago and along the Greenland shore.

They accomplished this in a decade that saw rapid political shifts concerning the Arctic, and it is hard not to see a correlation between that activity and the support given to explorers by the government of the United States. Canada started to explore its own Arctic. Two Dominion surveyors, Henry Youle Hind and his artist brother William, began an exploration of the Quebec-Labrador peninsula. Hind already had experience as a field geologist, and made important contributions to physical geography and ethnology.[20] Surveys and the scientific mapping of the land and its products contributed significantly to the emergence of a coherent Canadian

---

[14] G. C. Wallich, *Notes on the Presence of Animal Life at Vast Depths in the Sea; with Observations on the Nature of the Sea Bed, as Bearing on Submarine Telegraphy* (London, 1860); Wallich, *The North-Atlantic Sea-Bed: Comprising a Diary of the Voyage on Board H.M.S. Bulldog, in 1860; and Observations on the Presence of Animal Life, and the Formation and Nature of Organic Deposits, at Great Depths in the Ocean* (London, 1862).

[15] M. F. Maury, *The Physical Geography of the Sea* (New York, 1855).

[16] F. L. Williams, *Matthew Fontaine Maury, Scientist of the Sea* (New Brunswick, New Jersey, 1963).

[17] For mid-century debates in oceanography and hydrography, see M. Deacon, *Scientists and the Sea* (1971) pp. 276–331.

[18] Here used geographically not politically.        [19] See chap. 9.

[20] H. Y. Hind, "An Exploration of the Moisie River, to the Edge of the Table-Land of the Labrador Peninsula," *Journal of the Royal Geographical Society* 34 (1864) 82–7; *Explorations in the Interior of the Labrador Peninsula, the Country of the Montagnais and Nasquapee Indians*, 2 vols. (London, 1863). See also W. L. Morton, *Henry Youle Hind 1823–1908* (Toronto, 1980).

nationalism:[21] without seeking a causal connection between surveying and nationhood, one can note that the confederation of the Dominion of Canada in 1867 came not long after the first Canadian as opposed to the first British or American surveys of the North. The dream of "a Great Britannic Empire of the North," rivaling the United States, had been vigorous since the 1850s. "Thus, even prior to the birth of the new dominion, English- and French-speaking colonists had integrated a northern ethos into their nationalist rhetoric."[22] The further Canadian opening of that north land, extending into the continental Arctic but not yet significantly into the archipelago, was rapidly pursued in the decade following Confederation,[23] but that opening was for territorial sovereignty and the exploitation of resources, rather than for the prompt settlement of northern lands.

The American purchase of Alaska from Russia occurred in the same year as Canadian confederation, with territorial challenge accompanying national self-assertion.[24] The United States had conducted its negotiations with Russia in secrecy, perhaps to exclude the British. The British ambassador in Moscow, informed of the sale, visited the Russian Foreign Minister, Gorchakov, and relayed his comments as follows:

I said it might have been considered a friendly act on the part of the Russian Government if she had afforded Her Majesty's Government or the Government of Canada an opportunity of purchasing the territory which had been sold, but that this not having been so was materially unimportant, as I felt assured it would not have been bought.[25]

A further realignment of jurisdictions affecting the Canadian Arctic occurred in 1870, putting a seal to the decade with the transfer to Canada of the lands administered as a monopoly by the Hudson's Bay Company since its imperialist union with the North West Company in 1821.

---

21 S. Zeller, *Inventing Canada* (Toronto, 1987) p. 273.
22 Shelagh D. Grant, *Sovereignty or Security?* (Vancouver, 1988) p. 4.
23 Morris Zaslow, *The Opening of the Canadian North 1870–1914* (Toronto, 1971). For British and American tensions and dimensions of Canada's emergent nationalism, see Carl Berger, *The Sense of Power: Studies in the Ideas of Canadian Imperialism 1867–1914* (Toronto, 1970).
24 Alaska falls outside the geographic limits of this book. A useful study is James Alton James, *The First Scientific Exploration of Russian America and the Purchase of Alaska* (Evanston and Chicago, 1942).
25 Quoted in Trevor Lloyd, "Canada and the Circumpolar World – Comparisons and Challenges," in M. Zaslow, ed., *A Century of Canada's Arctic Islands 1880–1980* (Ottawa, 1981) p. 313, from Stuart R. Tompkins, *Alaska: From Promyshlenik to Sourdough* (Norman, Oklahoma, 1945) pp. 188–9.

Thus, in the course of the 1860s, Russia effectively removed itself as a presence in the North American Arctic; Canada achieved nationhood as a Dominion; Canadian surveyors began to explore the continental Arctic, and to consider more actively its wealth of raw materials; the United States of America achieved explicit sovereignty over the Alaskan part of the western continental Arctic; the Royal Navy kept out of the archipelago; and exploration of the archipelago, with its uncertain implications for sovereignty, was pursued mainly by Americans.

In 1860, the American zoologist Robert Kennicott was in the second year of an expedition to the Mackenzie River, collecting specimens for the Smithsonian Institution, while Constantin Drexler, taxidermist and ornithologist at the Smithsonian, was in James Bay, collecting birds for the museum.[26] Meanwhile, two additional American expeditions had set out for the Arctic, one aiming for the North Pole by way of Smith Sound, the other combining whaling with a search for Franklin relics in the archipelago.

## THE AMERICAN ROUTE: SCIENCE AND GOVERNMENT

*Isaac Israel Hayes and the United States North Polar Expedition 1860–1861.* Isaac Israel Hayes had already been to the Arctic when, straight from medical school, he had gone as ship's doctor on Kane's expedition in 1853–5. He had been one of the party of crewmen who, disenchanted with Kane's leadership, attempted to make their way south in a breakaway party and were rescued by the Inuit. Now, most improbably in the light of that experience, he was determined to lead his own expedition north, and to combine scientific observation with geographical exploration. But first he had to raise funds.

His account of the second Grinnell expedition, *An Arctic Boat Journey in the Autumn of 1854* (Boston and London, 1860) had brought in some money, but he needed a minimum of $20,000. He gave a series of lectures to scientific bodies and public lectures to obtain institutional and financial support but his efforts met with very limited success. He argued that a new expedition would extend Kane's supposititious discovery of an open polar sea, and would also settle the question about the existence of an undiscovered continent in the Arctic Ocean to the north of Asia. His first lecture was given to the American Geographical and Statistical Society in 1857,

[26] See chap. 9.

and perhaps his most crucial one was given at the meeting of the American Association for the Advancement of Science in Baltimore in the following year. Besides the open polar sea and the new arctic continent, he held out prospects of new knowledge in climatology, anthropology, and natural history.[27] The AAAS set up an arctic committee to confer with Hayes, but he was more concerned with material than with intellectual support. Alexander Dallas Bache encouraged him to lobby the government for funds, but U.S. government support had shrunk since Kane's expedition. Money came in principally from Henry Grinnell, as well as from other private donors and scientific societies in the United States and Europe. Hayes was able to acquire and strengthen a small vessel, with a crew of 14 officers and men. When they reached the Greenland shore, the addition of an interpreter, Inuit hunters, and Danish seamen brought the complement up to 19, a far cry from the larger vessels and 133 followers on Franklin's last expedition.[28] The government did provide some instruments; barometers, thermometers, and containers and spirits for zoological specimens came from the Smithsonian Institution; the United States Topographical Bureau provided two pocket sextants; the Coast Survey under Bache furnished the expedition with a combined transit and theodolite, a unifilar magnetometer, a reflecting circle, and other instruments. Spirit thermometers, a 4½" telescope, chronometers, and a seconds pendulum were also obtained, and Hayes was well satisfied with their scientific equipment.[29]

The expedition left Boston in July 1860, with Sonntag, also from Kane's expedition, and young Henry Radcliffe as assistant astronomer. They headed up the west coast of Greenland, and on into Smith Sound, where they met heavy ice. After hitting an iceberg and sustaining severe damage, they put in to what Hayes named Port Foulke, some twenty miles south of Kane's winter harbor. They used the pendulum for gravitational experiments in fall and spring, as well as astronomical observations. Game was plentiful, and an Inuit party remained with them through the winter.

In spring, they made sledging journeys with the dogs that survived from the previous year; too many had succumbed to a disease beginning with fits and ending in death. The dogs were not the only casualty. Sonntag had fallen through the ice and died of hypothermia. But the spring sledging did

[27] I. I. Hayes, "Observations upon the Practicability of Reaching the North Pole," *Proceedings of the American Association for the Advancement of Science* 12 (1858) 234–54; J. E. Caswell, *Arctic Frontiers: United States Explorations in the Far North* (Norman, Oklahoma, 1956) pp. 32–3.

[28] Berton, *The Arctic Grail* (Toronto, 1988) pp. 353–4.

[29] Caswell, *Arctic Frontiers*, pp. 35–6.

result in collections of plants and fossils, as well as the northward exploration of the coast of Ellesmere Island. Hayes's navigation was poor; he reckoned that he had come within 450 miles of the pole, whereas it subsequently emerged that he had greatly overstated his progress, and that the distance was nearer to 1,000 miles.[30]

He returned to Boston that summer, intending to return to the Arctic in the following year, but the Civil War frustrated that intention. The scientific results were more extensive than intensive but they included meteorological, magnetic, and astronomical observations (which were sent to the Smithsonian where Schott reduced them for publication),[31] as well as a measurement of the movement of a glacier during the winter, and natural history collections, some of considerable interest. The plant collection contained no startling surprises, and the northern limit at which specimens were collected was wrongly determined, but the southern limit was accurate, and the contribution to phytogeography was correspondingly valid.[32] The fossils, because of their northerly location, aroused considerable interest.[33]

Hayes had hoped that on his return, a fair measure of success in exploration, announced in lectures, a book,[34] and the sale of photographs, would pay the outstanding expenses of the expedition. The photographs, taken by Radcliffe, were a real novelty; they were the first from north Greenland, though not the first from the Arctic. The outbreak of civil war distracted both public and private interest, and Kane found himself putting up instruments, including those that had been lent through Bache, as security for a loan to pay his sailors; he was then much embarrassed when Bache pressed him for the return of almost $1,000 worth of instruments.[35]

[30] Berton, *The Arctic Grail*, p. 363.

[31] I. I. Hayes, *Physical Observations in the Arctic Seas, Made on the West Coast of North Greenland, the Vicinity of Smith Strait and the West Side of Kennedy Channel, during 1860 and 1861: Reduced and Discussed at the Expense of the Smithsonian Institution by Charles A. Schott, Smithsonian Contributions to Knowledge* vol. 15 art. 5. Smithsonian Institution Publication 196 (Washington, D.C., 1867).

[32] E. Durand et al., "Enumeration of the Arctic Plants Collected by Dr. I. I. Hayes in his Exploration of Smith's Sound between Parallels 78th and 82nd, during the Months of July, August and Beginning of September 1861," *Proc. Academy of Natural Sciences of Philadelphia* (1863, pub. 1864) 93–6.

[33] F. B. Meek, "Preliminary Notice of a Small Collection of Fossils Found by Dr. Hays [*sic*], on the West Shore of Kennedy Channel, at the Highest Northern Localities Ever Explored," *American Journal of Science* S2 40 (1865) 31–4.

[34] I. I. Hayes, *The Open Polar Sea: A Narrative of a Voyage of Discovery toward the North Pole, in the Schooner "United States"* (London, 1867).

[35] Caswell, *Arctic Frontiers*, p. 40.

*Charles Francis Hall 1860–1871*. In 1859, while Hayes was involved in raising funds for his expedition, and while McClintock was still away on the *Fox,* Charles Francis Hall, a newspaper proprietor and printer in Cincinnati, received, as he would later say, a call from God to go to the Arctic and search for Franklin. He was well prepared for the call. Kane was one of his heroes, and after 1857 Hall's notebooks took on an arctic focus. He sold his business, left his family, and set about preparing for an arctic voyage.[36] He soon found that he would not succeed in raising funds for his own vessel, but managed to purchase supplies and acquire a sextant, a pocket sextant, two self-registering thermometers, ordinary thermometers, and compasses.[37] Besides searching for Franklin, Hall's dream of reaching the North Pole for the United States was a powerful motive, albeit scarcely unique and inevitably frustrated. But in other ways his aims and achievements were unprecedented, resting on his determination and ability to learn from and to live with the Inuit.[38] Having taken passage on a whaler, which wintered on Baffin Island, he was befriended by an Inuit couple, Ebierbing and Tookoolitoo. He learned firsthand about their way of life, as well as something of the language; he was the first outsider to call them by their own name, Inuit, the people. The whaler spent an enforced second winter in the ice, and steamed home in 1862, with Hall and his Inuit friends on board, and not with Franklin but with Frobisher relics, to which he was led by Inuit accounts. Ebierbing and Tookoolitoo would return to the North with Hall on his next expedition, remaining his friends and companions until his death in 1871 on the Greenland shore.

Hall had collected a representative set of geological specimens around Frobisher Bay.[39] Otherwise there was not much formal science on Hall's first expedition. The innovation of living with the Inuit as nearly as possible as one of them was important, and would be followed by serious ethnologists in decades to come. Hall's experience, even more than Rae's travels, showed the superiority of the Inuit to naval ways of living in the high Arctic.

[36]  Chauncey C. Loomis, *Weird and Tragic Shores: The Story of Charles Francis Hall, Explorer* (New York, 1971).

[37]  C. F. Hall, *Arctic Researches and Life among the Esquimaux* (New York, 1865) pp. 585–7, a one-volume reprint of Hall, *Life with the Esquimaux,* (see note 38).

[38]  C. F. Hall, *Life with the Esquimaux: The Narrative of Captain Charles Francis Hall, of the Whaling Barque "George Henry," from the 29th May 1860, to the 13 September, 1862,* 2 vols. (London, 1864).

[39]  B. K. Emerson, "On the Geology of Frobisher Bay and Field Bay: Description of the Collections Made by C. F. Hall during his First Expedition in the Arctic Regions, 1860–1862," in J. E. Nourse, ed., *Narrative of the Second Arctic Expedition Made by Charles F. Hall . . . 1864–1869* (Washington, D.C., 1879) pp. 553–83.

More important for Hall was the fact that he had not only lived with "this noble people, . . . but enjoyed their society & mode of life. The happiest days of my life," he declared in a letter to John Barrow, "have been spent living with these iron sons and daughters of the North – sharing all their vicissitudes of life."[40]

Hall's next expedition was again aimed at discovering Franklin relics, and this time he was successful; once again he steamed north on a whaler, once again he lived with the Inuit, and once again formal science was not prominent. He had, however, made more of an effort to obtain apparatus, taking with him a somewhat fuller outfit of instruments, among them a dip circle supplied by the Coast Survey and broken before it could be used.[41] He had written in vain to the Smithsonian to ask if he could use the magnetic apparatus that they had furnished to Kane. Joseph Henry, secretary (that is, head) of the institution, was America's leading physicist and spokesman for science, and was also president of the American Academy of Science: his plurality of offices would have impressed an eighteenth-century English clergyman or a nineteenth-century French academic. He told Hall that the apparatus had been

lost in Mexico . . . , and we have no other to supply its place. Besides this unless some one were connected with the expedition properly educated for, and trained to the business of observation who should devote his whole time to the instruments, scarcely any results could be obtained that would add to what is already known in regard to the magnetism of the polar regions.

Here was a message that Hall would take to heart, and apply in selecting personnel for his third expedition in 1871. Natural history collections did not require such single-minded devotion, nor was it beyond the reach of competent amateurs. Hall's offer to make such collections met with a warmer response from Henry:

. . . we . . . will gladly receive whatever you can send us.[42] We are especially interested in the Natural History of the Arctic Regions. . . . The Frobisher relics too we will be much pleased to place on exhibition in our Hall, side by side with a gun formerly used by Sir John Franklin in his second journey, and a short sword, relic of his last fatal expedition, obtained on the lower Mackenzie through some Esquimaux.[43]

[40] Hall to Barrow, 3 Dec. 1862, Museum of American History, C. F. Hall papers.
[41] Nourse, *Narrative,* p. 85; Caswell, *Arctic Frontiers,* p. 49.
[42] For further discussion of the Smithsonian Institution's relations with arctic naturalists, see chap. 9.
[43] Joseph Henry to Hall, 24 March 1863, Museum of American History, C. F. Hall papers.

Henry's implicit dismissal of Hall as a magnetic observer must have disappointed him, while at the same time spurring him to make those regular observations, primarily in meteorology, that were within his competence.[44] Hayes, who was unimpressed, also attacked him by innuendo of his lack of formal scientific training. In April 1870, in what appears to have been an interview with the Committee on Foreign Affairs, Hall defended himself:

No, I am not a *scientific* man. Discoverers seldom have been. Arctic discoverers – all except Dr. Hayes – have not been scientific men. Neither Sir John Franklin nor Sir Edward Parry were of this class and yet they loved science and did much to enlarge her fruitful fields. Frobisher, Davis, Baffin, Bylot, Hudson, Fox, James, Kane, Back, McClintock, Osborn, Dease and Simpson, Rae, Ross and a host of other explorers were not *scientific* men.[45]

Hall's distinction between scientific men and amateurs who could still advance the sciences was increasingly sensitive as professionalization took hold of the sciences. We have repeatedly encountered tension between the demands of exploration and the demands of systematic scientific observation. The very habits of order and regularity required by the latter were unlikely to expedite the former. Leaving aside the acceptance of geography as a science, explorers and scientists were frequently impatient with one another. Halley and the entire institution of the Royal Navy, Banks and Captain Cook, and as we shall see, Stefansson and the Canadian community of scientists, all had simply different emphases and priorities. When the Royal Navy sent out arctic or other expeditions, ships' commanders generally received orders asserting the primacy of navigation over science. The Franklin searches form a particularly cogent example; but as the urgency of navigation receded, or even vanished as it did once a ship was iced in for the winter, science could assert itself, if only as a way of providing occupation through the long and tedious winter.

Hall's third expedition was the best sponsored and the most official of the three that he undertook. He asked Congress for $100,000, and received half that sum, as well as the pay of government employees who joined him. The scientific program was devised by the National Academy of Sciences. Joseph Henry, president of the academy, continued to be skeptical of Hall's scientific competence, and extended that skepticism to his officers and crew. Hall's choices had been made with a view to the Arctic rather than to sci-

---

[44] Nourse, *Narrative*, pp. 479–550, appendix II gives "Hall's Meteorological Journal, 1864–69." Notes on astronomical observations were presented on pp. 451–2.

[45] Quoted in Loomis, *Weird and Tragic Shores*, p. 239.

ence. "It is evident," Henry concluded, " . . . that the Expedition, except in its relation to geographical discovery, is not of a scientific character."[46] In spite of such reservations, Hall's expedition was presented with a detailed scientific program to follow,[47] with Henry giving instructions for meteorology, Simon Newcomb, Professor of Mathematics in the United States Navy,[48] advising on astronomy, the Pennsylvania naturalist and zoologist Spencer Fullerton Baird, then the Smithsonian's assistant secretary under Henry, advising on natural history, and Louis Agassiz handling glaciers, on which he was the world's leading authority.

Lady Franklin, interested in Hall because he had searched for records of her husband when few others thought there was anything more to be learned, had written suggesting that David Walker, former surgeon-naturalist on the *Fox* under McClintock, would be a good candidate as chief scientist. Henry and Baird at first agreed, but then changed their minds on the advice of August Petermann, the great German geographer. His candidate was Emil Bessels, a physicist, physician, and naturalist all in one. He had obtained a medical degree at Heidelberg, studied zoology at Jena and Stuttgart, and had been with Petermann's expedition in 1869 to the seas around Spitsbergen, where he had done distinguished work. Frederick Meyer, a German-born technician now in the U.S. Signal Corps, was appointed as meteorologist under Bessels, and R. W. D. Bryan, a recent graduate of Lafayette College, was a last-minute appointment as both astronomer and chaplain.[49]

The expedition had more than a set of instructions and a full-time experienced scientist; its complement of apparatus was also admirable. The Coast Survey, in spite of its earlier experience with Hall, lent astronomical instruments, which included a Würdemann transit, Gambey sextants, and artificial mercury horizons. Hayes, at least temporarily suspending his criticisms of Hall, lent a pendulum for gravitational work. They had ten chronometers, and there were magnetic instruments, chief among them a unifilar magnetometer and a dip circle. They had meteorological instruments, dredging equipment, geological hammers, a microscope,

---

[46] Quoted by Caswell, *Arctic Frontiers*, p. 56 from C. H. Davis, ed., United States Navy Department, *Narrative of the North Polar Expedition: U.S. Ship Polaris* (Washington, D.C., 1876) p. 637.

[47] The instructions were reprinted in Davis, *Narrative of the North Polar Expedition*, pp. 637–62.

[48] Newcomb was director of the American *Nautical Almanac* from 1877 until his death in 1897.

[49] Loomis, *Weird and Tragic Shores*, pp. 246–7, 257.

a camera and plates, containers and alcohol for preserving natural history specimens – and more besides.[50]

Hall was in overall command of the expedition, but he chose an experienced whaler, Sidney Buddington, to command his ship, the renamed and refitted *Polaris*. Hall's instructions were that he should head for the North Pole by way of what was now firmly established as the American route, along the west coast of Greenland. Because the pole was the first priority, the scientific program would, typically, be mostly concentrated in the winter months. *Polaris* steamed north to 82°11′ N, was stopped by ice, and, turning south in search of a suitable harbor for the winter, settled in at 81°38′ N, on the northwest coast of Greenland, in what the expedition named Thank God Harbour.

Bessels began organizing scientific observations. Meteorological observations were unprecedentedly comprehensive, including, besides the usual qualitative observations, and measurements of temperature, atmospheric pressure, wind velocity, and humidity, regular observations of atmospheric electricity, ozone concentrations, and solar radiation. An observatory was built and filled with the astronomical apparatus, the magnetic instruments were set up nearby in snow houses (although they were not used for the first half of the winter), and pendulum experiments were carried out.[51] Meanwhile, the expedition began to come apart. Hall, having returned late in October from a northward sledding trip, died on board ship early in November. It appears likely, following a recent autopsy, that he died of arsenic poisoning,[52] caused either by murder or by an accidental overdose in his medicine. Most of the crew's enthusiasm for reaching the pole died with Hall, although Bessels, disciplined in pursuing his scientific program as far as possible, also attempted further sledging to the north.

In the summer of 1872, *Polaris* was temporarily freed and started south, but was soon beset again. That October, nineteen members of the party were carried away from the ship on an ice floe, surviving the winter thanks to Joe Ebierbing and another of the Inuit, Hans Hendrik. They were to be rescued off the Labrador coast in 1873 by a Newfoundland sealer. Meanwhile, *Polaris*, having survived another winter, appeared next season to have sprung a leak; almost everything, including many of the scientific results and most of the apparatus, was thrown overboard prematurely and

[50]   Emil Bessels, *Die amerikanische Nordpol-Expedition* (Leipzig, 1879) pp. 10–12; Caswell, *Arctic Frontiers*, pp. 56–7; *Scientific Results of the United States Arctic Expedition: Steamer Polaris, C. F. Hall Commanding*, vol. I *Physical Observations by Emil Bessels* (Washington, D.C., 1876).

[51]   Caswell, *Arctic Frontiers*, pp. 60–1.        [52]   Loomis, *Weird and Tragic Shores*.

then lost when the ice floe broke up. Later, the crew set off in two boats built from the ship's timbers. Remarkably, they survived the ice fields, were picked up by a Scottish whaler, and, passing to other vessels, made their way home.[53]

Bessels went via Pond Inlet, where he studied the Baffin Island Inuit.[54] He had also managed to keep with him some of the scientific records of the expedition, and the first of two volumes of these was published in a hurry, partly so as to be of use to the new British Arctic Expedition, announced in 1874 for the following year. That volume was later suppressed for errors.[55]

This was not altogether propitious. The United States had, however, firmly entered the realms of northern science, and had joined, as it was later to command, the quest for the pole. There was a board of inquiry into the loss of the *Polaris*, and on it, representing the U.S. Signal Corps, was Captain H. W. Howgate, who would later plan an expedition to the Arctic;[56] so too would Howgate's junior in the Corps, Adolphus Washington Greely. The tragic outcome of Greely's expedition[57] would ensure its leader a prominent place in the arctic pantheon. Americans tended to glorify success more than the British, but like the British, they also had a taste for glorifying failure, especially tragic and hopeless failure.

## THE ARCTIC CRUSADE: LEARNED SOCIETIES TACKLE A RELUCTANT ADMIRALTY 1865–1874

By the mid-1860s, British advocates of arctic exploration were chafing at Admiralty inaction, and mortified by what looked like growing American

---

[53] E. Vale Blake, ed., *Arctic Experiences: Containing Capt. George E. Tyson's Wonderful Drift on the Ice-Floe, a History of the Polaris Expedition, the Cruise of the Tigress and Rescue of the Polaris Survivors: To Which is Added a General Arctic Chronology* (New York, 1874).

[54] Bessels, "Einige Worte über die Innuit des Smiths Island, nebst Bemerkungen über Innuit-Schädel," *Archiv für Anthropologie* 8 (1875) 107–22; "The Northenmost Inhabitants of the Earth," *American Naturalist* 18 (1884) 861–82.

[55] Caswell, *Arctic Frontiers*, p. 66. United States Navy Department, *Scientific Results of the United States Arctic Expedition*, vol. 1 (Washington, D.C., 1876) p. v, Bessels to Joseph Henry, 1 March 1875: "Some portions of the volume have been prepared in a somewhat hasty manner, in order to render the information collected immediately available for the use of the English expedition about to be dispatched to the same regions." Bessels's original observations are in the Smithsonian Institution Archives, RU 68.

[56] G. E. Tyson, *The Cruise of the Florence, or, Extracts from the Journal of the Preliminary Arctic Expedition of 1877–78. Edited by H. W. Howgate* (Washington, D.C., 1879).

[57] See chap. 8.

dominance in a field that was once almost a British monopoly. The Royal Geographical Society, under its president, Sir Roderick Murchison, decided to mount a thoroughgoing campaign, engaging the support of scientific societies at home and abroad, in a determined effort to awaken the Lords of the Admiralty from their post-Franklinian torpor, and persuade them to undertake renewed polar exploration.

Murchison first approached the senior society, the Royal Society of London, receiving in return a resolution that the proposed expedition, properly undertaken, would be "highly advantageous in the advancement of several branches of Physical Science." If the government sent out an expedition, then the Royal Society would propose a "detailed statement of scientific objects which might be prosecuted with advantage without interfering with the main geographical purposes of the Expedition, and to specify in particular the instruments and methods of research available for such objects."[58] The primacy of geographical exploration reflects the origin of the proposal, as well as the Admiralty's relegation of natural science to a subordinate role, except where it bore on navigation. The stress on physical science makes sense here, for natural history and the human sciences had their separate advocates.

The Linnean Society, devoted to natural history and zoology, was next in order of seniority, and next to respond,[59] and in spite of Hooker's conviction that there were probably no new species to be discovered in arctic North America, they were very keen on the idea. They argued not only that the expedition would be good for science, but that "maritime adventure and voyages of discovery in the pursuit of science have an excellent effect upon the Naval Service" in training officers to cultivate their powers of observation!

Drawing on their experience around the world, they disposed briskly of "the popular objection to North Polar Expeditions on account of the supposed danger": the council of the society was

convinced that it rests on a fallacy. The Linnean Society has, during the last half century, enrolled among its Members almost all the scientific Officers of the surveying and exploring expeditions of our Naval and other public services. . . . It has thus been deeply concerned in . . . judging of the comparative amount of loss and hardship incurred, the results showing a remarkable immunity from danger exemplified in the Polar voyages North and South, as compared with many others. With

[58] Secretary of the Royal Society to president of the Royal Geographical Society, 16 Feb. 1865, PRO MS Adm.1 5934.
[59] Secretary of the Linnean Society to secretary of the RGS, PRO MS Adm. 1 5934, n.d. but in response to letter of 2 March 1865. Founded in 1788, the Linnean Society received its charter in 1802.

the exception of Sir John Franklin's party, it is believed that not one Fellow of the Society has met with his death through Polar Discovery, whilst in those African Surveys and Explorations which are so warmly supported, there are very few of the numerous contributors to our publications who have not perished in the prosecution of their researches, and the numbers lost in, or in consequence of, scientific expeditions in India and other tropical countries, and in the interior of Australia, have been most deplorable.

Franklin's fate, like that of Mungo Park and many others, should be a guide to future expeditions rather than a warning against undertaking them. As for specific gains for science,

The most important results in natural history to be obtained from a voyage to the Arctic Ocean, are undoubtedly those that would extend our knowledge of the conditions of . . . life in those regions. It is now known that the Arctic Ocean teems with life, and that of the more minute organized beings, the multitude of kinds is prodigious . . .

The kinds of these animals, the relations they bear to one another, and to the larger Animals (such as whales, seals &c., towards whose food they so largely contribute), the conditions under which they live, the depths they inhabit, their changes of form, &c., at different seasons of the year, and at different stages of their lives, and lastly, their distribution according to geographical areas, warm and cold currents, &c., are all subjects in which very little is known.[60]

They also sought information about algae, diatoms, and other microscopic plant life of the polar seas. The skeletons of diatoms in particular could be correlated with fossil forms, and could thus contribute to dating rocks and to understanding ancient climatic changes. When it came to terrestrial plants, the most urgent questions were in biogeography, correlated with a study of currents and of the effects of climate. This would help to unravel anomalies; for example, "the Spitsbergen Flora contains American plants found neither in Greenland nor in Scandinavia."

The enormous increase in the sophistication of the theoretical framework of natural history in the half century since the Napoleonic Wars is abundantly clear from the Linnean Society's questions. Many developments had contributed to this increase, among them the evolutionary biology promulgated in Darwin's *On The Origin of Species* in 1859. Reinforced by Darwin's work was a keen awareness of the interdependence of different forms of life, and of their adaptation to specific environments. Also related, and indeed seminal for Darwin's work and for natural history in general, was historical geology, incorporating stratigraphy and paleontology, as well

---

[60] PRO MS Adm. 1 5934. M. P. Winsor, *Starfish, Jellyfish, and the Order of Life: Issues in Nineteenth-Century Science* (New Haven, 1976) discusses taxonomic and other problems in invertebrate marine biology.

as studies of climate; these all owed much to Lyell's synthesis in his *Principles of Geology* (1830–3). Biogeography, stemming from von Humboldt and others, had been advanced by Lyell, and was now passionately pursued by Joseph Hooker within a broadly Darwinian framework. Also underlying the Linnean Society's questions were issues in marine biology and oceanography, sciences which, owing something to arctic voyages, had been developing since John Ross's expedition in 1818. Oceanographic research had meanwhile gained greatly from significant improvements in sounding equipment.[61]

With Murchison, British geology's statesman, in the chair at the Royal Geographical Society, it was inevitable that the Geological Society of London would offer enthusiastic support, together with an earnest wish that someone "capable of making the required geological investigations" be part of the expedition. As for detailed proposals for geological work, these would be forthcoming as soon as the expedition was approved by the government.[62]

The Ethnological Society, founded in 1843 to study the physical and moral characteristics of the varieties of the human species, using archeological and perhaps even geological evidence as well as modern evidence, was similarly keen on polar exploration. Like the Linnean Society, they, in the person of their secretary, the eugenicist Galton, observed that "it is a matter of notoriety that as compared with geographical Expeditions in other directions, those to the Polar regions have, on the whole, been attended with remarkable little loss of life . . . " To be properly useful, such observations, they urged, should be carried out by competent observers. The stress on properly skilled observers is a reflection of two complementary trends in natural history, one toward its growth as a branch of the sciences needing its own professional core, the other toward the increasing coordination of amateur contributions by those professionals.[63] It was no longer enough, if it had ever been, simply to assume that naval officers could satisfy the needs of qualitative natural science, while quantitative studies, like those pursued by Sabine, required formal preparation. Certainly Parry, Sherard Osborn, and other officers had complained of their

[61]  A. McConnell, *No Sea Too Deep*, chaps. 4 and 5.
[62]  PRO MS Adm. 1 5934, 11 March 1865.
[63]  D. E. Allen, *The Naturalist in Britain: A Social History* (London, 1976) passim: chap. 14 especially discusses the combining of amateur and professional naturalists. Ruth Barton, "An Influential Set of Chaps," *BJHS* 23 (1990) 53–81, explores the limitations of the amateur-professional dichotomy. Frank M. Turner, "The Victorian Conflict between Science and Religion: A Professional Dimension," *Isis* 69 (1978) 356–76 also explores emergent professionalism in a way that goes beyond that dichotomy.

unpreparedness for systematic natural history; so had John Rae. As we shall see, recognition by serving officers of their scientific limitations, and criticism by scientists of those limitations, still did not persuade the Admiralty for some years to ensure the presence of adequate scientific complement on naval expeditions.

Most of the learned societies approached by the Royal Geographical Society were content to accept the latter's notion that the attainment of the North Pole was the first priority, and that other scientific work would be fitted in around it. But there were two respondents who expressed reservations, not about the proposed expedition but about its priorities. George Roberts, writing for the Anthropological Society, observed that getting to the pole by the shortest route in a matter of months would not allow for the collection of desired scientific information. "Indeed," he noted, "we should be inclined to regard the reaching of the North Pole in a single season rather a loss than a gain to general science inasmuch as it would be very difficult to persuade the general public (of whom we must not lose sight) that any more was to be done after the North Pole had been reached"; with the result that years would elapse before there was another chance to fill in gaps in the physical geography of the Arctic.[64]

The other critical comment came informally from St. Petersburg, where the Imperial Academy of Sciences hastened formally to support Murchison's proposal.[65] Count Feodor Petrovich Lütke, whose expedition in 1822–4 had obtained the first real scientific information about Novaya Zemlya, wrote with cheerful directness to Murchison, expressing the hope that academic and scientific support would succeed in "neutralizing the opposition of the adepts of the Cui bono principle." But he, in his character as self-styled arctic explorer emeritus, found that there was

too much prominence given to the hope or the desire of *reaching the Pole,* a thing that I consider of no importance compared to the great goal of exploring all the polar region, where the earth's pole – an abstract, mathematical point which doesn't even exist in reality – plays no part. The idea of planting the national flag there may smile on national self-respect; but one must not forget that to do this, one would have to take a mathematical line as a flagpole.[66]

One question to which the scientific societies responded with unanimity was about the best route to the pole. If getting to the pole was the main concern, then the Spitsbergen route might be fine; but for sciences other

[64]  8 May 1865, PRO MS Adm. 1 5934.
[65]  10 April 1865, Sec. perpetuel de l'Académie Impériale des Sciences, St. Petersbourg, PRO MS Adm. 1 5934.
[66]  Lütke to Murchison, 10/22 April 1865, PRO MS Adm. 1 5934.

than geography, and even for geography when it came to mapping the north shore of Greenland and the archipelago, the Smith Sound route, the one in danger of becoming the American route, was obviously superior, because many natural sciences cannot be studied in mid-ocean, and only the latter route had much land along it.

Armed with letters of support from major scientific societies, Murchison, writing as President of the Royal Geographical Society, wrote to the Duke of Somerset, First Lord of the Admiralty, urging the case for another naval expedition. As a geographer, he pointed out that the unknown polar area covered roughly two million square miles. Murchison believed that the pole was at the center of an ocean, in which case it followed "as a thermal law" that the climate around the geographical pole would be milder than that around the magnetic pole. That could have important implications for the navigation of the Northwest Passage. Magnetic data were important for science and navigation; obtaining them for the polar region would supply "a *desideratum* of real practical value," as Sabine, felicitously now President of the Royal Society of London, would be happy to elaborate.

Then Murchison became forward looking rather than retrospective in his argument. In 1882 there would be an opportunity in the Antarctic of observing the transit of Venus over the sun. This, although Murchison did not explain as much to the noble duke, was important because, as Edmond Halley had pointed out, the parallax of the sun could be obtained by observing transits of Venus from places on the earth where the displacement by parallax was greatest. The parallax of the sun was important because it was the key to the accurate determination of the sun's distance from the earth, and because that distance connected terrestrial measurements with celestial ones, navigational science was potentially a great beneficiary of good transit observations. Now transits of Venus were infrequent phenomena, occurring only four times every 243 years; the transit of 1882 would be the second and last in the nineteenth century.[67] So Murchison had a point, which probably escaped the duke. In order for the transit observations to stand a reasonable chance of being accurate and therefore useful, there would in 1882 be a need for officers "who have *already* been trained to ice navigation by service in the Polar Regions," so the expedition now proposed would be an investment for the subsequent crucial observations. The Admiralty was not much impressed, and Britain was one of the most sluggish nations to become involved in the whole program of observations that

[67]  The previous transit was in 1874.

clustered around the transit observations; but here in 1865 was an argument that should have encouraged British participation in the International Polar Year of 1882–3.[68]

The argument based on the transit of Venus was the strongest one scientifically, but Murchison's presumed failure to explain it must have weakened his case with a First Lord who was far from being an astronomer or navigator. More powerful was the memory of the Franklin disaster, and full awareness of the cost of the Franklin searches. The Admiralty, thus bolstered, turned itself for the time being into a branch of the Circumlocution Office, devoted to the art of how *not* to do it.

Within the Admiralty, response was entirely critical of the Royal Geographical Society's proposal. Weaknesses were immediately identified by the Second Sea Lord, Admiral Sir Frederick Grey: "to attain the objects aimed at by the various scientific bodies a mere rapid voyage into the supposed polar basin would not suffice & . . . one if not two winters would be necessary nor could one expedition accomplish the whole of the objects sought." Here was an argument for doing much, cooperatively, as was the case in the International Polar Year, or for doing nothing, which was the Admiralty's current choice. The lack of unanimity about the best route, via Spitsbergen or Smith Sound, was also noted. In any case, whether or not the scientific inducements were sufficient for government to support the proposal, "there is no advantage whatever to the Naval Service from these expeditions." The First Lord observed that he had seen a deputation on this subject a year ago, and had given them no encouragement. There would be no support from within the Admiralty.[69]

## MARKING TIME

The Admiralty's refusal was for the time being definitive. Arctic expeditions would be left to other nations in Europe, whose preferred route was via Spitsbergen or east Greenland. The expeditions would also be left to those whose interest was more commercial than scientific, including whalers and fur traders,[70] to private initiatives, such as the mountaineering explorer Edward Whymper's in Greenland, or to an extension of the travels of artists and photographers.

[68] See chap. 8.
[69] Admiralty Memorandum, "North Polar Region: Proposed Naval Expedition," 1 June 1865, PRO MS Adm. 1 5934. Draft of letter, Somerset to Murchison, 29 Nov. 1865, PRO MS Adm. 1 5934.
[70] For the HBC's support of scientific work, see chap. 9.

In matters scientific, German arctic expeditions were by now the best supported.[71] They explored Greenland, and they attempted to reach the pole by way of Spitsbergen, now called Svalbard. The second German North Polar Expedition, led by Carl Koldewey, had thirty-one members, including six scientists equipped with the best modern instruments, which they used "in every spare moment."[72] The Austro-Hungarian Empire made its presence felt in exploration of the Eurasian Arctic.[73] Sweden was also active, notably in the expeditions of Baron A. E. Nordenskiöld.[74] But none of these scientific expeditions in the late 1860s and early 1870s was concerned with the archipelago to the west of Greenland. Instead, Greenland itself became the focus of exploration and study, for example in the geological, oceanographic, and botanical work by the arctic geographer Robert Brown of Campster, who accompanied Whymper to Greenland.[75] The Danish presence in Greenland, and especially the work of Hinrich Johannes Rink, accelerated knowledge of the island.[76] But scientific knowledge of the Canadian Arctic, except through American expeditions, was put on hold. Especially galling to British advocates of scientific exploration must have been the expeditions of William Bradford, the American painter, who in 1864 had organized a voyage to the Labrador coast, to paint icebergs and scenery. He returned in 1869 to Greenland and Davis Strait, with camera, sketchbook, and palette, in an expedition inspired by Kane's writings and made solely for the purpose of art.[77] Isaac Israel Hayes was among his com-

71    A. Petermann, "Die Deutsche Nordpol-Expedition 1868," *Mitt. aus Justus Perthes' Geogr. Anstalt 14* (1868) 207–28; C. Koldewey et al., " Die zweite Deutsche Nordpolarexpedition," *Zeitschrift für Geschichte der Erdkunde 6* (1871) 1–45. G. Stäblein, "Traditionen und aktuelle Aufgaben der Polarforschung," *Die Erde 109* (1978) 229–67; "Historische Aspekte der deutschen geowissenschaftlichen Polarforschung," *Polarforschung 51* (1981) 219–25.

72    F. L. McClintock, "Resumé of the Recent German Expedition, from the Reports of Captain Koldewey and Dr. Laube," *Proceedings of the Royal Geographical Society 15* (1871) 102–11.

73    See, e. g., K. Weyprecht, "Scientific Work of the Second Austro-Hungarian Polar Expedition, 1872–74" (translated from the *Geographische Mittheilungen 21* p. 65), *Journal of the Royal Geographical Society 45* pp. 19–33.

74    Kirwan, *The White Road* (London, 1959) pp. 190 et seq.

75    R. Brown, Journal, Greenland Expeditions 1867, SPRI MS 441/2/1; Brown, "Notes on the Fauna & Flora of Greenland 1867," SPRI MS 441/4. This Robert Brown was the son of Thomas Brown; the names are common in Scotland, and I have not found any connection between the Browns of Campster and the great botanist Robert Brown, the subject of Mabberley, *Jupiter Botanicus.*

76    Henry Rink, *Danish Greenland: Its People and Its Products,* Robert Brown, ed. (London, 1871).

77    William Bradford, *The Arctic Regions Illustrated with Photographs Taken on an Art Expedition to Greenland by William Bradford: With Descriptive Narrative by the Artist* (London, 1873); Alpheus Spring Packard, *The Labrador Coast: A Journal of Two Summer*

pany. The photographs taken as they steamed north along the Greenland coast were crisp, of strong contrast, but lacked a sense of space or atmosphere. Nevertheless, Bradford took precise notes about light, colors, mutability, and refractions. The combination of photographs, notes, and sketchbooks was to serve him well: "Why, my photographs have saved me eight or ten voyages to the Arctic regions, and now I gather my inspirations from photographic subjects."[78] The arctic sublime[79] was becoming too readily available, and Britain was losing primacy in exploring the Arctic.

It was not, as we have already noted, that Britain had opted out of polar science altogether. H.M.S. *Challenger*, commanded by George Strong Nares, and with Wyville Thomson, a leading zoologist, spent four years from 1872 carrying out an ambitious program covering most of the sciences and every ocean except the Arctic, with special importance attached to the physical geography and biology of the Antarctic seas. The latest improvements in apparatus were adopted, so that the sounding and dredging gear, water bottles, thermometers, and the rest were of the best and latest design.[80] This was very well, indeed magnificent, but it was no salve to the frustration of arctic officers and scientists, who were anxious to reassert Britain's lapsed northern glory.

The Royal Geographical Society remained the center of the arctic crusade. They had long had a standing arctic committee, which constantly sought to engage the support of other scientific bodies. In December 1872, the Chancellor of the Exchequer and the First Lord of the Admiralty received a delegation arguing yet again for an expedition in the direction of Smith Sound: members of the delegation were the distinguished assyriologist Sir Henry Rawlinson,[81] President of the Royal Geographical Society; the fashionable physician Sir Henry Holland, President of the Royal

*Cruises to that Region: With Notes on its Early Discovery, on the Eskimo, on its Physical Geography, Geology, and Natural History* (New York, 1891).

[78] *Philadelphia Photographer*, 21 Jan. 1884 p. 8, quoted in N. Spassky, *American Paintings in the Metropolitan Museum of Art* vol. II (1985) p. 165. See also the paintings of Frederic Edwin Church who, also inspired by Kane and Hayes, painted icebergs with a style independent but reminiscent of Turner's.

[79] C. C. Loomis, "The Arctic Sublime," in U. C. Knoepflmacher and G. B. Tennyson, eds., *Nature and the Victorian Imagination* (Berkeley and Los Angeles, 1977) pp. 95–112.

[80] McConnell, *No Sea Too Deep*, pp. 106–16. There is a vast literature from and about the *Challenger* expedition. Starting points include T. H. Tizard, *Narrative of the Cruise of H.M.S. Challenger: With a General Account of the Scientific Results of the Expedition* (Edinburgh, 1885); John Murray, *A Summary of the Scientific Results Obtained at the Sounding, Dredging, and Trawling Stations of H.M.S. Challenger* (Edinburgh, 1895); Eric Linklater, *The Voyage of the Challenger* (London, 1972).

[81] Rawlinson deciphered the Persian cuneiform vowel system; he also had a distinguished military career in Persia and Afghanistan.

Institution; Professor William Carpenter, physiologist, oceanographer, complete naturalist, and President of the British Association for the Advancement of Science; Joseph Hooker, President-elect of the Royal Society of London; and a group of prominent arctic officers.

The arguments were more formally proposed than before, while echoing or building upon those advanced by Murchison. The failure of German and Swedish expeditions to penetrate far into the ice between Greenland, Spitsbergen, and Novaya Zemlya supported the British (and American) preference for Smith Sound. Two ships would be needed, one to forge as far north as possible, the other to be held in support near the entrance to Smith Sound. Exploration would aim at Greenland and at the icy seas believed to be to the north of it. The scientific program was more detailed than anything the RGS had formulated before. Geographically, the polar ocean and the north Greenland shore offered exciting prospects. In hydrography, Carpenter and Hooker argued that exploration in northern waters, preferably in the "northern Greenland seas," was a necessary complement to the *Challenger*'s southern researches. In botany, phytogeography would be "a most important contribution to the history of palaeontology, botany, and terrestrial physics." Oceanography would explore the teeming organisms of the northern seas, in their environmental, ecological, and even geological contexts, relating paleontology to modern studies. The isolation of human communities in the far North, and the discovery of their artifacts well above presently inhabited latitudes, indicated that the expedition would be valuable for ethnology and archeology alike.

Such arguments failed to persuade the First Lord, or the Chancellor, whose reply, written on the last day of the year, insisted that he could not justify a second scientific expedition while paying for the *Challenger*.[82] Nothing happened to change the picture during 1873, although the Royal Society appointed its own arctic committee to confer with the RGS arctic committee, "which," as Clements Robert Markham noted with grudging satisfaction, "is a step in the right direction."[83] Markham, prolific secretary and then president of the Hakluyt Society and a major force in the Royal Geographical Society, was to be a back-room and a front-bench spokesman for polar exploration, arguing persuasively, blustering, and ma-

---

[82]   H. C. Rawlinson, "Address," *Proceedings of the Royal Geographical Society* 17 (1873) 266; *The New Arctic Expedition: Correspondence between the Royal Geographical Society and the Government* (London, 1873).

[83]   C. R. Markham to [? Robert Brown of Campster], [1873], SPRI MS 441/9/33; Royal Geographical Society Committee Minute Book Sept. 1872 – Oct. 1877, entry 3 July 1873 p. 57; Royal Society of London MS MM 15.45.

neuvering to achieve his ends. It was he who was to be largely responsible for the creation of the British National Antarctic Expedition of 1901–4, and for the selection of Captain Scott as its leader.[84] Markham was also to succeed in getting his British Arctic Expedition sent north via the Smith Sound route, but not quite yet.

## 1874: THE TIME HAS COME

Foreign competition was undoubtedly a factor in piquing British pride, and the steady pressure from geographers, scientists, and arctic officers may have had some effect. But I believe that two events in 1874 were decisive. One was a formal request from American interests to be granted mineral rights on Baffin Island. The immediate although initially invisible consequence was the beginning of negotiations to transfer the arctic archipelago to Canada. A Colonial Office official remarked that the main reason for the transfer of the islands was "to prevent the United States from claiming them, not from the likelihood of their being any value to Canada."[85] The transfer was not effected until 1880. The other development in 1874 was a change of government in Britain. Gladstone, whose government had sent the *Challenger* expedition south, was defeated over the Irish University Education Bill, and resigned; Queen Victoria sent for Gladstone's archrival, Benjamin Disraeli, inviting him to form a government. The opportunity to show the flag in the archipelago, thus helping to frustrate the Americans, while at the same time trumping Gladstone's southern scientific expedition with a new northern one, would prove too much to resist.

[84] C. R. Markham, *Antarctic Obsession: A Personal Narrative of the Origins of the British National Antarctic Expedition 1901–1904*, ed. and intro. Clive Holland (Alburgh, Norfolk, 1986).

[85] Quoted in Grant, *Sovereignty or Security?*, p. 5, quoting in turn from Gordon W. Smith, *Territorial Sovereignty in the Canadian North: A Historical Outline of the Problem* (Ottawa, 1963) p. 5. See also Zaslow, ed., *A Century of Canada's Arctic Islands 1880–1980* pp. xiii–xvii.

# 7

# Science North: The British Arctic Expedition 1875–1876

There were in 1874 no significant scientific reasons for renewed arctic exploration via Smith Sound, over and above those already advanced in the previous decade.[1] Lobbying by the Royal Geographical Society had continued. On 6 December 1873, Sir Henry Bartle Edward Frere,[2] then President of the RGS, had sought an interview with Gladstone. The interview never took place: that very month, Gladstone's government was defeated, and in January 1874 Disraeli became prime minister. The new political context changed everything.

In October 1874, Rawlinson, once again President of the RGS, wrote to Disraeli, drawing attention to the success of Weyprecht's Austro-Hungarian expedition. Mere days later, the new hydrographer of the Admiralty, Captain Sir Frederick Evans, prepared a confidential memorandum about a North Polar expedition.

Now Evans, who had previously, in common with the Lords of the Admiralty, resisted proposals for further arctic exploration, suddenly advanced arguments identical with those emanating from the RGS.

The retrospect of the extended exertions of Great Britain in the field of Arctic research, the important results that have been obtained therefrom, their cessation in past years, and in this interval the persevering efforts of other nations to supple-

---

[1] There were, however, more articulate and theoretically focused statements of the problems; see, e. g., Georg von Neumayer, "Die geographischen Probleme innerhalb der Polarzonen in ihrem inneren Zusammenhang beleuchtet," *Hydrographische Mittheilungen: Herausgegeben von dem Hydrographischen Bureau der Kaiserlichen Admiralität 2* (1874) 52–4, 63–8, 75–82. For background, see W. Kerz, "Georg von Neumayer und die Polarforschung," *Polarforschung 53* (1983) 91–8.

[2] Diplomat and statesman, president of the RGS 1873–4. See F. V. Emery, "Geography and Imperialism: The Role of Sir Bartle Frere (1815–84)," *The Geographical Journal 150* (1984) 342–50.

ment, and if possible, eclipse the century of exertions made by this Country, form, it must be allowed, reasonable grounds, so far as a worthy emulation is concerned, for all interested in Geographical Science, and especially for Arctic travellers, to urge another trial.[3]

The Admiralty accepted the arguments. Disraeli was keen on the idea, and wrote in November that "Her Majesty's Government have determined to lose no time in organizing a suitable expedition."[4]

Selecting the leader of the expedition was the Admiralty's responsibility, but their decision must have had a pleasurably ironic relish for Disraeli. Who could be better, in every sense, than the Aberdonian commander of H.M.S. *Challenger,* George Strong Nares, a man who was even now leading that great scientific expedition? He was at fifty-three the youngest surviving officer with sledging and sailing experience from the Franklin searches.[5] He was met at Hong Kong by a telegraphic message calling him off the ship, and instructing him to return to England, there to make ready an expedition toward the North Pole.

The Admiralty informed the RGS and the Royal Society,[6] whose arctic committees were galvanized. The RGS undertook to prepare a selection of papers on arctic geography and ethnology for the use of the expedition.[7] Clements Markham explained to Joseph Hooker, now President of the Royal Society, that his aim was for the two societies to furnish the expedition with "all that is known in the various branches."[8] The Royal Society undertook the preparation of a "Greenland Manual" covering natural

---

3   PRO MS Adm. 1 6313.
4   Disraeli to Rawlinson, 17 Nov. 1874, reprinted in *The Navy,* 22 May 1875 p. 482; G. Hattersley-Smith, "The British Arctic Expedition, 1875–76," *Polar Record* 18 (1976) 17–26.
5   He had been on H.M.S. *Resolute* under Henry Kellett on Kellett and McClintock's Franklin search expedition 1852–4. Margaret Deacon and Ann Savours, "Sir George Strong Nares (1831–1915)," *Polar Record* 18 (1976) 1–15.
6   RSL Council Minutes 17 Dec. 1874; RGS Committee Minute Book Sept. 1872–Oct. 1877, 18 Jan. 1875.
7   *Arctic Geography and Ethnology: A Selection of Papers on Arctic Geography and Ethnology, Reprinted, and Presented to the Arctic Expedition of 1875, by the President, Council, and Fellows of the Royal Geographical Society* (London, 1875). The RGS was willing, through its most vocal advocate Clements Markham, to take full credit for persuading other scientific bodies and the government to mount this new expedition: see C. R. Markham, "The Arctic Expedition of 1875–76," *Proc. RGS* 21 (1877) 536–55; *The Threshold of the Unknown Region* (London, 1876); and *The Fifty Years Work of the Royal Geographical Society* (London, 1881). See also J. E. Caswell, "The RGS and the British Arctic Expedition, 1875–76," *Geographical Journal* 143 (1977) 200–10.
8   C. R. Markham to J. D. Hooker, 11 Jan. 1875, MS Royal Botanic Gardens, Kew [hereafter RBG] J. D. Hooker papers, Voyage of H.M.S. Alert and Discovery, Letters &c. [hereafter Hooker papers, Voyage] f. 174.

history, geology, and physics, under the superintendency of Thomas Rupert Jones, Professor of Geology at the Royal Military College. The expedition was to leave at the end of May 1875, just four months after Hooker wrote to Jones.[9] Habits of organization at the Royal Military College may have helped; the *Manual* was ready in time.[10]

The instructions covered sciences from the old *Admiralty Manual*, with additions for new branches of observation made possible by new developments in instruments, notably in spectroscopy, and for new categories of information required by theoretical advances. There were orders concerning eclipses of the sun and occultations, tidal observations, pendulum observations, and the detection of meteoric or cosmic dust in arctic snow, as reported in Greenland by Professor A. E. Nordenskiöld, who was then Sweden's greatest scientific explorer of the Arctic.[11] Of tidal observations, Samuel Haughton wrote that a lot of time had been wasted by careful but wrongly made observations, whereas a single month's observations, properly made, could be most valuable. The most useful results could be obtained at "the times of solstice and equinox," and times of observation should be according to "*mean solar time*, not . . . apparent solar time,"[12] the former being time by the clock or chronometer, corresponding to the fiction of days of equal duration through the year, the latter being time by the sun, as it appears on a sundial.

There were no novelties in the magnetic instructions, with the only advances since Franklin's day being modest ones in instrumentation. There

---

[9]   J. D. Hooker to T. R. Jones, 22 [Jan.] 1785, SPRI MS 336.
[10]   T. Rupert Jones, ed., *Manual of the Natural History, Geology, and Physics of Greenland and the Neighbouring Regions: Prepared for the Use of the Arctic Expedition of 1875, under the Direction of the Arctic Committee of the Royal Society . . . together with Instructions Suggested by the Arctic Committee of the Royal Society for the Use of the Expedition: Published by Authority of the Lords Commissioners of the Admiralty* (London, 1875). Royal Society of London CMB 2 p. 89.
[11]   For Nordenskiöld's detection in Arctic snow of particles consisting mainly of iron but containing cobalt, which proves their nonterrestrial character, see Poggendorf's *Annalen 151* p. 154. Nordenskiöld had been to Spitsbergen in 1858, led an expedition there in 1864, organized the Swedish North Polar expeditions of 1868 and 1872, and led an expedition to Greenland in 1870. This last expedition was a source of scientific data important for the British Arctic Expedition of 1875–6 (A. E. Nordenskiöld, "Account of an Expedition to Greenland in the Year 1870," Greenland *Manual*, Jones, ed., pp. 389–447, reprinted from the *Geological Magazine 9* [1872] 289, 355, 409, 449, 516). Nordenskiöld is best known for his triumphal navigation of the Northeast Passage (Nordenskiöld, *Vegas färd kring Asien och Europa*, 2 vols. (Stockholm, 1880–1); George Kish, *North-East Passage: Adolf Erik Nordenskiöld, His Life and Times* (Amsterdam, 1973). For a survey, see Tore Frängsmyr, "Swedish Polar Exploration," in T. Frängsmyr, ed., *Science in Sweden: The Royal Swedish Academy of Sciences 1739–1989* (Canton, Mass., 1989) pp. 177–98.
[12]   Royal Society's *Instructions*, p. 8.

was, for each of Nares's two screw steam ships, *Alert* and *Discovery*, a set of instruments containing a portable unifilar magnetometer, for determining the absolute horizontal force at a fixed station; a Barrow's or Kew pattern dip circle, for determining inclination, and fitted with special extra magnetic needles for determining the total force; a portable declination magnetometer, for differential observations at a fixed station; azimuth and sledge compasses for traveling; and a Fox dip circle, for observing the inclination and intensity in sledging or traveling parties.[13] Sabine's article in the *Admiralty Manual* remained authoritative for naval expeditions.

The meteorological instructions were also standard, except for a request by the physicist George Gabriel Stokes that observers should carefully note any correlation between magnetic vectors and the alignment of auroral streamers and arches.[14] The usual investigations into atmospheric electricity were indicated, especially in connection with aurora; but a new degree of accuracy was introduced by William Thomson's electrometer, which was insulated and carefully screened from external interference.[15]

Spectra had for some time been used as a tool for chemical analysis. Work begun in the 1850s and significantly developed in the 1860s had by the 1870s established spectrum analysis as valuable to chemists, as well as challenging to physicists.[16] Now Stokes drew the expedition's attention to the need for spectroscopic studies of aurora, and of solar radiation "with a view to terrestrial absorption." It was known that when the sun was near the horizon, additional lines and bands appeared in its spectrum, and that most of these lines, at any rate, were due to absorption by water vapor in the atmosphere. The arctic atmosphere was not only unusually cold, but also unusually dry, so that the lines and bands due to water vapor would be much reduced. If there were other chemical species not hitherto detected in the atmosphere, their absorption spectra might then be detectable. Stokes gave formal instructions that were supplemented by Norman Lockyer, a major contributor to the field.[17] Lockyer had a meeting with Lieutenant Alfred A. Chase Parr, who would be responsible for the spectroscopic work;

[13]  For details, see Levere, "Magnetic Instruments in the Canadian Arctic Expeditions of Franklin, Lefroy, and Nares," *Annals of Science 43* (1986) 57–76 at 68–9.
[14]  A. J. Meadows and J. E. Kennedy, "The Origin of Solar-Terrestrial Studies," *Vistas in Astronomy 25* (1982) 419–26 looks at auroral variations and their correlation with other phenomena.
[15]  G. L'E. Turner, *Nineteenth-Century Scientific Instruments* (1983) pp. 199–200.
[16]  See, e. g., J. Norman Lockyer, *Studies in Spectrum Analysis* (New York, 1878).
[17]  Frederick Evans, Hydrographer, to Nares, 27 May 1875, SPRI MSS 665/2/1–2. Lockyer's official instructions in the use of spectroscopes were published in the Royal Society's *Instructions*, pp. 28–32.

he wrote a letter discussing the scales and spectrometers, and then raised another question related to the atmosphere's absorption of solar radiation, while omitting any consideration of refraction:

> You get the whole spectrum from the sun at mid-day & [sunlight] is white. You get the blue end absorbed at sunrise & sunset & the sunlight is red. Now it may be that in consequence of the extreme cold in the North the number of particles in the atmosphere which so absorb the blue light may be reduced, if so the intensity of the red may be reduced & the tint of red may not be the [result].
>
> ... a green sky comes from the absorption of both ends of the spectrum. Are there green skies? You will be doing great good to science & a personal kindness to myself if you will note all curious atmospheric colours.[18]

There were also instructions for observing the polarization of light, including that from solar halos, and a miscellany of other problems was indicated by the physicist John Tyndall, who had recently been described as one of the pioneers of the kingdom of science.[19] Tyndall wanted to know about the height of icebergs, the rate of advance of glaciers, the presence or absence of germs, the range of a dog whistle, an organ pipe, and a pistol in arctic air, the formation of snow crystals, and a dozen more items, all tied to theoretical debates in science, but sounding as if they had come from a catalogue of questions by a seventeenth-century virtuoso.[20]

The physical sciences were to be handled in a manner that was by now traditional in the navy, with naval officers receiving instructions, instruments, and where needed minimal instruction from specialists. The officers would then carry out the program of observations laid down for them, and the results, where reliable, and where money allowed, would be reduced following the expedition's return. There would be no dedicated astronomer or magneticist on this expedition.

The Admiralty's arctic committee had originally taken the line that natural history could also be handled by naval personnel. This seemed not unreasonable, since many British gentlemen, naval and otherwise, were amateur naturalists, sometimes of no mean order. In spite of the barrier of Latin names, it generally seemed to require less training to be, say, an amateur botanist or ornithologist than to be an amateur magneticist or astronomer; experience in the field was more important for natural history. The Admiralty's arctic committee recommended that the expedition should employ only naval officers, and no civilian scientists[21] – a remarkable contrast

---

[18]    Lockyer to A. A. C. Parr, 28 May 1875, SPRI MS 666/3. Parr failed to observe green skies.
[19]    *Vanity Fair*, April 1872.        [20]   Tyndall in the Royal Society's *Instructions*, pp. 34–5.
[21]    PRO Adm. 1 6328, 4 Dec. 1874; Royal Society of London CMB 2 pp. 55–6 re 4 Dec. 1874.

with the arrangements for the *Challenger* expedition, and one bringing into question the navy's understanding of the needs of science. The general view of the scientific societies was that there should at least be full-time naturalists on this expedition; the arctic committee was persuaded by this view, and its members reversed their earlier recommendation.[22] The Royal Society was so informed, and was asked to suggest appropriate candidates. Their first recommendation was Chichester Hart of Trinity College, Dublin, a young man whose principal distinctions were that he had "worked up the flora of parts of the west of Ireland in the field," and that he had been awarded a "pedestrian prize" in athletic sports at Dublin. The other naturalist, chosen from the two who had applied, was Henry Wemyss Feilden, a distant connection of the magistrate and novelist Henry Fielding. He was an army officer,[23] born in barracks in Ireland and subsequently to be engaged in most of the major wars in his lifetime, in the United States, China, India, and South Africa.[24] Feilden had spent a brief spell in the Faeroe Islands studying ornithology, thereby gaining some small reputation as a naturalist. He was scarcely over-qualified, or even an obvious candidate. His letter of application was suitably modest:

Dear Sir, By the advice of Prof. Newton, I write you a line to offer my services. . . . I am afraid my qualifications may be somewhat limited for such an undertaking. . . . I believe, however, that I am well acquainted with the birds that an Arctic expedition, even reaching the Pole is likely to meet with, I am fairly acquainted with the Mediterranean & Northern forms of mollusca and can make observations on the Geology of the countries I pass through.[25]

---

[22]  Memorandum 6 January 1875, PRO Adm. 1 6367.

[23]  The Parliamentary Papers relating to the expedition refer to him as Captain Feilden, R.A. [i. e., Royal Artillery]. The army list of 1875, however, shows that he was a member of the 4th (the King's Own Royal) Regiment. I am grateful to Mr. C. L. Gardner for pointing out this discrepancy. C. R. Markham, *The Threshold of the Unknown Region*, 4th ed. (London, 1876) p. 401 summarizes Feilden's military career, and provides an explanation: ensign in the 42nd Highlanders 1856, Indian Mutinies at Lucknow (medal and clasp); staff officer, 1st Gwalior Infantry 1858, served against rebels in Bundelcund 1859; transferred to 8th Punjab Infantry, served with them at the Taku forts (medal and clasp); lieutenant in the 44th and returned to England 1861; American Civil War, captain and assistant adjutant-general on General Beauregard's staff, then senior officer on the staff at the siege of Charleston; on General Hardee's staff when opposing Sherman's march; returned to England 1866, adjutant to the Lancashire Rifle-Volunteers; paymaster of the 18th Hussars 1868, served with them in India; 1868–73, paymaster of the 4th, and in 1873 of the brigade of Royal Artillery at Malta – hence, presumably, his characterization in 1875 as captain, R.A.

[24]  Royal Society of London MSS CMB 2 pp. 55–6, 61–4. C. R. Markham, *The Threshold of the Unknown Region*, pp. 401–2. Levere, "Henry Wemyss Feilden, Naturalist on HMS *Alert* 1875–1876," *Polar Record* 24 (1988) 307–12.

[25]  Feilden to Sclater, 25 Nov. 1874, RSL MS M C.10.172.

There had been talk about taking a Scot, the geologist and naturalist Robert Brown of Campster, who argued publicly that one of the two civilian naturalists should be a geologist: "for . . . the geological questions to be solved are not the least important of all those which await the labours of these gentlemen."[26] Brown himself was not selected, because of his "great inaccuracy." But there was considerable pressure to appoint a geologist. Clements Markham, who was no scientist but certainly not one to remain silent through ignorance, rated the work of a geologist as

the most important of all. Medical and other officers can carefully collect all plants and animals, and note localities and habits; but collecting is the very smallest part of geology; and it requires a thoroughly skilled man, well acquainted with all that is known, to bring back the information. Of course this branch of the work has the most interest for geographers, and I venture to think that a good geologist should certainly go.[27]

Markham combined the perspective of a geographer with an acceptance of Darwin's warnings in the *Admiralty Manual* about the need for study in the library and experience in the field. Nares was asked whether he could take a geologist, and replied that the ships were so small that each would have room for only one scientific officer, once the officers, crew, and stores were accounted for.[28] To take an additional scientist meant leaving a sailor behind, a poor idea since the crews were already minimal. "I would point out," he continued,

that two of the Medical Officers have been appointed to the ships in consequence of their scientific accomplishments and knowledge of Natural History [in the old tradition of surgeon-naturalists], also that a gentleman specifically appointed for geological observations would have a very limited field for his Examination, as our number will render it impossible for him to travel any distance away from the ship by sledges.

. . . Regarding the subject of geological observations, great advantages would be gained if the Naval Officers who will be employed on sledging Expeditions, which if the [land] is [continued], must necessarily take them along two hundred, or three hundred miles of previously unexamined coast line, were to obtain such geological instruction as to make them at all events good collectors.[29]

26   R. Brown, "The Arctic Expedition: Its Scientific Aims," *Popular Science Review* (1875) 154–5.
27   C. R. Markham to J. D. Hooker, 11 Jan. 1875, RBG Hooker papers, Voyage . . . , f. 174.
28   For a list of Nares's officers, see C. R. Markham, *The Arctic Navy List: Or, a Century of Arctic and Antarctic Officers, 1773–1873: Together with a List of Officers of the 1875 Expedition, and Their Services* (London and Portsmouth, 1875).
29   PAC MG 29 B12 vol. 3, Nares *Letterbook*, Nares to c.-i.-c. 24 Feb. 1875.

In spite of his experience as commander of the *Challenger* expedition, Nares underestimated the preparation necessary for effective collecting. He was unable to put scientific priorities to the fore when they conflicted with what he still saw as his primary mission: geographical exploration toward the pole, and the maintenance of naval discipline and efficiency. Nor could he see the cogency of an argument for experienced rather than minimally trained observers. In the event, he would be proved both right and wrong; naval officers contributed little to geology, but the geological work of Feilden, an amateur but official naturalist, was to stand unsurpassed until the middle of this century.

The Admiralty accepted Nares's advice, and no geologist was appointed. *The Scotsman*, perhaps shielding the rejected Brown, expressed indignation at this exclusion of a geologist. What the Admiralty should do was refuse the nomination of the botanist, put a geologist in his place, and turn over "what little botanical work is required" to Edward Lawton Moss,[30] one of the surgeons, who was quite up to it.[31] Here was the beginning of friction between surgeon and naturalist that would spread beyond the expedition.

The naturalists, like the officers responsible for physical observations, received instructions through the Royal Society. One of their main tasks was "to ascertain all facts bearing upon the distribution or possibly gradual disappearance of Mammalian life in the direction towards the Pole."[32] Geographical distribution and its change through time were important in zoology just as they were in botany, because of the dominance of Darwinian theories of evolution. For the same reason, variations between southern and northern forms of the same species were to be carefully noted. There were a number of specific questions about arctic species, including polar bear, musk-ox, and arctic fox; also of interest, should it occur, was the sickness among "Eskimo dogs," which had so depleted Hayes's teams. Standard instructions were also given for observing cetaceans, and for collecting and studying birds and fishes, with due attention to distribution.

The instructions for observing and collecting molluscs made clear the importance of dredging, for paleontology as well as zoology:

---

[30] Moss was born in Ireland in 1843, educated first at Dublin, and graduated in medicine from St. Andrews in 1862. He entered the navy in 1864. By 1875 he had contributed several papers to the Linnean Society, the Zoological Society of London, etc. This information is from the biological sketch in C. R. Markham, *The Threshold of the Unknown Region*, pp. 407–8.

[31] *The Scotsman* [Feb. 1875], cutting in RBG, Hooker papers, Voyage . . . , p. 38.

[32] Albert Günther, "Instructions for Making Observations on, and Collecting Specimens of, the Mammalia [not including *Cetacea*] of Greenland," Royal Society *Instructions*, p. 36.

The palaeontological basis of the glacial epoch consists mainly in the identification of certain species of Mollusca, which inhabit the Polar seas and are fossil in Great Britain and even as far south as Sicily. But such species may owe their present habitat and position to other than climatal causes, viz., to the action of marine currents. Certain small Spitzbergen species . . . have lately been found everywhere in the depths of the North Atlantic as well as in the Mediterranean; and the question naturally arises what is the home of these species . . . ? That question cannot be answered for want of sufficient information. . . .

It is hoped that each of the vessels to be fitted out for the Polar expedition will have a donkey engine, by which the dredges can be lifted; and that a sufficient supply of necessary apparatus will be provided, regard being of course had to the limited space allowed for such a secondary object. The great experience of Capt. Nares renders any suggestions as to dredging quite superfluous.[33]

It is striking that even a scientific advisor to a scientific expedition could describe work in marine biology as secondary, presumably to geographical work.

Fossil shells were to be collected, and their position and height above sea level noted. "The former conditions and climate of the Polar region may be thus ascertained, and a new chapter opened in the history of our globe."[34]

Then came instructions for collecting and preserving hydroids and polyzoa, on the construction and use of the towing net, and notes on the animals that might thereby be obtained. These notes manifested the considerable advance during the preceding decades in an understanding of the life cycles of some very curious creatures.[35]

Joseph Hooker gave instructions for botany. He still thought it probable that there were no new species to be discovered in the Arctic, at least in North America. But there was still much to learn about their distribution, the conditions under which they grew, and their life cycles. There were questions of hybridization (how else to account for the number of intermediate forms of willow and saxifrage?); a need for the careful collection, hitherto lacking, of arctic lichens and mosses; a study of the transportation of seeds by icebergs or other ice formations, and the resistance of those seeds to cold and to saline immersion, all central to any historical account

---

[33]  J. Gwyn Jeffreys, "Instructions for Making Observations on, and Collecting the Mollusca of, the Arctic Regions," Royal Society *Instructions*, p. 49. Jeffreys was a lawyer turned conchologist, active in several scientific organizations, including the Royal Society (and the Royal Society club), the BAAS, and the Linnean Society.

[34]  Ibid., p. 50.

[35]  Ibid., G. J. Allman, pp. 52–60. M. P. Winsor, *Starfish, Jellyfish, and the Order of Life* describes these advances.

of distribution, and therefore crucial for an evolutionary interpretation; and a study of marine algae and diatoms, including the depths at which they were found.[36]

Then there were instructions for geological work, given by Professor A. C. Ramsay, Director of the Geological Survey, and John Evans, President of the Geological Society. The naturalists should tackle the identification of sedimentary rocks, recording their sequence, alignment, fossil content, and lithological character; the identification of igneous rocks, formed by upwelling through the earth's crust, and their sequence, cleavage, and penetration by dykes; and they should make a similar study of metamorphic rocks, such as slate formed from clay. They should seek to identify major formations that could be compared lithologically or paleontologically with such known American or European formations as the Silurian, Devonian, or Carboniferous. They should look for and record mineral lodes in fractures and faults. They should carefully record the fossil floras in north Greenland and Ellesmere Island. The Miocene flora was particularly rich and attractive, because at that time, as Scandinavian expeditions had shown, Greenland and other circumpolar lands had "a luxurious flora of ever-green trees and shrubs, oaks, magnolias, chestnuts, cypresses, redwoods (*Sequoia*), ebony, &c."[37] In geology even more than in zoology, information brought back by numerous British expeditions was being integrated into a circumpolar picture. That picture was emerging from studies by several nations, notably in recent Swedish, Austrian, and, to a lesser extent, American polar work. The Austrians and the Swedes had worked largely to the east of Greenland, although Nordenskiöld had led an expedition to the west; the Americans had concentrated their energies along the Smith Sound route. Here were the beginnings of a true international enterprise. Clements Markham and the Royal Geographical Society might echo John Barrow in stressing the national glory accruing from polar exploration, but science, excluding geographical discovery, was increasingly dependent on international exchange and cooperation.

Distinct from the observations of rocks and fossils, but contributing to an understanding of erosion and other aspects of geological change, was the study of glaciers and icebergs: Ramsay and Evans proposed a more comprehensive set of questions and observations than had been given to previous British expeditions.

---

[36] Hooker, in Royal Society *Instructions*, pp. 62–7.
[37] R. Brown, "The Arctic Expedition: Its Scientific Aims," *Popular Science Review* (1875) p. 159.

Mineralogy, once independent and then integrated with Wernerian geology, now enjoyed a semiautonomous existence, using chemical tools, such as the blowpipe, to identify specimens.[38] Much the most exciting prospect in mineralogy was the extension of Nordenskiöld's reported discovery of large masses of meteoric iron in Miocene rocks on the west coast of Greenland at Ovifak. Was the iron meteoric or not?[39] It would be good if the naturalists could visit the site again and explore the point. Feilden, scrutinizing Nordenskiöld's account, clearly had some reservations. At Ovifak, the cliffs of basalt and red wacke (a sandstonelike rock derived from the breakdown of basalt) rise 2,000 feet above sea level. The ironstones lay together at the foot of the cliffs, so if they were meteoric they must have fallen before the cliffs were formed, and then been released by the action of the sea from the overlying cliffs. Perhaps Nordenskiöld's alternative explanation, of a terrestrial origin of the iron, was more probable.[40] When Feilden got to Ovifak, a hand lens showed him that pieces of basalt were embedded in the iron, which argued against the latter's meteoric origin.[41]

Clearly, with responsibility for geology, mineralogy, and the full range of the life sciences, the naturalists would have their work cut out, because there would be only one of them on each ship. Equally clearly, neither of the naturalists selected possessed the kinds of qualifications, based on formal study and experience, to be able to achieve all that was required of them. Robert Brown had stated reasonably enough that the naturalists should be qualified, or else "for the credit of English science they had better be left at home."[42] They were neither qualified nor left at home; but they were overloaded with work. The contrast with the scientific staffing of the *Challenger* expedition is striking, as is the lack of volunteers from the

---

[38]   For the changing interrelations between mineralogy and the newer science of geology, see Rachel Laudan, *From Mineralogy to Geology: The Foundations of a Science, 1650–1830* (Chicago, 1987); W. R. Albury and D. R. Oldroyd, "From Renaissance Mineral Studies to Historical Geology, in the Light of Michel Foucault's *The Order of Things*," *BJHS* 10 (1977) 187–215; Roy Porter, *The Making of Geology: Earth Science in Britain 1660–1815* (Cambridge, 1977) pp. 175–6.

[39]   For Nordenskiöld's account of his expedition to Greenland, including the discovery of the problematic iron, see the report reprinted in the Greenland *Manual* pp. 389–447. For a more detailed account of the iron, see Walter Flight, "On Meteoric Iron[s] Found in Greenland," Greenland *Manual*, pp. 447–67, reprinted from the *Geological Magazine* NS 2 (1875) 115, 152; N. Story Maskelyne, Royal Society *Instructions*, pp. 80–1.

[40]   Feilden, note enclosed in "Journal HMS Alert 1875–76," RGS Library MS 387B.

[41]   Feilden, "Notes from an Arctic Journal," *The Zoologist* 3rd series 2 (1878) 375.

[42]   Brown, *Popular Science Review* (1875) 163.

scientific community for Nares's expedition. The navy's attitude to non-naval personnel speaks volumes for the dominance of old models of control in the Arctic.

The spokesmen of the RGS, it will be recalled, had reserved for themselves the provision of advice about arctic geography and ethnography. Among their offerings was a report from the arctic committee of the Anthropological Institute, a group including Sherard Osborn and Clements Markham, and which, therefore, was vigorously in favor of the expedition. They offered the opinion, on less than conclusive evidence, "that there are or have been inhabitants in the unexplored region to the north of the known parts of Greenland. If this be the case, the study of a people who have lived for generations in a state of complete isolation, would be of the highest scientific interest."[43] They listed general questions, including the diligent collection of skulls, and the collection of as complete a vocabulary as possible. There were inquiries about the "Religion, Mythology, and Sociology of Eskimo Tribes"; questions about the remains of ancient races, and about Eskimo warfare, art and ornamentation; and questions about stature, intelligence, marriage and funeral customs, food, and kayaks and other forms of transportation.

Apart from this plethora of directions, there were instructions from Dr. Armstrong, the Medical Director-General of the navy, for the maintenance of health and the avoidance of scurvy. Armstrong proposed a liberal diet of meat (two pounds daily) with a "proportionate quantity of vegetables and antiscorbutics.... [W]ith a scale of diet smaller than this I consider that debility of a scorbutic character must ensue, and that at an early period, if the men are much exposed to hard work and intense cold." He recommended an alternation of fresh preserved meat with salt meat, and cabbage and preserved potato as the bulk of the vegetable food. Armstrong attached "the greatest possible importance to the daily administration of Lemon juice," to begin the day after fresh vegetables ceased: "but this must be carried out on the most rigid principles ... without one day's interruption." To ensure that the men took their daily ration, it should be administered, mixed with sugar in water, in the presence of the officer of the watch. Away from the ship, when traveling by sledge, the same ration of lemon juice should be administered.[44] Nares

---

43 "Report of the Anthropological Institute" in *Arctic Geography and Ethnology* ... (London, 1875) p. 279.

44 PRO Adm. 1 6361. For details of theories of the cause of scurvy, see K. J. Carpenter, *The History of Scurvy and Vitamin C* (Cambridge, 1986), especially chap. 6, pp. 133 et seq., "Problems in the Arctic and the Ptomaine Theory (1850–1915)."

copied Armstrong's recommendations into his "Remark Book," which accompanied him on the voyage.[45]

The ships were refitted and outfitted, the crews assembled, and late in May the Admiralty sent formal instructions[46] to Nares on board H.M.S. *Alert* at Portsmouth. His ships would be accompanied by the *Valorous* a little beyond Disco, Greenland, for the provision of extra coal and stores, and to bring back Nares's first reports; *Valorous* would also be available thus far for towing, if needed, and would perform dredging, sounding, and other scientific work before returning home.[47] *Alert* and *Discovery* would continue north, picking up interpreters and dogs along the way, making vigorously for winter quarters. *Discovery* was not to winter beyond the 82nd parallel. This would enable her to serve as a back-up ship for *Alert*'s crew, if need be so that the men could escape to the relief ship that would go to Smith Sound in the summer of 1877. *Alert*, no matter how far north she steamed, was not to winter more than 200 miles north of *Discovery*. "[I]t is not desirable, under any circumstances, that a single ship should be left to winter in the Arctic regions." This, with instructions about marker cairns, supply depots, and more, showed a proper post-Franklin concern with safety, allied to geographical ambition.

The expedition was one of "exploration and discovery," and its object was "the advancement of science and natural knowledge," or, as *Punch* put it, men went to the Arctic because of a desire "to look into the works of Creation, to comprehend the economy of our planet, and to grow wiser and better by knowledge!"[48] The most approved instruments had been supplied, several officers had been instructed in their use, and Nares was to give them every fair opportunity of contributing to science. Similar support was to be given to the naturalists.

The departure was, for the public, one long celebration. The mayor of Portsmouth gave a dinner for the officers, and among the crowds examining the ships were the Prince of Wales and the emperor of France. One of the last items to come on board was a gift for *Alert*'s naturalist Feilden, a pair of snowshoes of unfamiliar pattern, and the cause of much hilarity. In the event, they would prove most serviceable, which was no surprise since John Rae had sent them.[49]

[45]	PAC MG 29 B12 vol. 3, pp. 1 et seq.			[46]	PRO Adm. 1 6367.
[47]	Sailing orders for *Valorous*, PRO Adm. 1 6337. Wm. B. Carpenter, "Report on the Physical Investigations carried on by P. Herbert Carpenter, B.A., in H.M.S. 'Valorous' during her Return Voyage from Disco Island in August 1875," *Proc. Roy. Soc.* 173 (1876) 230–7.
[48]	*Punch*, 3 July 1875.
[49]	Feilden to Rae, 26 May 1875, RGS MS Correspondence Block 1871–80.

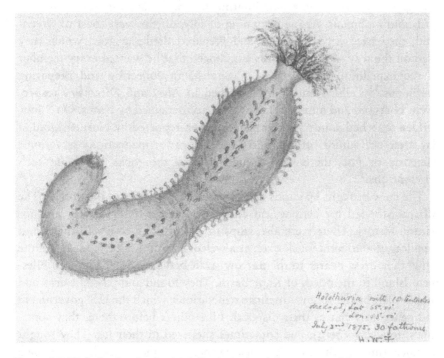

Figure 27. Holothurian dredged 2 July 1875, sketch by H. W. Feilden. © Royal Geographical Society.

The expedition set off at the end of May,[50] with cheering crowds all along the waterfront and piers, the local regimental bands playing, the rigging of every ship in port manned, loud cheers from all, whistles and horns blowing, and a telegram of encouragement from Queen Victoria.[51] Henry Frederick Stephenson, commander of H.M.S. *Discovery,* said that he could only compare the crowd witnessing their departure to "the excitement caused by the Shah's visit to the fleet at Spithead."[52]

The issue of lime juice (not lemon) began on 6 June. Four days later they hit a gale; as soon as the wind abated, the towing net was put overboard.

[50] The principal published narrative of the expedition was Nares, *Narrative of a Voyage to the Polar Sea during 1875–6 in H.M. Ships 'Alert' and 'Discovery'* . . . *with Notes on the Natural History edited by H. W. Feilden, F.G.S., C.M.Z.S., F.R.G.S. Naturalist of the Expedition,* 2 vols. (London, 1878). For a briefer account, see R. Johnston, *The Arctic Expedition of 1875–6 Compiled from Official Sources with a Summary of Previous Adventures in the Arctic Seas* (London [1877]).

[51] Feilden, "Journal HMS Alert 1875–76," RGS Library MS 387B.

[52] H. F. Stephenson, *Discovery* Journal, May–Nov. 1875, entry 29 May 1875, National Maritime Museum, Greenwich, MS STP 4a.

The gale soon renewed itself, limiting the naturalists to observing sea-birds and mammals. At the beginning of July, off the west coast of Greenland, they tried for soundings, and prepared dredging gear, which they used on the way to Disco. Nares announced that he wanted every member of the expedition to assist the naturalists in collecting and preparing specimens.[53] Feilden and Hart, attached to *Alert* and *Discovery* respectively, observed and hunted ashore, often accompanied by Nares. On 9 July, Feilden searched vainly for the meteoric iron reported by Nordenskiöld, a day after two junior officers had gone ashore for magnetic observations. Meteoric or not, there was enough iron in the rocks to vitiate such observations.[54]

The crew brought up corals and other unlikely species in the dredge. The officers stopped for geophysical and astronomical observations, and for natural history. There were also stops to collect dogs, to let off their Inuit ice pilots, and to hunt musk-oxen as a welcome source of fresh meat. All the while they drew nearer to the narrow strait between Greenland and Ellesmere Island to the north of Kane Basin. They found and passed cairns and supplies left by the recent American expeditions, which the U.S. government had generously put at their disposal. Like others before them, they sometimes dodged icebergs, and sometimes sheltered in their lee. They sought channels through the pack, or, lacking them, steamed, sailed, and crashed through the ice fields, balancing the need to forge northward against the dangers of being crushed between masses of ice, and avoiding being frozen in until they had chosen their winter quarters.

### MAGNETIC OBSERVATORIES ON THE SHORES OF
### THE POLAR SEA

By late August, *Discovery* was settled for the winter off the north shore of Lady Franklin Bay, off a coast quite different from Greenland's with its cliffs capped by a sea of ice. Grinnell Land, as that part of Ellesmere Island was known, appeared "as a series of peaked mountains rising to an altitude of 2000 to 3000 feet, with deep valleys intervening between them," and no glaciers in the valleys.[55] *Alert* forged north until iced in for the winter at

---

[53]   Nares to Commander Markham et al., 13 July 1875, letter book PAC MG 29 B 12 vol. 3.
[54]   June–July 1875: Feilden, "Journal"; Nares, *Voyage;* Nares, Captain's Log, *Alert,* PAC MG 29 B12 vol. 2; Stephenson, *Discovery* Journal.
[55]   Feilden, "Notes from an Arctic Journal," *The Zoologist* 3rd series 3 (1879) 16.

Figure 28. *Discovery* in winter quarters, with the depot and observatories on shore, September 1875. National Archives of Canada C-25921.

INSIDE THE UNIFILER HOUSE.

Figure 29. "Inside the Unifiler [*sic*] House," E. L. Moss, *Shores of the Polar Sea.* Fisher Library, University of Toronto.

Floeberg Beach, at latitude 82°27′ N, close by the modern station named after the ship. It was the most northerly point hitherto reached in the Canadian Arctic. Near each ship, observatories were soon erected: astronomical observatories, magnetic observatories, and pendulum houses built in England and of which, "strange to say," not a single piece was missing.[56] *Alert*'s observatories were in place a month later.[57] On *Discovery*, Lieutenant Fulford, in charge of that ship's magnetic work, "gave a regular house warming in magnetic house" on 11 September. Over the following weeks, he worked with the Barrow dip instrument, took some "most unsatisfactory measurements" of dip with Lloyd's needles, and set about building another house for the declinometer. In mid-October he learned that there were iron bolts in the chocks under the pedestal for one of his instruments, so all the results for the previous month had to be scrapped.[58]

Meanwhile, up at Floeberg Beach, *Alert*'s magnetic officer, Lieutenant Reginald Giffard, was rhapsodizing about his magnificent magnetic observatory, a dome 14′ across and 10′6″ high. Another smaller observatory was to be built nearby with a connecting tunnel between them.[59] Eventually, there were three ice houses for the observatory, one for the unifilar instrument for measuring horizontal intensity, another for the Barrow dip circle for dip and total force, and the third for the declinometer, for measuring declination or variation.[60]

Magnetic observations continued through the winter at both stations, hampered by the difficulty of handling small metal screws in intense cold, and reading measurements when one's breath froze over the instrument. These snow houses at least had the advantage of taking their temperature from the earth rather than the air, so that sometimes the internal temperature was as much as 40°F warmer than the outside air. Under these conditions, "an observer well muffled in furs could remain for four or five hours at a time watching the swinging magnetic needle."[61] In the spring of 1876, Fulford wrote to Giffard to give him an outline of the winter's magnetic work:

---

[56] Stephenson, *Discovery* Journal, 2–10 September.
[57] Nares, Captain's Log, 14–18 October.
[58] Reginald Baldwin Fulford, private journal, H.M.S. *Discovery*, PAC MG 29 B37, microfilm A1176.
[59] G. A. Giffard, *Alert* Journal, 16 Oct. 1875, SPRI MS 41.
[60] A. H. Markham and Giffard to Nares, 16 March 1877, PRO MS Adm. 1 6431. Markham shared magnetic responsibilities with Giffard.
[61] E. L. Moss, *Shores of the Polar Sea . . .* (London, 1878) p. 37.

We have been most unfortunate with our instruments smashing the Fox the jewels where the axle bears being forced out; but we have had it most admirably repaired by our armourer with steel planes like the Barrow & it works very well indeed.[62] The weighted needle of the Barrow had its axle broke as well, so we use a weight on one of the Fox needles for finding the total force; the same needle for deflecting the No. 3 of the Barrow screwed on the face and also for that long experiment of deflections of Lloyds.

We have taken weekly a dip and total force; only missing twice I think; two or three times we have continued the dip & total force observations for 24 hrs. Two or three times a month unifilar observations & bearings with our azimuth of a mark noting the Differential Declinometer at the same time.

The Declinometer has been noted hourly since 22nd October, & any unusual disturbances 5 minutely observations were noted; the declinometer house being [100] yards from the ship on the floe.[63]

They had indeed been as assiduous in using their instruments as they had been unlucky with them, from the bolts under Fulford's pedestal to the breakages that he reported to Giffard. They effected ingenious repairs and substitutions, which unfortunately meant that they were effectively working with different instruments for the same series of observations. Steel planes or new needles would certainly affect the performance of a dip circle, and failure to calibrate the modified apparatus, and to note in the observations just when the change in an instrument occurred, meant that most of those laboriously obtained magnetic results were useless, in spite of subsequent efforts to recalibrate the apparatus at the Royal Observatories.[64] But even where the observations were otherwise wasted, they served to counteract boredom, and gave a sense of purpose when travel was impossible.

Stephenson, writing to Nares to report on *Discovery*'s activities in winter quarters, seemed unaware of problems with the magnetic instruments, and was well content with the range of magnetic, tidal, meteorological, and astronomical observations; only when it came to navigational instruments was he critical. The sextants supplied, he observed, were "decidedly of a very common order."[65]

[62]  Cf. H. F. Stephenson, *Discovery* Journal 1875–6, National Maritime Museum MS STP 4b.
[63]  Fulford to Giffard, 27 March 1876, SPRI MS 249. Fulford gave a series of lectures on magnetism, beginning on 1 Oct. 1875: see his journal.
[64]  Nares to Admiral George Elliott, H.M.S. *Alert*, Portsmouth, 4 Dec. 1876, Nares letter book, PAC MG29 B12 vol. 3.
[65]  Stephenson to Nares, H.M.S. *Discovery* at Bellot Harbour, 26 March 1876, PAC MG29 B12.

GEOLOGY[66]

As light returned, the expedition's emphasis shifted from work in the observatories or on the ships to exploration in geography and natural history, including geology. The dog teams and the men were given more vigorous exercise; tents, sledges, and the rest of the expeditionary paraphernalia were put in order. Sledging recommenced, in spite of severe cold in early March, when thermometers registered temperatures below −70°F. Christian Petersen, interpreter and dog driver, came back to *Alert* severely frostbitten from an abortive attempt to reach *Discovery;* he never recovered, and died at Floeberg Beach. Days after his return, another attempt to reach *Discovery* succeeded, and soon sledges were taking off at a great rate, laying down food depots, exploring the nearby coast, and generally preparing for northward exploration.

Feilden had occupied himself through the winter with materials collected in the fall, with short geologizing trips near the ship, with the ship's scientific library, and with the sickness of sledge dogs, which took its toll, and increased the expedition's dependence upon man-hauling. The sun returned on 2 March, and Feilden was soon busy with fieldwork. He took part in sledging expeditions, ranging from day excursions to journeys of a fortnight's duration in April and May, when his geological observations recorded an expedition to the United States Range on northern Ellesmere Island.

Floeberg Beach was protected by ice floes, and overlooked by cliffs rising to a plateau dotted with lakes. Feilden found on the plateau fossil shells of species similar or identical to those lying on exposed beaches, concluding

[66] Levere, "Henry Wemyss Feilden (1838–1921) and the Geology of the Nares Strait Region: with a Note on Per Schei (1875–1905)," *Earth Sciences History* 10 (1991) 213–18. Geological and paleontological results of the expedition are: C. E. De Rance and H. W. Feilden, "Appendix No. XV. GEOLOGY," Nares *Narrative* (1878) vol. 2 pp. 327–45; Feilden and De Rance, "Geology of the Coasts of the Arctic Lands Visited by the Late British Expedition under Captain Sir George Nares," *Quarterly Journal of the Geological Society of London 34* (1878) 556–67; De Rance, "The Geology of the Arctic Regions," *Transactions of the Manchester Geological Society 14* (1877–8; published 1878) 441–7; R. Etheridge, "Palaeontology of the Coasts of the Arctic Lands Visited by the Late British Expedition under Captain Sir George Nares," *Quarterly Journal Geological Society of London 34* (1878) 568–639; Feilden, "The Post-Tertiary Beds of Grinnell Land and North Greenland," *Annals and Magazine of Natural History* series 4, 20 (1878) 493–4; Feilden and J. G. Jeffreys, "The Post-Tertiary beds of Grinnell Land and North Greenland," *loc. cit.* series 4, 20 (1877) 483–93; Jeffreys, "The Post-Tertiary Fossils Procured by the Late Arctic Expedition, with Notes on Some of the Recent or Living Mollusca from the Same Expedition," *loc. cit.* series 4, 20 (1877) 229–42.

that Ellesmere Island had begun to rise in relatively recent times. He observed the action of ice in all its forms, explaining the formation of bays and elevated lakes from such action. "The surface of our plateau land shows evidences of intense glacial action; the almost vertical strata of the hard limestones, slates, grits [and] schists of the district have <been> ground . . . down to a level, or rather, plane off with, a gentle slope toward the sea[;] subjected to aerial degradation, the slaty cleavage of the rocks, assisted by the frost and snow, has been unable to resist the various agencies of destruction arrayed against it."[67]

In an order established by Feilden's collection of more than two thousand fossils,[68] the oldest rocks along the strait between Greenland and Ellesmere Island[69] were Paleozoic. First were "the ancient fundamental gneiss and crystalline rocks, that have been described by so many observers as fringing the coasts of Greenland." These rocks, "underlying the synclinal of Paleozoic rocks of the Parry Archipelago, continue northwards, and form the shores of Smith Sound on either side, occupying the entire coast of Ellesmere Land from Cape Isabella to Cape Sabine, rising to a height of 2,000 feet." On top of these, between Scoresby Bay and Cape Creswell, in latitude 82°40' N, were the Cape Rawson Beds, so named by Feilden, and the first in the region to receive a formal stratigraphic name.[70] These were "a vast series of azoic rocks, newer than the fundamental gneiss, and probably unconformable to it, but older than the fossiliferous Silurians." At Cape Rawson itself, near Floeberg Beach, the strata "are abruptly terminated in sea-cliffs, . . . and exhibit fine sections of jet-black slates . . . " They

exhibit the wildest tokens of contortion and disturbance, for miles you sail along them, the strata almost vertical, they appear to have been crumpled up like a puckered ribbon, and then some tremendous denuding agent has shaved them off, and after that the glacier and ice action has grooved them into peaks and hollowed them out into valleys.[71]

Feilden was delighted whenever he had the opportunity to collect or to acquire fossils. At the end of May 1876, he, Captain Nares, and others were on a sledge trip south. Throughout the whole night of 30 May,

67  Feilden, Alert Journal, 20 March 1876.
68  Feilden, MS journal 1875–6, list of fossils, British Museum, Natural History, Palaeontology Library; Feilden, MS "Post-Pliocene and Recent Specimens from Arctic Localities," British Museum, Natural History, Palaeontology Library.
69  This is now known as Nares Strait.
70  P. R. Dawes and R. L. Christie, "History of Exploration and Geology in the Nares Strait Region," Meddelelser om Grønland, Geoscience 8 (1982) 19–36 at 25.
71  Feilden to J. D. Hooker, 12 Nov. 1876, RBG, Hooker papers, Voyage . . . , ff. 54–8.

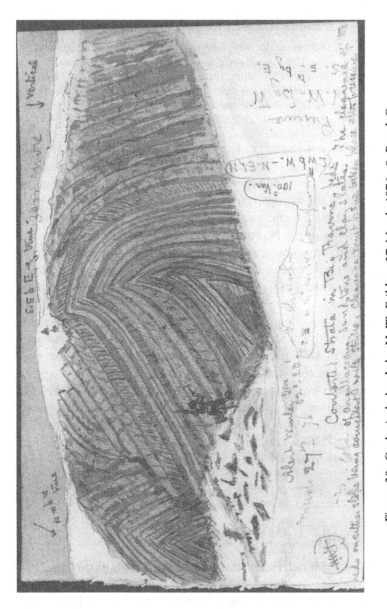

Figure 30. Geological sketch by H. W. Feilden, 27 March 1876. © Royal Geographical Society.

Captain Nares and I, and two of the men, laboured in the ravines collecting fossils, and by the morning of the 31st a goodly pile was stacked in front of the tent. . . . we gave ourselves barely sufficient time for meals, and hardly any to sleep. . . . From the large collection of carboniferous fossils that we made near Cape Joseph Henry we were only able to take away a selection; and a very large collection, ready for transportation, is now lying on the coast of Grinnell Land, in latitude 82°45′ N.[72]

Giffard brought back a fragment of calcareous limestone from Crozier Island, which Feilden found contained a fragment of a *Spirifer*, a genus of fossil brachiopods. Others had pointed out the fossiliferous limestone formation elsewhere in the archipelago,[73] so Feilden was more delighted than surprised. Nonetheless, he remarked, "this discovery of Carboniferous fossils within 400 miles of the N. Pole, will open up a series of speculations, amongst geologists, of profound interest."[74] He noted that such evidence of lush forest growth pointed to a formerly much milder climate.[75] Carboniferous limestones occurred in the Feilden and Parry Peninsulas on the north coast of Grinnell Land: "assuming the same strike continued over the Polar area, a prolongation of these limestones would pass through Spitsbergen, where this formation has been recognized, and contain some identical species." These rocks in the Nares Strait region contain marine fossils that also argued that the arctic seas had once been much warmer.

On a northbound expedition in May, camping at Depot Point, Feilden expressed his frustration with the oldest rocks, which were relatively uninteresting to him because of their lack of fossils – "I am getting very tired of these Azoic rocks."[76] He was to find compensation on the homeward journey that summer, when he explored a Miocene coal bed at *Discovery*'s winter quarters.[77]

### NATURAL HISTORY

Across the Atlantic and into Davis Strait, Feilden had made notes on pelagic birds, and on the occasional whale. Ashore, at Godhavn and elsewhere on the coast of Greenland, land birds were noted, as well as the local flora. Ascending to the highlands of Disco, Feilden observed the rush of arctic

72  Feilden, "Notes from an Arctic Journal," *The Zoologist* 3rd series 2 (1878) 313–20, 372–84, 407–18, 445–51; 3 (1879) 16–24, 50–8, 89–108, 162–70, 200–2, at 99–100.
73  E. g., Samuel Haughton, "On Fossils Brought Home from the Arctic Regions in 1859, by Capt. Sir F. L. M'Clintock," *Journal of the Royal Dublin Society 3* (1860) 53–8.
74  *Alert* Journal, 1875–6, 7 May 1876.
75  De Rance and Feilden, "Geology," p. 332 underlines this point.
76  *Alert* Journal, 13 May 1876.       77  See notes 138–41.

spring: "it was curious to notice that the blossoming of these [alpine] plants appeared to keep pace with the retreat of the snow, and just as soon as the snow dissolved the flowers appeared."[78]

As they sailed north, Feilden noticed a falling off in the abundance of flora and bird life. "The sea, however, was more prolific in life: a small dredge, let down in thirteen and a half fathoms, brought up many Mollusca, Star-fishes, and Crustaceans. By dipping buckets, we captured hundreds of *Clio borealis* [a mollusc that swims using lobes on its feet like wings or flippers] and *Limacina arctica* [a kind of sea slug]; when the two species were placed in the same vessel the *Clios* eagerly devoured the *Limacina*."[79] In Melville Bay, "the bugbear of Arctic voyagers in the days of sailing vessels," the bird life matched the multitudinous marine life: "Myriads of Little Auks swarmed round us, busily employed fishing . . . , flocks of them diving just in time to avoid the ship's stem."[80] As they rounded Cape York and steamed north between the islands and into Smith Sound, Feilden, like the Americans who had gone before him, was overwhelmed by the sheer numbers of birds:

There is a large breeding place of Looms [thick-billed murres, also known as Brünnich's guillemots] on the north-eastern face of Hakluyt Island, and myriads of Little Auks were flying up and down to their nesting haunts in the talus of the cliffs. The breeding places of the sea-fowl along the shores of this region appear to be continuous, and are occupied by incredible numbers of Looms and Little Auks. Dr. Kane, in those pathetic chapters of his charming work [*Arctic Explorations*], which relate how he and his worn-out companions escaped from Smith Sound, and traversed Melville Bay in frail open boats during the summer of 1855, tells us how his party subsisted almost entirely on the spoils of these aukeries and loomeries. . . .

[Along Fulke Fiord, near the settlement Etah,] the air above was filled with thousands upon thousands of specks, which were Little Auks passing from their breeding-places to the sea.[81]

Further north, ashore near Hayes's winter harbor of 1860–1, Feilden found arctic poppies in full bloom, and a party of knots, birds that migrated through Greenland and Iceland, but whose breeding grounds had hitherto been unknown, a matter for mere speculation.[82] He argued that their presence, like that of all creatures, was tied to food supplies, which accounted for the rapid decrease in numbers of individuals and of species

---

[78]   Feilden, "Notes from an Arctic Journal," 373.          [79]   Ibid., 379–80.
[80]   Ibid., 382–3.          [81]   Ibid., 407, 410.
[82]   Ibid., 408. See Alfred Newton in Jones, ed., Greenland *Manual*, p. 102; R. Brown, "The Arctic Expedition: Its Scientific Aims," 161–2. For a recent study of knots on Ellesmere Island, see D. N. Nettleship, "The Breeding of the Knot *Calidris canutus* at Hazen Camp, Ellesmere Island, N.W.T.," *Polarforschung* 44 (1974) 8–26.

north of the 82nd parallel. Knots, turnstones, and sanderlings included saxifrage and insects in their diet. Abundant small crustacea, in the cracks between the sea ice and the land, helped to sustain the arctic tern, while the brant's range was probably coincident with that of *Saxifraga oppositifolia*.[83] Feilden was startled to find butterflies and a bee, as well as other more numerous and unwelcome insects.

*Discovery's* winter harbor was in a zoologically richer environment than *Alert's*. As soon as the ships reached Lady Franklin Bay, they found a herd of musk-ox – and promptly slaughtered the lot, thereby acquiring about 2,000 pounds of fresh meat, and contributing to the navy's traditional role as a threat to that species throughout the Canadian Arctic. The environs of Floeberg Beach were much less productive. That fall, *Alert's* complement saw one seal and killed another; caught a few hares; saw tracks of foxes and lemmings, but not the animals themselves; and Feilden found the skeleton of a musk-ox. It was late in the season, and bird life was very sparse.[84]

Migratory birds began to arrive early in March, rock ptarmigans being the first that Feilden noted. A ptarmigan killed near Cape Joseph Henry on 29 May had the unhappy distinction of being "the most northern ornithological specimen ever secured."[85] Snowy owls and snow buntings were the other early arrivals; but in June, knots, sanderlings, turnstones, and long-tailed jaegers piled in. Other species observed near Floeberg Beach included ravens, red phalaropes (which Feilden called by their alternate name, grey phalaropes), arctic terns, and ivory gulls, which arrived in mid-June, having been absent since 1 September 1875. Long-tailed ducks, kind eiders, brant geese, and an unidentified loon (probably red-throated) completed the far northern list.[86] Many other species, including auks, kittiwakes, and murres, were noted further south along Smith Sound.

Feilden worked diligently with gun and notebook in pursuit and study of arctic birds, including their nests and eggs. None proved as frustrating in this respect as the knot:

Night after night I passed out on the hills trying to find the nest of the Knot. Not a day passed without my seeing them feeding in small flocks; but they were very wild, rising with shrill cries when one approached within a quarter of a mile on the mud flats on which they were feeding. It is very extraordinary, considering the hundreds of miles traversed by myself and my companions, – all of us on the look-out for this

[83]  Feilden, "Appendix No. III. Ornithology," in Nares, *Narrative*, vol. 2 pp. 206–17 at 207.
[84]  Feilden, "Notes from an Arctic Journal," 21.        [85]  Ibid., 99.
[86]  Feilden, "List of Birds Observed in Smith Sound and in the Polar Basin during the Arctic Expedition of 1875–76," *Ibis* 4th series 1 (1877) 401–12. A modern reference with good distribution maps is L. L. Snyder, *Arctic Birds of Canada* (Toronto, 1957).

bird's eggs, and several of us experienced bird's nesters, – that we found no trace of its breeding until the young in down were discovered [on 30 July 1876, just one day before *Alert* broke out of its winter quarters].[87]

Hart, the naturalist on *Discovery*, saw his first knots on 31 May, and, in a region richer in bird life, succeeded in finding nests.[88] His zoological and geological notes are thinner than Feilden's, but he made a good botanical collection.[89] The expedition's botanical work was based on collections at thirteen places, including the winter harbors of each ship.[90] That was more than sufficient to characterize regions botanically. The reduction in the number of species as one went further north, and a corresponding diminution in height, were familiar and expected. It was only in the more southerly parts of Greenland along their route that there was anything approaching uniform ground cover, consisting first of

small tufted perennials of low matted growth, through which the Arctic willows and *Ericaceae* [heaths] trail and extend their branches, the first alone rarely rising under the shelter of a cliff to a height of three or four feet. Through this brownish green carpet, which is about the hue of an Irish mountain bog, conspicuous and beautiful blossoms of *Rhododendron, Azalea, Diapensia, Pyrola,* and other ericaceous plants, are lavishly scattered; while the cream-coloured *Dryas,* the snowy-white *Cerastium* and *Stellaria,* the pink *Silene,* and the gorgeous red-purple *Saxifraga,* often form luxuriant sheets of colour, the latter being comparable to our Scotch heather, though richer in its effects.[91]

There were sometimes brilliant carpets of pink or white, but for the most part the individual flowers were brilliant miniatures, far removed in scale from the vegetation of temperate climes. Other differences were those of color: blues were rare, true reds were lacking, "and most of all is felt the absence of a green sward, such as the eyes are accustomed to at home." Further north, vegetation rapidly diminished; and although Discovery Bay was botanically among the richest stations explored, plants were found only in patches.[92] As Feilden's journal shows, those patches persisted richly into the northernmost parts of Ellesmere Island.

---

[87] Feilden, *Ibis* (1877) 407–8; "Notes from an Arctic Journal," 107. This was the first evidence of the knot's breeding. See Richard Vaughan, *In Search of Arctic Birds* (London, 1992) pp. 158–63.

[88] H. C. Hart, "Diary *(Discovery)* 1876," PAC MG 29 B30.

[89] Hart, "On the Botany of the British Polar Expedition of 1875–6," *Journal of Botany, British and Foreign* (1880) 52–6, 70–9, 111–15, 141–5, 176–82, 204–8, 235–42, 303–6.

[90] Sverdrup's expeditions added substantially to the high arctic flora: see chap. 9.

[91] Hart, "Botany of the British Polar Expedition," 52.     [92] Ibid.

Hooker had stressed the importance of establishing the distribution of plants. Hart drew up a flora arranged according to latitude, commencing with plants whose northern limits were nearest the pole. Most northerly of all were a saxifrage (*S. oppositifolia*) and the arctic poppy (*Papaver nudicaule*).[93]

Among Hart's more significant observations was that the arctic flora almost entirely lacked annuals producing seeds and seedlings:[94] arctic plants generally need more than one season to complete their cycle of growth and reproduction.

*Scurvy, and return to England.* Spring was a time of vigorous sledging, exhausting to men and to dogs depleted by sickness. Even before they had set up winter quarters, everyone on *Alert* realized that the open polar sea that so many had dreamed of was a myth. Instead, they were faced with a sea of ice that never melted, and that the officers soon christened the Palaeocrystic Sea.[95] A chorus, composed by the Rev. H. W. Pullen, chaplain on *Alert,* was cheerful:

> Not very long ago
> On the six-foot floe
> Of the palaeocrystic sea,
> Two ships did ride
> Mid the crashing of the tide –
> The 'ALERT' and the 'DISCOVERY'.
> The sun never shone
> Their gallant crews upon
> For a hundred and forty-two days;
> But no darkness and no hummocks
> Their merry hearts could flummox:
> So they set to work and acted Plays.[96]

The reality of the frozen sea was far from cheering; traversed by pressure ridges and hummocks, it was appalling terrain for sledging. As Nares's fleet-surgeon, Dr. Thomas Colan, reported to Armstrong: "The passage

[93]   Ibid., 141–4. William Hooker's *Flora* was the primary flora for arctic explorers, and Joseph Hooker's paper on arctic flora was also important, as were other contributions reprinted in Jones, Greenland *Manual.* A most useful modern guide is A. E. Porsild, *Illustrated Flora of the Canadian Arctic Archipelago,* National Museum of Canada Bulletin No. 146, Biological Series No. 50 (Ottawa, 1957).

[94]   Hart, "Botany of the British Polar Expedition," 53.

[95]   Clements Markham wrote to Robert Brown on 27 Nov. [1876]: "This palaeocrystic sea has upset all my cherished theories; but at the same time it has opened out new ideas and speculations which are equally interesting" [SPRI MS 441/9/35].

[96]   2 March 1876, enclosure in Feilden's *Alert* journal, RGS.

over the frozen sea was of the heaviest description. Roads had to be made with pickaxes ere the sledges could be drawn through, a mile or a mile and a half made good being considered a good day's work."[97] The geographical pole was further off than Hayes had thought, and beyond reach for Nares and his crews. Nevertheless, they would try. Expeditions went out, between the ships, inland, and along the northern shores of Ellesmere Island and Greenland. Ignorant of the ice drift that would inevitably frustrate their efforts, they also tried to force their way over the sea ice, with boats and sledges, to attain the highest northern latitude possible.[98] Most of their progress would be by man-hauling, in the tradition eulogized by Clements Markham and found to be definitively disastrous in the Antarctic expedition of Captain Scott.[99] That tradition saw man-hauling as superior to the use of dogs, and certainly there were some conditions that dogs, even with excellent drivers, could not handle.[100] This northward expedition, led by Clements Markham's cousin Albert Hastings Markham, was to be among the most demanding of all. Their departure in early April, on sledges with heraldic pennants flying, was filled with optimism, boisterous rivalry, and determination, all at sorry odds with their task. First, woolen blankets, canvas tents, and wool and canvas outer garments all became steadily heavier as they absorbed water vapor, so that as the trip progressed and the men tired, they found themselves dragging an ever increasing weight. Man-hauling heavily laden sledges was brutal, punishing work. Time and again, to go around obstacles, they marched three times as far as they actually progressed. Another problem was produced by their daily rations, which included two ounces of rum for each man, but no lime juice. The weight of fuel needed to melt ice to dissolve sugar and render the juice palatable was considered excessive on a journey where weight was to be pared to a minimum; they would, after all, be hauling or carrying everything themselves. Previous expeditions had undertaken sledging trips, some of considerable duration, without encountering scurvy, but they had either supplemented their diet with fresh meat,[101] or else they had not traveled across such

---

[97] Colan to Armstrong, 27 Oct. 1876, PRO MS Adm. 1 6431, printed in Blue Books C. 1636, *Journals and Proceedings of the Arctic Expedition, 1875–6* (London, 1877) p. 462.

[98] Nares, *Narrative*, vol. i p. 345.

[99] Markham's traditional bent appears with total and devastating clarity in Clements Markham, *Antarctic Obsession: A Personal Narrative of the Origins of the British National Antarctic Expedition 1901–1904*, ed. and intro. Clive Holland (Alburgh, Norfolk, 1986).

[100] A. H. Markham, "On sledge travelling," *Proc. RGS 21* (1876) 110–20.

[101] Raw meat was better than cooked or preserved meat as an antiscorbutic.

demanding terrain. Nares's decision not to require that lime juice be used during sledging expeditions was wrong, however reasonable his arguments.

For all the need to minimize weight, Markham's party carried magnetic apparatus, and used it until they collapsed. On 7 May, Markham noted: "I fear we have almost reached our highest latitude . . . ," and then immediately reported: "Built up a snow pedestal on which I fixed the Fox [dip circle] and was employed the whole afternoon in taking observations but I fear with an unsatisfactory result, I think my magnetic variation was out." On the following "beautiful warm sunny day," he remained behind "and made a complete, and satisfactory, series of magnetic observations." Two days later, he had to face facts:

I have at length arrived at the conclusion, although with a great deal of reluctance, that our sick men are really suffering from scurvy, and that in no mild form: the discolouration of their limbs, their utter prostration and helplessness, low spirits, loss of appetite, and other symptoms are I think decidedly scorbutic. If this should really be the case, I may consider "my little game as played out," for I can hardly expect to see them again fit for any work until they can be supplied with fresh meat and vegetables. I am unwilling for the men to suspect that they are really suffering from this terrible disease, but at the same time I shall serve out to them some lime juice, which I shall tell them will be in lieu of their grog, being a better blood purifier. I only wish I had taken more of this excellent antiscorbutic with me – it was extremely fortunate I thought of bringing some but we only have a couple of bottles on each sledge.[102]

By this time, one third of Markham's men were invalids, there seemed to be more soft snow to slow their retreat, and the hummocks, some more than forty feet high, began to seem like hostile fortresses. One of the lime juice bottles cracked in thawing, and Markham used his body's warmth to thaw the other one more slowly.[103]

Before returning, he decided to let his sick and exhausted crew rest for two days, and to use that time for observations of magnetic force, dip, and variation, accurate determination of their position, and the measurement of deep-sea temperatures with a Casella's thermometer.[104] In order to make

[102]   A. H. Markham, "Notebook and Sledging Journal 3 April–8 May 1876," SPRI MS 396/2. Edited and abridged extracts from this notebook are in Nares, *Narrative*, vol. i pp. 350 et seq. See also Great Britain, Parliament, Blue Book (1877) c. 1636.

[103]   A. H. Markham, *The Great Frozen Sea: A Personal Narrative of the Voyage of the "Alert" during the Arctic Expedition of 1875–6*, 3rd. ed. (London, 1878) p. 343.

[104]   See A. McConnell, "Historical Methods of Temperature Measurement in Arctic and Antarctic Waters," *Polar Record* 19 (1978) 217–31; "Six's Thermometer: A Century of Use in Oceanography," in M. Sears and D. Merriman, *Oceanography: The Past* (New York, 1980) pp. 252–65; *The Construction and Use of a Thermometer by James Six F. R. S: Prefaced by an Account of his Life and Works and the Use of his Thermometer over Two*

the last measurement, they cut through sixty-four inches of ice, a dangerous expenditure of their depleted energies, and found the bottom was only seventy-one fathoms beneath them. They took temperature readings at ten-fathom intervals, let down a bread bag and hauled it up full of small crustaceans and other creatures. Scientific work by men in such a state was positively heroic. Then they headed south, with the sledges dragging more heavily, with more and more of the men unable even to walk, and those dragging the sledges moving their legs with great difficulty and in intense pain. When they drank to Queen Victoria's health on her birthday, 24 May, "we could muster only four and a half good and sound pair of legs!" Three days later, five men were utterly unable to move, five others were nearly as bad but managed to hobble after the sledges, three others had all the signs of incipient scurvy, and only two officers and two men could be considered effective. "This was, it must be acknowledged, a very deplorable state of affairs."[105]

Lieutenant Parr left Markham's exhausted crew just twenty-seven miles from the *Alert,* made good time to the ship, roused a rescue party, and had them swiftly on their way, preceded by Dr. Moss, *Alert's* surgeon, and another officer on snowshoes accompanying a dog-drawn sledge with medical and other relief. They made a forced march, reaching Markham's crew fifty hours after Parr had set out, but too late for one man, who died hours before the rescue arrived. The main rescue party, headed by Nares, arrived the next morning. Back at the ship, lime juice, rest, and nourishment contributed to the gradual recovery of the survivors.

Other sledging expeditions were similarly afflicted, so that in May and June a majority of the crew had scurvy. *Discovery's* crew were in almost as bad shape. Albert Markham concluded, and Nares's actions evince the same belief, that the principal cause of scurvy was the absence of fresh animal and vegetable food, to which all arctic expeditions had been exposed, rather than the specific lack of lime or lemon juice. When fresh food was lacking, then darkness, damp, cold, and poor ventilation were predisposing causes, acting as triggers.

A more thorough break-up of a healthy and strong body of men it would be difficult to conceive. Not only had the men engaged in the extended party under my command been attacked with scurvy, but also those who had been absent from the

*Hundred Years* by Jill Austin and Anita McConnell (London, 1980). See also Wolfgang Matthäus, "The Historical Development of Methods and Instruments for the Determination of Depth-Temperatures in the Sea *in situ,*" *Premier Congrès Internationale d'Histoire de l'Océanographie* (Monaco, 1966) pp. 35–47.

[105]  A. H. Markham, *The Great Frozen Sea,* pp. 346–70.

ship only for short periods, and some, who may be said never to have left the ship at all, or if they did, only for two or three days! The disease then could not be attributed to any special circumstance connected with sledge travelling.

The seeds must have been sown during the time, nearly five months, that the sun was absent, and we were in darkness. . . . To . . . the predisposing causes our expedition was exposed for a very much longer period than any other which sent out extended travelling parties. For this reason other expeditions were exempt from scurvy while we were attacked. In short the different result was caused by the difference in latitude.

This might seem a little too simple, but Markham was convinced that neither diet nor winter routine sufficed to explain the outbreak. As for the much-vaunted value of lime juice as an antiscorbutic, it was simply not enough:[106]

Lime-juice, though most useful in warding off for a time and delaying an attack of scurvy, and as a cure, will not, with other circumstances unfavourable, prevent an outbreak. . . . Some of our men had scurvy who never left the ship and never ceased to take their daily rations of lime-juice, and others were attacked who went away travelling at a time when daily rations of lime-juice formed a part of the sledge dietary.[107]

The matter was controversial, and was to be the subject of a formal naval inquiry when Nares returned to England. But meanwhile the fact of scurvy was undeniable, rendering insupportable the prospect of a second winter in the ice. On 16 June Nares announced that, in order to guard against a repetition of the attack of scurvy, he had determined to give up all further northward exploration. Instead, both ships would proceed south to a region where game was plentiful, thereby solving the problem of fresh meat; one of them would stay there for the winter, in order to explore Hayes Sound in the following spring, while the other returned home.[108] The return of Markham's crews in desperate condition determined Nares before the end of the month to head for home with both ships, a decision, as Albert Markham remarked, showing great moral courage. He was obviously right, because only ten men out of his whole ship's company were by then fit for work, and some of these were still convalescent. Freeing *Alert* from

[106]  K. J. Carpenter, *The History of Scurvy and Vitamin C*. Note that lime juice is considerably less rich in vitamin C, and is therefore a less effective antiscorbutic, than lemon juice. See also A. H. Smith, "A Historical Enquiry into the Efficiency of Lime Juice for the Prevention and Cure of Scurvy," *Journal of the Royal Army Medical Corps* 32 (1919) 93–116, 188–208.

[107]  A. H. Markham, *The Great Frozen Sea*, pp. 370–1.

[108]  Nares letter book, 16 June 1876, PAC MG29 B12 vol. 3.

the ice was difficult; not until 31 July and after much labor did they succeed in blasting a path ahead of the ship to open water.[109]

The ships were together again on 11 August, and it took another ten days before they were free from the ice and moving south together. During those ten days, Hart showed Feilden a thick seam of coal in a nearby valley. Hart had already found some fossilized leaf impressions, and he and Feilden discovered a few more, apparently conspecific with those found in Miocene deposits in Greenland and Spitsbergen. Feilden made a return visit, and formed "a very considerable collection of these leaf-impressions."[110] They were to prove quite as interesting as the specimens of living plants,[111] constituting the last significant scientific fruit of the expedition. Both ships sailed home to Portsmouth, and to controversy.

WELCOME HOME: THE SCURVY COMMITTEE

The expedition returned to a very mixed welcome. Arctic geographers were impressed. August Petermann eulogized the expedition: "It is this pure interest for scientific progress that cannot be too much commended . . . to conduct two vessels through this most dangerous ice alley and safely back again, has never been done before."[112] Clements Markham publicly and privately, and always aggressively, celebrated the achievements of Nares, his officers and crew.[113] The expedition had sledged and mapped three hundred miles of new coastline and had shed light on the nature of a large section of the Polar Ocean.[114] One party, under Albert Markham, had reached the highest latitude hitherto attained. There had been extensive magnetic, meteorological, and tidal observations, and first-rate geological fieldwork.[115] The naturalists had

[109]  Nares, Narrative; A. H. Markham, The Great Frozen Sea.
[110]  Feilden, "Notes from an Arctic Journal," 165.
[111]  See Heer, Flora Fossilis Arctica (1878), discussed in note 141.
[112]  Petermann to President of RGS, Gotha, 8 Dec. 1876.
[113]  C. R. Markham, The Royal Geographical Society and the Arctic Expedition of 1875–76: A Report (London, 1877); "The Arctic Expedition of 1875–76," Proc. RGS 21 (1877) 536–55; "Arctic Expedition 1875–76," MS vol. RGS CRM 65.
[114]  J. R. Lotz, "Northern Ellesmere Island: A Study in the History of Geographical Discovery," Canadian Geographer 6 (1962) 151–61.
[115]  Nares, The Official Report of the Recent Arctic Expedition (London, 1876) gives an account of auroral, magnetic, spectroscopic, polarimetric, electrical (atmospheric), meteorological, oceanographic (including measurement of currents, tides, and ocean temperatures), and astronomical observations.

besides made comprehensive collections and studies of the flora and
fauna.[116].

Nares met enthusiasm, even adulation from the public, but he ran into
professional criticism. He had failed to reach the pole, failed to avoid
scurvy, and failed to endure a second winter in the ice. *The Navy*, an influential periodical, attacked him: "no less than nine banquets; the honour
of dining with Her Majesty; the dignity of a K.C.B.; Admiralty approval;
and other advantages have been received by Captain Sir George Nares for a
failure quite as glaring as that of the unhappy Admiral Byng."[117] Instead of
receiving such rewards, *The Navy* fulminated, Nares should be court-
martialed. The main ground for such prosecution would be Nares's "dis-
obedience of orders in not concentrating the strength of the Expedition on
'its primary object, to attain the highest northern latitude, and if possible
reach the North Pole.' "[118] The subordination of natural and geophysical
science to geography should have been absolute. Then there was Nares's
neglect of duty in not personally leading the sledging parties; his neglect of
the sanitary arrangements laid down by Armstrong; and his improper and
unjustified declaration that the pole was impracticable.[119] Most of those
criticisms were ill-judged, as the Admiralty recognized. Nares vainly requested a court-martial to clear himself, and Clements Markham drafted a
furious list of the "lies and slanders" respecting the expedition, invented after its return.[120]

Unhappily there was one accusation that did need looking into: the Admiralty set up an inquiry into the causes of the outbreak of scurvy.[121] The
inquiry rejected the notion that conditions on board the ships during the
winter caused the subsequent outbreak of scurvy, and reaffirmed the value
of lime juice as an antiscorbutic. They found that Nares had been reminded
by his fleet-surgeon of the desirability of taking lime juice on sledging expeditions, and had decided, primarily because of the weight of fuel in-

[116]  Nares, *Narrative*, vol. 2 pp. 352–3.
[117]  Byng was executed on his own deck for not having more vigorously prosecuted a fight
with the French; as Voltaire wryly noted, the execution was "pour encourager les autres."
[118]  On 29 Oct. 1876 Sir J. Biddulph transmitted the Queen's congratulations to Nares, adding that "Her Majesty desires me to write to say that she trusts that now the question [of
reaching the pole] is settled, and that no more lives will be risked in what appears to be
a hopeless attempt" (PRO Adm. 1 6390).
[119]  *The Navy*, 9 Dec. 1876 p. 562.      [120]  C. R. Markham, MS RGS CRM 65 p. 11.
[121]  *Report to the Lords Commissioners of the Admiralty on the Cause of the Outbreak of
Scurvy in the Recent Arctic Expedition; on the Adequacy of the Provision Made in the
Way of Food and Medicine; and on the Propriety of the Orders for Provisioning the
Sledge Parties* (London, 1877). This was promptly met by C. R. Markham, *A Refutation
of the Report of the Scurvy Committee* (Portsmouth, 1877).

volved, that this could not be done without leaving out essential stores. As partial compensation, a double ration of lime juice was issued for a month before sledge parties set out, and lime juice was left at those depots that would be reached about the end of May, together with fuel to melt it. Once the weather was warm enough to dispense with fuel, lime juice was issued to all sledging parties. Previous expeditions, as well as some members of Nares's expedition, had been afflicted by scurvy even when lime juice was regularly administered, and there was some suspicion that bad meat might cause scurvy. The committee of inquiry found that the lack of lime juice among the sledging parties was the main cause of scurvy, but admitted to enough qualifications based on prior arctic experience to avoid disciplining Nares. No one was happy with the result.[122]

### HENRY FEILDEN AND JOSEPH HOOKER

Nares was not the only one to encounter trouble on his return. Feilden, the more active of the naturalists on the expedition, found himself caught between his notions of a soldier's duty, and the notions of Joseph Hooker, President of the Royal Society, about what was due to science and to himself as its personification.

Things began well. Nares was generous and public in his praise of Feilden, stating in his official report that "no one moment has been lost by this indefatigable collector and observer. . . . I am only doing him justice when I state that he has been to this Expedition, what Sabine was to that under the command of Sir Edward Parry."[123] Feilden made appropriately modest disclaimers to Hooker, and then started preparing his reports. He was working on "a small typical collection" that he had brought from his cabin, with the bulk of the specimens remaining on board. As soon as he had something ready, he would send it to Hooker, but this would have to be

---

122 C. R. Markham, *A Refutation;* Nares to Admiralty, draft 5 June 1877, PAC MG29 B12 vol. 3; Nares to the Secretary of the Admiralty, 10 Nov. 1876, PRO Adm. 1 6390; Nares to Armstrong, n.d., Remark Book PAC MG29 B12 vol. 3 ff. 12–13; Admiralty to Nares, 5 June 1877, PAC MG29 B12 vol. 3; Armstrong, letter of 14 Nov. 1876, in Nares remark book, PAC MG 29 B12 vol. 3; Patrick Black, *Scurvy in High Latitudes: An Attempt to Explain the Cause of the "Medical Failure" of the Arctic Expedition* (London, 1876).

123 Nares *Official Report,* p. 47. Nares later nominated Feilden as a candidate for F.R.S., but Feilden was not elected (Nares to H. Woodward, 8 March 18[8/9]9, Ellen S. Woodward Collection, Blacker Wood Library, McGill University, and RSL election certificate records).

through Nares, "as I am still on full pay of the Admiralty."[124] He gave
Hooker a quick preview. There were invertebrates from Smith Sound,
echinoderms whose distribution confirmed that Greenland was an island, a
flora that seemed Greenlandic in character, with between twenty and thirty
species of flowering plants between 82° and 83° north. There were no
whales in Smith Sound because of over-hunting by whalers. There were terrestrial mammals, with lemmings being the most numerous, providing food
for other creatures. There were evidences of old Inuit settlement, indicative
of a previous milder climate; pieces of driftwood arguing for an oceanic
drift from Siberia and Bering Strait; and best of all, geological observations
and specimens: "I think the Geology is our strongest point."

In the following month, Feilden reported to Nares that he had removed
the natural history collections to his home near Woolwich from the ship,
which was too damp for them. The geological specimens were either at his
home, or in the Museum of Practical Geology in Jermyn Street, where officers of the Geological Survey were helping with the arrangement of the
collection. Robert Etheridge, paleontologist to the Geological Survey, was
working on the fossils, and other help was available – Hooker had offered
to examine the plants. Feilden was separating marine specimens brought up
by the dredge from the crowded stew in which they had traveled to London;
he had sent a musk-ox carcass to the British Museum because it was dete-
riorating rapidly; and generally, he sought to put the collections in order.
He hoped that it would be possible for him, under Nares's supervision, to
work up the natural history of the expedition, and that appropriate arrangements could be made for his salary and out-of-pocket expenses. As he had
told Hooker, he considered himself still under Nares's command. Nares
supported his request.[125]

Hooker was displeased, and he let it be known. He felt that the reports
should have come directly to him, and that the specimens should have
been at his disposal, not Feilden's. Nares sought to smooth things over,
explaining that Feilden was still under Admiralty orders, and reporting to
his superior officer. "[A]ll his information is at the disposal of the Royal
Society and no one else – but it must be sent through the Admiralty until
he is released."[126] Hooker was not satisfied with that, and said so. Feilden
told him:

[124]  Feilden to Hooker, 12 Nov. 1876, RBG, Hooker Papers, Voyage pp. 54–8.
[125]  Feilden to Nares, 7 Dec. 1876, RBG, Hooker Papers, Voyage, 222–7, Nares Remark
       Book 15 Dec. 1876, PAC MG29 B12 vol. 3.
[126]  Nares, 27 Dec. 1876, RBG, Hooker Papers, Voyage.

I have been brought up from boyhood as a soldier. I was employed by the Admiralty, and consequently waited for orders from them. Who ever heard of a soldier leaving his post until duly relieved. Now that you tell me that the council of the R.S. is the body to which I am *wholly* and *entirely* responsible you will find me as willing a servant as ever you met with.[127]

Feilden began to send Hooker plants, and to ask his advice about other specimens. But Hooker had jumped the gun, as Nares told him, while agreeing that the sooner Feilden and Hart were under Royal Society orders, the better. Then the hydrographer, Evans, had to poison the waters by expressing astonishment that the naturalists were so wanting in the proprieties as to turn their backs on the Royal Society.[128]

Hooker, now thoroughly irritated, waited a month and then wrote icily to Feilden:

Let me remind you, in the spirit of true friendship that the expedition is widely regarded as a failure & waste of money. Viz it appears that the *two* paid naturalists (the first ever paid on an arctic Expedition) cannot bring out their *one year's* collections from the "limits of life" without paid aid, there will be the most disagreeable comments made on the Royal Society's choice of Naturalists and on the Naturalists themselves. The world still looks to the scientific results of the Expedition for the salvation of its credit.[129]

Feilden responded by writing to the Treasury, with a copy to Hooker, announcing that he would no longer undertake any work under the direction of the council of the Royal Society, but that he would within three months submit a report to the Treasury, on his own responsibility and without remuneration, on leave from his regimental duties.[130] Hooker realized that he had gone too far, and wrote to amend the affair. Feilden, while willing to accept a friendly gesture, justified himself, explained that arrangements with the Treasury would no longer allow him to work under the original terms, and offered to meet Hooker and return his letter if a friend of Hooker's and a member of the Royal Society's council would read both party's letters. The meeting took place.[131]

---

[127]  Feilden to Hooker, 27 Dec. 1876, RBG, Hooker Papers, Voyage, f. 65.

[128]  Feilden to Hooker, 2 and 8 Jan. 1877, RBG, Hooker Papers, ff. 66–8; Nares to Hooker, 20 Jan. 1877, ff. 192–3; F. J. Evans to Hooker, 5 Jan. 1877, f. 53.

[129]  Hooker to Feilden, 10 March 1877, RBG, Hooker Papers, ff. 139–41.

[130]  Feilden to Lingen, Permanent under-secretary to Treasury, cc. Hooker, 14 March 1877, RBG, Hooker Papers, f. 81.

[131]  Feilden to Hooker, 14 March 1877, RBG, Hooker Papers, f. 82. Hooker's humiliating letter is in the archive with the rest of the correspondence.

300         *Science and the Canadian Arctic*

A new arrangement was struck. Hart was not producing much. Feilden undertook to write up the ethnology, mammalia, and ornithology of the expedition, while specialists would write up other subjects. Some of the specialists, predictably, published the results of their work independently, to Nares's and Feilden's chagrin. Relations between Hooker and Feilden nevertheless regained cordiality.[132]

BOTANY AND PALEOBOTANY

Hooker himself wrote up the botany. He was delighted at the thoroughness of the collection, which included sixty-nine identifiable flowering plants from north of 82°, as well as nearly as many again from the Greenland coast to the south – ten more than had been obtained by all previous explorers of Melville Island 5° to the south. That fall he told Feilden:

I am busy with your [phenogams] & more interested than ever. There are sixty-nine flowering plants and ferns from 80°[−83°] . . . No fewer than 15[133] are neither in Spitzbergen or Melville Island, & of these 2 are not even in Greenland! On the other hand there are only 12 common to Melville & Spitzbergen (both so much farther South) that are not in your collection . . . and of these 5 are not Greenland plants at all. Of Spitzbergen plants not found in 80–83°, there are only 5, & none of these are Greenlandic. Of Melville Island plants not in 80–83 there are 10, and only two of them are Greenlandic and these two are confined to E. Greenland!

The flora possessed Greenland plants lacking in other islands of the Canadian archipelago to the west, and in Spitsbergen to the east, whereas it lacked plants that either or both of those regions possessed, but which were also lacking in Greenland:[134] "Thus your Flora of 80–83 is a northward extension of the Greenlandic, & with two plants added that are not Greenlandic or Melville Isd. or Spitzbergen! – and which are not found anywhere in the Arctic circle for many degrees further south. . . . "

---

[132] Nares to Hooker, 6 May 1877, Feilden to Hooker, 15 May 1877, "Instructions to Naturalists," RBG, Hooker Papers, ff. 96–7, 101–3, 196–7, 211–13; Hooker to Feilden, 2 Nov. 1877, RGS correspondence block 1871–80.
[133] This number was subsequently revised downward to 12. See "Appendix No. XIV. Botany. By Sir Joseph D. Hooker, C.B., K.C.S.I., President Royal Society: With Lists of Flowering Plants, by Professor D. Oliver; Musci, by W. Mitten; Fungi, by Rev. W. J. Berkeley; Algae and Diatomaceae, by Professor George Dickie," in Nares, *Narrative*, vol. 2 pp. 301–26 at 301.
[134] Ibid., p. 302. For Hooker on Greenland's flora, see chap. 5.

Distribution[135] and its causes in climate, winds, and currents were vitally important to Hooker: "To my mind all this indicates occasional warm winds or warmth of ocean currents which other polar lands do not enjoy – also . . . that the interior of Greenland enjoys a climate & vegetation that its coasts do not [betray]. Will you dine with me at the R.S. . . . ?"[136]

*Oswald Heer.* As Robert Brown of Campster had pointed out, parts of the Arctic had once enjoyed a much warmer climate, and a correspondingly luxuriant flora. The distinguished Swiss paleobotanist Oswald Heer had worked on arctic fossil collections made from the Canadian archipelago to Spitsbergen, and was working on a magisterial *Flora Fossilis Arctica*. The Greenland *Manual* prepared for Nares's expedition made sure that the naturalists were properly alerted to the existence of Miocene and other fossil plants.[137]

The exposed coal seam, twenty-five to thirty feet deep, lay four miles north of *Discovery's* winter quarters. Fossilized impressions of plants were collected by both naturalists, and by Dr. Edward Lawton Moss, surgeon in the *Alert*, who sketched and painted careful records of the expedition.[138] Moss was jealous of Feilden's status as official naturalist, judging him too ignorant for his duties,[139] but they managed to cooperate in collecting fossil

---

[135] Phytogeography was for geographers as well as botanists. See W. J. Thistleton-Dyer, "Lecture on Plant-Distribution as a Field for Geographical Research," *Proc. RGS* 22 (1878) 412–45; and note the great botanist Robert Brown's prominence in the RGS (he was on the society's committee).

[136] Hooker to Feilden, 20 Nov. 1877, RGS correspondence block 1871–80.

[137] Reproduced therein were: pp. 368–73, Oswald Heer, trans. Robert H. Scott, "On the Miocene Flora of North Greenland," from the *Report of the Thirty-Sixth Meeting of the British Association for the Advancement of Science . . . 1866* (1867) 53–5; pp. 374–7, "Notice of Heer's 'Flora Fossilis Arctica' (Carboniferous Fossils of Bear Island and Spitzbergen, and Cretaceous and Miocene Plants of Spitzbergen and Greenland) communicated by Robert H. Scott," from the *Geological Magazine* 9 (1872) 69–72; pp. 378–85, Heer, "The Miocene Flora and Fauna of the Arctic Regions," from *Flora Fossilis Arctica* vol. 3 (Zürich, 1875); pp. 386–9, Heer, "The Cretaceous Flora and Fauna of Greenland," from *Kungliga Svenska Vetenskapsakademien Handligar* 12 (1874) 5–7, 16–18.

[138] See his sketches in the SPRI, and his *Shores of the Polar Sea: A Narrative of the Arctic Expedition of 1875–76 . . . illustrated by Sixteen Chromo-Lithographs and Numerous Engravings from Drawings Made on the Spot by the Author* (London, 1878). His field sketches are better than his paintings. The finest paintings from the expedition were made by Thomas Mitchell, who also took a series of photographs: see Michael Bell, "Thomas Mitchell, Photographer and Artist in the High Arctic, 1875–76," *Image* 15 (1972) 12–21.

[139] The friction between Moss and Feilden is clear from numerous passages in the latter's *Alert* Journal 1875–6.

302         *Science and the Canadian Arctic*

plants. Hart seems to have done nothing with his collection, but Moss and Feilden independently sent their specimens to Heer. Feilden sent twenty-six species to Moss's fourteen; three of them were new to Heer.[140] The result was another volume in Heer's series on arctic flora that compared these collections with others made in Greenland and in Spitsbergen.[141] Heer also compared the fossils with the modern range of species around the "coal mine." Not only were the species different, but in all but two cases (willows and sedges), even the genera were different. There was a general coincidence between the Miocene fossil flora of Grinnell Land and that of Greenland, just as there was between their living flora. The fossil flora was as luxuriant as Robert Brown had hoped, with horse tails, yews, cypresses, beech trees, elms, viburnums, and many more. There were the usual problems of taxonomy. Heer renamed a genus of yew after Feilden, and four species within it became *Feildenia rigida H[ee]r., F. major Hr., F. bifida Hr.,* and *F. Mossiana Hr.;* the last embodied a form of subordination that Moss could scarcely have welcomed.

Before this volume was published, Heer received the Royal Society's Royal Medal for his paleobotanical researches. He was working in a difficult field, and was one of the few whose results were trustworthy. As Hooker wrote to J. W. Dawson of McGill University in January 1879:

I agree with you that the Veg. Palaeontology of the Arctic regions has to be redone: but I do not limit this redoing to those regions, I would make it include all [known] regions, for that we have as yet no approximate conception of the extent, or composition, or affinities, or chronology, or geographical distribution, of the vegetation of any one past epoch. I believe that we are in the dark as to the affinities of perhaps 90 per cent of the [enumerated] fossil plants; & that [our methods] of determining the ages of beds, by these means are very feebly tentative. In saying this I do not [undervalue] the definite results obtained by Heer . . . and others.[142]

Classifying fossil plants was and is a difficult discipline. Feilden and the others rightly did not try to sort them out; sending them to Heer showed good judgment.

---

[140]  Heer to Feilden, 10 Nov. 1877, RGS correspondence block 1871–80.
[141]  O. Heer, *Flora Fossilis Arctica: Die fossile Flora der Polarländer* vol. 5, pt. i, *Die miocene Flora des Grinnell-Landes gegründet auf die von Capitän H. W. Feilden und Dr. E. Moss in der nähe des Kap Murchison gesammelten fossilen Pflanzen* (Zürich, 1878); for a summary, see Heer, "Notes on Fossil Plants Discovered in Grinnell Land by Captain H. W. Feilden, Naturalist of the English North Polar Expedition," *Quarterly Journal of the Geological Society of London* 34 (1878) 66–72.
[142]  Hooker to J. W. Dawson, 4 Jan. 1879, McGill University Archives MG 1022 Accession No. 2211/65/38.

ZOOLOGY

Feilden was competent in zoology. We have already noted that he wrote up the ornithology himself, and he did the same for the mammals.[143] There were no surprises, but many fine observations, and confirmation of the suggestions of others. He noted the profusion of flowers immediately around the den of an arctic fox, whose presence had fertilized the soil. He found several dead lemmings, whose skulls had been penetrated by the canine teeth of foxes, as well as two ermines killed in the same way:

Then to our surprise we discovered numerous deposits of dead lemmings: in one hidden nook under a rock we pulled out a heap of over fifty. We disturbed numerous "caches" of twenty and thirty, and the ground was honeycombed with holes each of which contained several bodies of these little animals, a small quantity of earth being placed over them.

Here was confirmation of the suggestion that foxes laid up supplies of food for the winter. Feilden also found remains of a hare, and wings of young brants from the previous season – evidence that the foxes used the same den in successive years.[144]

They saw no living polar bears around Floeberg Beach, there being little there to tempt them away from the richer hunting grounds of Baffin Bay's open north water,[145] although ringed seals penetrated as far as the icy polar sea. Feilden was quite clear that whales could not inhabit the frozen polar ocean to the north of Grinnell Land; there was little hope that arctic discoveries in that region could further extend the range of whaling.

His extensive observations on the musk-ox were accurate. So too was his perception that "the number of musk-oxen in Grinnell Land is extremely limited, whilst the means of subsistence can only supply the wants of a fixed number; consequently, after an invasion such as ours, when every animal obtainable was slaughtered for food, it must take some years to re-stock the ground."[146]

Ten species of fish were collected between latitudes 78° and 83° N, of which some were previously known from the western Arctic, others from

---

[143] Feilden, Appendix No. II. Mammalia, in Nares, *Narrative*, vol. 2 pp. 192–205; "On the Mammalia of North Greenland and Grinnell Land," *Zoologist* Series 3 1 (1877) 313–21, 353–61.
[144] Feilden in Nares, vol. 2 p. 194.
[145] The north water is a "recurring polynya"; see Moira Dunbar and M. J. Dunbar, "The History of the North Water," *Proc. Royal Society of Edinburgh (B)* 72 (1972) 231–41.
[146] Feilden in Nares, vol. 2 p. 201.

Spitsbergen, and one new species, which Albert Günther of the British Museum named after Nares, *Salmo naresii.*[147]

There was comprehensive coverage of the other zoological collections, to which Feilden had contributed most of the specimens: molluscs, insects, crustacea, echinoderms (including holothurians, starfish, and sea urchins), sponges, and more besides.[148]

Beyond his surprise at finding butterflies on the shores of the polar ocean, Feilden had been little impressed at his collection of insects. Robert McLachlan, however, who wrote up the insects and spiders,[149] had no hesitation in stating that the entomological collections were the most valuable of all the zoological ones, because they proved the existence "of a comparatively rich insect fauna, and even of several showy butterflies, in very high latitudes." Their presence posed a puzzle, because "[o]ne month in the year is the longest period in which they can appear in the perfect state, and six weeks is the period in each year in which phytophagous [i. e., plant-eating] larvae can feed; so it appears probable that more than one season is necessary, in most cases, for their full development . . . " Also remarkable was the paucity of beetles, elsewhere so numerous as to make it clear that the Creator loved them. The incidence and distribution of species was, here as

---

[147]   A. Günther, "Appendix No. IV. Ichthyology" in Nares, *Narrative,* vol. 2 pp. 218–22.

[148]   Nares, *Narrative,* vol. 2; E. A. Smith, "Appendix No. V. Mollusca," pp. 223–33; R. McLachlan, "Appendix No. VI. Insecta," pp. 234–9; E. J. Miers, "Appendix No. VII. Crustacea," pp. 240–56; W. C. McIntosh, "Appendix No. VIII. Annelida," pp. 257–59; W. P. Sladen, "Appendix No. IX. Echinodermata," pp. 260–82; G. Busk, "Appendix No. X. Polyzoa," pp. 283–9; G. J. Allman, "Appendix No. XI. Hydrozoa," pp. 290–2; "Appendix No. XII. Spongida," pp. 293–5; H. B. Brady, "Appendix No. XIII. Rhizopoda Reticularia," pp. 295–300. See also O. P. Cambridge, "On Some New and Little-Known Spiders from the Arctic Regions," *Annals and Magazine of Natural History* Series 4, *20* (1877) 273–85; E. J. Miers, "Report on the Crustacea Collected by the Naturalists of the Arctic Expedition in 1875–76," *ANH* Series 4, *20* 273–85; P. M. Duncan and W. P. Sladen, "Report on the Echinodermata Collected during the Arctic Expedition, 1875–76," *ANH* Series 4, *20* (1877) 449–70; H. J. Carter, "Arctic and Antarctic Sponges," *ANH* Series 4, *20* (1877) 38–42; H. Brady, "On the Reticularian and Radiolarian Rhizopoda (Foraminifera and Polycistina) of the North Polar Expedition of 1875–76," *ANH* Series 5, *1* (1878) 425–40; G. Busk, "List of Polyzoa Collected by Captain H. W. Feilden in the North Polar Expedition; with Descriptions of New Species," *Journal of the Linnean Society of London: Zoology* 15 (1880) 231–41; W. C. M'Intosh, "On the Annelids of the British North-Polar Expedition," *Zoology* 14 (1879) 126–34; P. M. Duncan and W. P. Sladen, *A Memoir on the Echinodermata of the Arctic Sea to the West of Greenland* (London, 1881); W. H. Feilden, "Arctic Molluscan Fauna," *Zoologist* Series 3, *1* (1877) 435–40; E. A. Smith, "On the Mollusca Collected during the Arctic Expedition of 1875–76," *ANH* Series 4, *20* (1877) 131–46.

[149]   McLachlan, "Appendix VI" in Nares, *Narrative.*

elsewhere, a major preoccupation of naturalists; and once more, Edward Forbes's account of the common origin of arctic and alpine flora and fauna was invoked.

In like manner, the historical dimension of distribution patterns was repeatedly addressed throughout the zoological essays, so that, for example, similarities were noted between the arctic crustacea and the fauna of "the Post-tertiary glacial beds of Scotland, and also, of course, to that of the North British seas."[150] Geology's imperialism could find an ally in historical zoology.

## CONCLUSION

The British Arctic Expedition of 1875–6 had achieved a good deal, especially in geology and natural history. The theoretical development of these sciences had progressed rapidly in recent decades, even more rapidly than geophysical and atmospheric sciences. The united front of geographers, physical scientists, and life scientists in lobbying government for renewed scientific exploration of the Arctic reflected an increasingly interconnected view of the sciences, one more manifestation of the realization of von Humboldt's vision. The Greenland *Manual,* which was so rapidly and efficiently assembled by the Royal Society, and the elaborate scientific instructions given by that body and by the Royal Geographical Society to the members of the expedition, reflected a growing body of knowledge, and growing sophistication in its interpretation.

The sciences in 1875 had come a very long way from their state in 1818, when John Ross sailed on the first post-Napoleonic search for the Northwest Passage. In the earth sciences, Charles Lyell had made the model of history part of the definition of geology.[151] The emphasis upon historical process and development came to characterize not only the life and earth sciences, but also the physical sciences, for example in astronomy and cosmology, leading Lord Kelvin to deny that Darwin was entitled to the time he needed for his theory to work.[152]

---

[150] Miers, "Appendix VII" p. 255.
[151] C. Lyell, *Principles of Geology,* 3 vols. (London, 1830–3; reprinted Chicago, 1989–91) vol. 1 pp. 1, 3, 4.
[152] For an elaboration of this point, see Levere, "Elements in the Structure of Victorian Science, or Cannon Revisited," in J. North and J. Roche, eds., *The Light of Nature: Essays in the History and Philosophy of Science Presented to A. C. Crombie* (Dordrecht, 1985) pp. 433–49.

The study of processes over time and space needed extensive, coordinated, and cooperative study. The very successes of Nares's expedition
pointed to the limitations of its model, a single expedition to essentially one
locality. The argument of cooperation among the practitioners of different
branches of science, and between the scientists of different nations, was also
implicit in the composition of the Greenland *Manual*, with its cross-
referencing between disciplines formerly distinct, and with contributions
from scientific observers of several different nationalities. The British Arctic
Expedition should have represented, and in some ways did come to represent, a watershed between old and new styles of scientific exploration of the
Arctic. If science had been the only goal, and national glory no part of the
expedition's mandate, there would have been no temptation to repeat such
scientifically insignificant feats as an attempt to reach the North Pole. Old
Count Lütke in St. Petersburg had spoken for science when he decried the
passion for seeking to plant national flags at the pole. Scientific geographers might agree; but there were always geographers like Clements
Markham for whom national glory came first, and for whom the pole was
one version of the arctic grail.[153] The tension between their kind of exploration, and the kind advocated by the Royal Society, was repeatedly to come
into play, continuing to frustrate and bedevil arctic expeditions. But for a
while, cooperative international circumpolar science came to enjoy greater
favor than striving for the pole. How it did so, in the first International Polar Year of 1882–3, is the subject of the next chapter.

[153] The Northwest Passage was the other version.

# 8

# From Nationalism to Internationalism in Science: The International Polar Year 1882–1883

The arctic archipelago lay to the north of British North America, which since 1867 was partly united within a confederated Dominion of Canada. Historically, the Royal Navy had been the most frequent southern presence, and the British government had come to think of the arctic islands as its private fiefdom. In the years following the resolution of the Franklin searches, other nations began to explore the archipelago. There were competitive aspects to arctic exploration and associated scientific work, but there were also signs of nascent international cooperation. Most evident was the transnational complexion of magnetic investigations, where term-days of intense systematic measurement were kept at observatories around the world, using moderately although not wholly standardized apparatus.[1] Von Humboldt's ideal of a worldwide survey was more nearly realized in magnetism than in other realms of terrestrial science: magnetic observations in the polar regions were especially important.

There were other signs of cooperation. Bessels had rushed to publish the results of the United States North Polar expedition of 1871–3, so that the impending British Arctic Expedition under Nares could make use of this data. Similarly, the United States had made food caches from the *Polaris* expedition available to the British.

The Netherlands had been active in exploring the Arctic to the east of Greenland in the sixteenth and seventeenth centuries, and a revival of their interest in the north was heralded by the foundation of the Dutch Geographical Society in 1873. A Dutch officer, Laurens Beijnen, sailed in *Pandora* on a private British expedition to the Arctic, where science was very

---

[1] A brief account of later observations is G. A. Good, "The Study of Geomagnetism in the Late 19th Century," *Eos* 69 (1988) 218–28.

307

much a sideline. With that experience, Beijnen campaigned for Dutch scientific research in the Arctic. A Dutch expedition sailed to the eastern Arctic in 1878. Its mandate was scientific, and its sounding apparatus came from the *Challenger*'s equipment, which had been lent by the British government.[2] The expedition took an English photographer, received advice from the British hydrographer Evans about apparatus for physical oceanography, and sent an officer to the Meteorological Office,[3] and to Kew Observatory for instruction in the use of magnetic instruments. Their research program was based on plans devised by Austrian officers. Here was a national expedition with an international flavor.[4]

In the same manner, Nordenskiöld's results had been important for Nares's expedition in 1875–6. When Nordenskiöld announced his plans to tackle the Northeast Passage, the president of the Royal Geographical Society proposed that Britain assist the Swedes.[5]

## THE UNITED STATES, SCIENCE, AND POLAR COLONIZATION

International cooperation was on the rise in the Arctic. So, however, were what looked like territorial ambitions, buttressed by the use of science as a state activity. The United States began to consider an expedition of which the ultimate objective was the establishment of a permanent scientific colony[6] at Lady Franklin Bay on Ellesmere Island, along the "American route"; initial proposals were, however, merely for a temporary colony. Plans were drawn up by Captain H. W. Howgate of the U.S. Army. An important point in the debate was the reversal of the traditional priority of geographic over scientific work:

> Geographic discovery has hitherto been the objective point. . . . An absolute change of operations must be had. [What is needed is] a long stay, which will give ample opportunity for observations and the conduct of scientific enquiry under the most

2  PRO Adm. 1 6455.
3  The Meteorological Office kept comprehensive records, as did the Admiralty: see *Contributions to our Knowledge of the Meteorology of the Arctic Regions*, vol. 1 (London, HMSO, 1885), listing the observations of thirty-six expeditions since 1819.
4  W. F. J. Mörzer Bruyns, "The Dutch in the Arctic in the Late 19th Century," *Polar Record* 23 (1986) 15–26.
5  PRO Adm. 1 6514, 11 Feb. 1879. Tore Frängsmyr, "Swedish Polar Exploration," in T. Frängsmyr, ed., *Science in Sweden: The Royal Swedish Academy of Sciences 1739–1989* (Canton, Mass., 1989) pp. 177–98.
6  Cooke and Holland, *The Exploration of Northern Canada* (Toronto, 1978) p. 242.

favorable conditions. . . . Ordinarily, the expeditions have been so conducted as to actually preclude scientific discovery – all appliances left at home, and almost continuous locomotion.[7]

Given such prospects, the scientific community was understandably keen on the expedition. Joseph Henry wrote in support of it, urging the adoption of a program of pendulum experiments, geomagnetic observations, studies of tides and winds, and the collection of natural history specimens. Henry was in favor of a colony that would endure for several years. If Congress voted funds, the National Academy would provide scientific instructions.[8]

In like vein, Elias Loomis of Yale wrote enthusiastically about the importance of a study of polar phenomena for every question of terrestrial physics. "If the information which has been acquired upon the various subjects in the numerous Polar expeditions of the last half century were annihilated, it would leave an immense chasm which would greatly impair the value of the researches which have been made in other parts of the world." Commercial benefits had accrued, and more could be expected, from studies of weather, tides, and geomagnetism: "we may confidently anticipate that any advance in our scientific knowledge respecting . . . the physics of the globe will impart increased security to commerce." A whaling fleet of a dozen vessels had been wrecked in the Arctic in 1876, with property worth $500,000 destroyed, "all because of a lack of proper knowledge of climatic and tidal observations."[9] As the Cincinnati Chamber of Commerce put it, science, commerce, and trade were "inseparably linked together."

R. W. D. Bryan of the U.S. Naval Observatory was "opposed to all spasmodic efforts to reach the Pole," but very much in favor of sustained observations and explorations from a fixed base. Scientific, commercial, and military opinion coincided in supporting Howgate's proposal.[10]

In the first stage of the expedition, George Tyson, who was widely experienced in whaling and polar navigation,[11] wintered on Baffin Island, collecting supplies and hiring Inuit, superintending scientific observations and collections, and engaging in whaling to support the voyage. Pelagic

[7]  House of Representatives, *Report No. 181: Expedition to the Arctic Seas*, Washington, D.C., 22 Feb. 1877.

[8]  *Proposed Legislation, Correspondence, and Action of Scientific and Commercial Associations in Reference to Polar Colonization* (Washington, D.C., 1877).

[9]  House of Representatives, *Report No. 181*, 22 Feb. 1877.

[10]  *Proposed Legislation* (1877). See also Senate Report No. 94, 13 Feb. 1878, and PRO Adm. 1 6455 and 6456.

[11]  Tyson had been with Charles Francis Hall on his last polar expedition, and led the party that drifted south on an ice floe in 1872–3, and were kept alive thanks to Ebierbing and Hendrik, and were rescued off the Labrador coast: see chapter 6.

birds were caught and skinned, meteorological records were kept, and altogether, "[d]uring the winter the scientific work was carried steadily on,
never stopping day or night.[12] . . . Mr. Kumlien[13] employed his time on the
seals and on other animal life."[14] Kumlien did far more than zoological
work; as Spencer Fullerton Baird, secretary of the Smithsonian, remarked,
Kumlien's ethnological data offered "one of the most complete and finished
descriptions ever given of" the Inuit.[15] This efficient beginning was also the
end. Tyson learned in August 1878 that the main expedition had been deferred. It never took place,[16] although the United States did not give up the
idea of polar colonization. Thus by the late 1870s, two models of polar science were in play, one based on international cooperation, the other based
on national colonization. As it turned out, these models were not necessarily mutually exclusive, but the emphasis was increasingly on the former.
When proposals for the International Polar Year were made to the United
States, Congress was already convinced and was able to move rapidly in
sending out an expedition along the lines proposed by Howgate.


PRELUDE TO THE INTERNATIONAL POLAR YEAR


*Weyprecht and company.* Plans for the IPY derive especially from the
activities of three figures: the geographer August Petermann, the geophysicist and naval officer Georg von Neumayer, and above all Karl
Weyprecht, a lieutenant in the Austro-Hungarian Navy, and a geophysicist
and explorer.[17] Petermann had urged further exploration of the region

12    Tyson credited Orray Taft Sherman for this activity, which resulted in his *Meteoro-
      logical and Physical Observations on the East Coast of British America* (Washington,
      D.C., 1883).
13    Ludwig Kumlien, "Contributions to the Natural History of Arctic America, Made in Con-
      nection with the Howgate Polar Expedition, 1877–78," *Bulletin of the United States Na-
      tional Museum*, No. 15 (Washington, D.C., 1879). Kumlien wrote reports on mammals,
      birds, and ethnology; twelve other fields were handled through the National Museum.
14    George E. Tyson papers, expedition notes, diaries, journals 1871–8, National Archives
      (Washington, D.C.) RG 401 (33^A). See also G. E. Tyson, *The Cruise of the Florence, or,
      Extracts from the Journal of the Preliminary Arctic Expedition of 1877–78*, H. W. How-
      gate, ed. (Washington, D.C., 1879).
15    Quoted in Caswell, *Arctic Frontiers* (Norman, Oklahoma, 1956) p. 93.
16    The end of the story for Howgate was his arrest in 1881 for defrauding the Signal Service
      of large sums.
17    For Petermann, see chap. 7. J. Georgi, "Georg von Neumayer (1826 bis 1909) und das 1:
      Internationale Polarjahr 1882/83," *Deutscher hydrographische Zeitschrift 17* (1964)
      249–71; W. Kertz, "Georg von Neumayer und die Polarforschung," *Polarforschung 53*
      (1983) 91–8. H. Wild, "History of Weyprecht's Proposal for International Polar Scientific

Figure 31. Karl Weyprecht. University of Toronto Library.

between Spitsbergen and Novaya Zemlya. Such advice contributed to the creation of the Austro-Hungarian North Pole Expedition of 1872–4. Weyprecht was one of its leaders; he returned from the expedition convinced that there must be a change in the style of polar research. Instead of geographic competition, what was needed was scientific cooperation. Instead of geographic exploration, systematic coordinated observations were needed with each expedition staying in place in the Arctic for a year: no longer could exploration allow "the somewhat too partially topographical perception of its aims."[18] Weyprecht's first announcement of his hopes for polar exploration was made before the Academy of Sciences in Vienna in January 1875, while Nares was putting his expedition together.

Meanwhile, quite independently, Neumayer had been setting out ideas for the geophysical exploration of the Antarctic. As director of the German naval observatory, and since 1872 a member of the Hydrographic Office of the Imperial Admirality in Berlin, he was an influential figure. He had been advocating polar work for several years, including southern observations of the next transit of Venus. In 1874 he published a statement about geographical problems in the polar regions.[19] He was seeking comprehensive coordinated information, very much of the kind sought by Weyprecht.

Eighteen seventy-four was a busy year for making plans for polar exploration. In that year, the Bremen Association for the German North Polar Passage applied for an Imperial grant in support of another expedi-

Research," *Mittheilung der Internationalen Polar-Commission* (St. Petersburg, 1882) part 1:1–12; E. Ihne, "Carl Weyprecht, der Nordpolarforscher: Ein Beitrag zur Geschichte der Polarforschung," *Archiv für die Geschichte der Naturwissenschaften und der Technik 5* (1913) 1–29 and *Carl Weyprecht der österreichische Nordpolfahrer: Erinnerungen und Briefe gesammelt und zusammengestellt von Heinrich von Littrow* (Wien, Pest, Leipzig, 1881).

[18]  J. Payer and K. Weyprecht, "The Austro-Hungarian Polar Expedition of 1872–74," *Journal of the Royal Geographical Society*, 45 (1875), 1–33; W. Barr, *The Expeditions of the First International Polar Year: The Arctic Institute of North America. Technical Paper No. 29* (Calgary, 1985); N. H. de V. Heathcote and A. Armitage, "The First International Polar Year," in *Annals of the International Geophysical Year 1957–1958*, vol. 1, *The Histories of the International Polar Years and the Inception and Development of the International Geophysical Year* (London, 1959) pp. 6–104; F. W. G. Baker, "The First International Polar Year, 1882–83," *Polar Record 21* (1982) 275–85; Weyprecht, *Die Nordpol-Expeditionen der Zukunft und deren sicheres Ergebniss, vergleichen mit den bisherigen Forschungen auf dem arktischen Gebiete* (Leipzig, 1876). The quotation is from Weyprecht's letter to General Albert Myer, Chief of the U.S. Signal Office, Trieste, 20 May 1879, National Archives (Washington, D.C.) RG 27, Records of the U.S. Weather Bureau, Meteorological Correspondence of the Signal Office, Misc. [1879] #1349.

[19]  Neumayer, "Die geographischen Probleme innerhalb der Polarzonen in ihrem inneren Zussamenhange beleuchtet," *Hydrographische Mittheilungen: Herausgegeben von dem Hydrographischen Bureau der Kaiserlichen Admiralität 2* (1874) 51–3, 63–8, 75–82.

tion. The response was to temporize, and to refer the question to a commission, which reported in Berlin in October. The commission was a distinguished one, socially and scientifically.[20] They advised against any support for the Bremen proposal, precisely on the grounds that Weyprecht was advocating. They were against isolated expeditions. They were, however, in favor of a comprehensive scheme involving a circle of stations, established as far as possible continuously around the North Pole, and pursuing a carefully coordinated program of observation and research. They made not merely administrative, but also scientific proposals: thirty-one closely printed pages of them. They identified questions in meteorology, hydrography – including what became glaciology – magnetism and electricity (both geophysical and auroral), geodesy, geography, botany, zoology, and anthropology. They made it clear that the goal was to understand land, sea, and air. In hydrography, for example, they required observations from the surface of the sea to its bottom, noting temperature, density and salt content, substances present, currents, the physical, chemical, and geological properties of bottom sediments, and such special phenomena as the plasticity of ice. They were also concerned with the details of marine biology, specifying observations ranging from microscopic organisms to whales. They paid attention to instrumentation and to the current state of theory, and altogether did their job admirably. The report[21] reads more professionally than the advice put together by the Royal Society of London for the Nares expedition. Clearly Weyprecht's ideas had found swift support within the German Empire.

The commission adopted Weyprecht's notion of international cooperation, and recommended that the government approach other governments with arctic interests to explore the feasibility of an international polar year. The Imperial government acted on the report, and initially letters were sent from the Chancery Office of the empire to the British Foreign Office, and also to the governments of Sweden, Norway, Russia, and the United States. The British response was to propose consulting the principal British scientific bodies, but in any case to wait until the British Arctic Expedition under Nares returned, which was expected to be in 1877.[22] It would not be unreasonable to conclude that there was a lack of urgency and of enthusiasm among the Lords of the Admiralty.

---

[20] Among its fourteen members were Baron von Richthoven, the physicist and electrical engineer W. Siemens, and Neumayer.

[21] *Bericht der Kommission zur Begutachtung von Fragen der Polarforschung* (Berlin, 1875).

[22] PRO Adm. 1 6392.

Weyprecht, undeterred or perhaps unaware of British tepidity and of American nationalism,[23] had addressed the Association of German Naturalists and Physicists at Gratz, and had received the vigorous support of his friend Count Wilczek, who had financed his previous expedition, and now committed himself to support a station in Novaya Zemlya for a year.[24] Weyprecht and Wilczek prepared detailed proposals to be submitted to the International Meteorological Congress scheduled for Rome in 1877 but delayed until 1879 by war in the Balkans.

The Meteorological Congress was enthusiastic, and, as a result, the first International Polar Congress was held in Hamburg in October, with Neumayer as the opening chairman.[25] Issues to be decided were the number of observations and the locations of stations; when to start, and how long to continue; and what instruments were to be used for the observations, since standardized instrumentation was necessary for the coherence of the project. Most delegates at Hamburg lacked governmental authority, but thanks to Wilczek, Austria could promise the Novaya Zemlya expedition. There was a possibility that Russia might establish a station at the mouth of the Lena. Prussia talked of the East coast of Greenland, and of one or two antarctic expeditions. Norway proposed to tackle Finmarken (its northernmost region), Denmark West Greenland, America Cape Barrow in Alaska, Sweden hoped for a station at Spitsbergen, and France, while making no promises, contemplated an expedition to the Antarctic. Conspicuous by its absence was Great Britain, which, together with Portugal, sent apologies rather than delegates.

That winter, the Swedish and Norwegian minister in London inquired of the British government how far it wanted to support the project. The British response, which was either disingenuous or remarkably inefficient, was that this was the first Her Majesty's government had heard of the idea of an international polar year, that it therefore had no plans to participate, and that there was no estimate of cost; the Admiralty could therefore not consider the proposal, and thoughts of cooperation were therefore premature.[26] It would in any case be necessary to consult the Royal Society of London to see if they encouraged the project. One official noted in an attached memorandum that spokesmen for the Royal Society needed to be handled with care because they were likely to encourage anything that

[23]   The proposed Howgate expedition would share results with other nations, but the scheme of polar colonization was resolutely American.
[24]   Heathcote and Armitage, "First International Polar Year."
[25]   H. Wild, ed., *Mittheilungen der internationalen Polar-Commission*, pt. 1: 2.
[26]   PRO Adm. 1 6507, 20 Dec 1879, Swedish and Norwegian Minister in London.

would advance science, without much regard for cost. Perhaps part of *their* funds might be used for the purpose.[27] Dickens's Circumlocution Office was alive and well.

A different kind of opposition came from Sir Clements Robert Markham, die-hard supporter of old-style exploration. He wrote indignantly:

I look upon Weyprecht's scheme as unpractical and of course most injurious to geographical research. He wants men to sit down for a course of years to register observations at one spot, and not to explore. Such a scheme could only have been proposed by the most unpractical of specialists. But I must say that it is rather cool of some of the newspapers to ask the Geographical Society to advocate a scheme which is avowedly opposed to its main object, geographical discovery.[28]

Markham's opposition was rear-guard action. Weyprecht's scheme won increasing international support. It had been determined at the Hamburg Conference in 1879 that commitments for a minimum of eight arctic stations were needed for success. By August 1880, when the International Polar Commission met in Bern, Austria, Denmark, Norway, and Russia had engaged themselves. At their next meeting, in St. Petersburg in August 1881, detailed plans were made.[29]

AMERICAN, BRITISH, AND GERMAN EXPEDITIONS

Fourteen major stations were set up for the year's observations. Two were in the southern hemisphere, a German station in South Georgia and a French one near Cape Horn. The other twelve stations were distributed around the North Pole, at Point Barrow (United States), the mouth of the Lena (Russia), Novaya Zemlya (Russia), Spitsbergen (Sweden), Bossekop on the Norwegian arctic coast (Norway), Sodankylä in Lapland (Finland), Jan Mayen (Austria), Godthaab on the west coast of Greenland (Denmark), Fort Rae in the North-West Territories (Britain), Lady Franklin Bay (United States), and Kingua Fiord on Baffin Island (Germany). The last two just listed, or with a slight geographical stretch, the last three, fell within the Canadian Arctic. The Dutch set out for Dikson at the mouth of the Yenisey on the Russian arctic shore, but were trapped in the ice, and drifted for a year. There were also a number of auxiliary expeditions both north and south.[30]

[27]   PRO Adm. 1 6507, 23 Dec 1879.
[28]   C. R. Markham to Robert Brown, 7 Dec 1880, SPRI MS 441/9/20.
[29]   H. Wild, ed., *Mittheilungen der internationalen Polar-Commission*, pt. 1: 6–13 states the program for the IPY expeditions.
[30]   Barr, *Expeditions*, pp. 4–5; *Nature* (1882) 294–7; (1883) 423–4.

Here I shall consider only major expeditions within the geographical limits of the Canadian Arctic.[31]

*The Lady Franklin Bay expedition.*[32] When the British Admiralty was informed of American plans for a station at Lady Franklin Bay, the hydrographer, Captain Sir Frederick Evans, was unimpressed: "This . . . appears to be a renewal of the Howgate Expedition of 1880 . . . which was unsuccessful. There is now engrafted on the Howgate Expedition, the taking part in a scheme (not yet matured) for various nations to found stations in the Arctic Circle" for meteorological and magnetic observations. Evans was doubtful of success. Indeed, looking at all the proposals, he was convinced that "the time available, the money voted, the means proposed, all appear equally inadequate for the contemplated purpose."[33]

The hydrographer's response was in line with Britain's lack of enthusiasm for the IPY in general. In fact, the American proposal did combine the aims of the defunct Howgate expedition with those of the IPY, thus conferring on America's scheme for polar colonization the dignity of contributing to an international scientific enterprise. The expedition was commanded by Second Lieutenant Adolphus Greely, an acting signals officer in the cavalry, who had volunteered, and had been appointed with presidential approval. There were to be two other officers, one surgeon, one photographer, one astronomer, and nineteen enlisted men, as well as Inuit hunters and dog drivers. One of the officers would take charge of exploration, while the other would have charge of the scientific work: "One commissioned officer and three enlisted men are expected from the Signal Office, to form, with the astronomer, the scientific corps of the expedition, and to work under instruction from the Chief Signal Officer of the Army." The surgeon would double as naturalist.[34]

---

[31]  For the other expeditions, see Barr, *Expeditions*, and Heathcote and Armitage, "First International Polar Year," as well as the separate national reports of the contributing expeditions.

[32]  There is a considerable literature about this expedition. See A. L. Todd, *Abandoned: The Story of the Greely Arctic Expedition 1881–1884* (New York, 1961); T. Powell, *The Long Rescue: The Story of the Tragic Greely Expedition* (London, 1961). A good and brief recent account is in Berton's *Arctic Grail* (Toronto, 1988) pp. 435–86. Adolphus Greely's own account is *Three Years of Arctic Service: An Account of the Lady Franklin Bay Expedition of 1881–84 and the Attainment of the Farthest North*, 2 vols. (London, 1886). His formal report is *International Polar Expedition: Report of the Proceedings of the United States Expedition to Lady Franklin Bay, Grinnell Land, House of Representatives Miscellaneous Document No. 393*, 2 vols. (Washington, D.C., 1888).

[33]  PRO MSS Admirality 1/6593, 17 May 1881.

[34]  Memorandum, Washington, D.C., 27 April 1880, Library of Congress, A. W. Greely Pa-

They were to be taken by steamer to Lady Franklin Bay, there to be left to establish a permanent station near the coal seam discovered during Nares's expedition. It was intended that the steamer should return to Lady Franklin Bay in 1882 and 1883 with supplies. If ice prevented the supply vessel from reaching the station, that vessel would stay in Smith Sound as long as ice conditions permitted, and before leaving would leave a cache of its supplies on Littleton Island, together with a relief party. Greely was to move south by September 1883.[35]

The scientific instructions[36] were those issued by the International Polar Conference at Hamburg in 1879. Meteorological observations were to include air and sea temperatures, barometric pressure, humidity, wind, clouds, and precipitation.[37]

Comprehensive magnetic observations would be made; variation would be measured using instruments[38] with small needles, in contrast to the large magnetic needles or bars used in Gauss's apparatus. The conference had recommended hourly observations of variation, with readings made just before and just after the hour; Weyprecht had indicated that this was not good enough, and proposed that such measurements be made at two and one minutes before and after the hour, as well as on the hour. Greely was instructed to adopt Weyprecht's proposal, and to follow Göttingen time in doing so. Auroral observations would be made simultaneously with the magnetic ones.

There was also a range of elective magnetic and meteorological observations, and observations in hydrography, oceanography, natural history and geology.

The expedition looked promising, with clear scientific instructions, seemingly generous material resources, and well-considered plans for supplies and relief. For a change, the expedition would be an army rather than a navy one, something that seemed sensible in view of the plans for a protracted stay at a "permanent" station on land. Much of their establishment

---

pers Box 72. This represents a shift from the proposal of 12 April, printed in Greely, *International Polar Expedition*, vol. 1, p. 98.

[35] Ibid., pp. 97–100.          [36] Ibid., vol. 1, pp. 100–6.

[37] Meteorological apparatus is listed in ibid., p. 107, together with other apparatus (geographical, astronomical, magnetic, and pendulum apparatus; deep-sea sounding apparatus "will be left to the United States Coast Survey"). See also Signal Office, War Department, *Signal Service Notes No. X*, Ernest A. Garlington, *Report on Lady Franklin Bay Expedition of 1883* (Washington, D.C., 1883), a copy in Library of Congress, A. W. Greely Papers Box 71 Folder 4.

[38] The magnetic apparatus listed (Greely, *International Polar Expedition*, vol. 1 p. 107), was: "One complete magnetometer – Fauth & Co. – unifilar declinometer – catalogue No. 70, price $400, extra light needles and mirror for auroral disturbances."

was prefabricated, for assembly on arrival. For example, the magnetic observatory, a wooden building secured with copper nails, went with them on the steamer.

In June 1881 they set out from New York in the sealing steamship *Proteus*, made a good passage, and arrived at Lady Franklin Bay in August. The ship returned home, leaving the expedition's members to build Fort Conger at Discovery Harbour. Greely was humorless, insecure, and a rigid disciplinarian, a combination that was guaranteed to create friction, and likely to lead to disaster in the protracted close-quartered isolation of two arctic winters. One of his lieutenants, Kislingbury, had an exchange with Greely leading to his resignation when it was too late to rejoin *Proteus* on its departure from Fort Conger. It took months before they could work together. More serious was the clash between Greely and the expedition's surgeon, Octave Pavy,[39] a civilian who had joined the ship at Greenland and who saw no reason to accept military discipline.

In spring there were sledging expeditions that achieved new records to north, west, and east. Back at the base, scientific observations and collections went ahead, sometimes with more energy than exactitude. On 10 June Greely noted that Jens, one of their Greenland Inuit, shot a bird "which is either a knot *tringa canutus* or [a purple sandpiper] *tringa maritima* . . . Jens who knows the bird well gives it the eskimo name for the latter while Dr. Pavy thinks it is the former . . . Later this bird proved to be a knot; wherever the purple sandpiper is spoken of 'knot' should be substituted in this year's journal."[40] The bird was presumably a knot, because they were just beyond the northern limit of the purple sandpiper's range, and squarely within the knot's range.

Until June 1882, magnetic observations were made only on term-days by Sergeant Edward Israel of the Signal Corps, the expedition's astronomer and magnetic observer, who had received training before joining *Proteus* en route in Newfoundland.[41] That meant that the expedition had been collecting data for almost a year before the International Polar Year began. The American expedition to Lady Franklin Bay was early because it was planned and approved by Congress before it was incorporated in the scien-

---

[39]  See, e.g., Greely, 24 Sept. 1882, Library of Congress, Greely papers, Military Papers, Box 69, Journal 1 May 1882 – 28 Feb. 1883. Pavy was also on hostile terms with Greely's other lieutenant, Lockwood: see Greely, Journal, 10 June 1882. Pavy found Greely vain and authoritarian: see Pavy, notes, National Archives (Washington, D.C.) NARS RG 27 Box 12.

[40]  Greely Journal, 10 June 1882.

[41]  Greely, *International Polar Expedition*, vol. 2 pp. 480–1.

tific programs of the IPY. Overall plans for the IPY had aimed for observations in 1881–2, but the International Polar Commission had found it necessary to postpone the concerted fieldwork for a year, "in order that more time might be given to the instrumental outfit and better organization of the various parties."[42] So the Americans were there first, with a relatively light schedule of magnetic observations until the IPY began. At that point, far more frequent observations were required. These began on 19 June; and since the program was too demanding for a single observer, one of the officers, Lieutenant Lockwood, learned to carry out the routine magnetic and meteorological observations.[43]

In November, a spectacular auroral display was accompanied by marked magnetic disturbances. The event was particularly striking since the expedition was too far north for frequent impressive auroras. Greely ordered observations every five minutes over the twenty-nine hour period that the disturbances lasted, with the observers taking two-hour shifts.

A relief expedition was sent out in 1882, but failed to reach Greely. The expedition survived a second winter, but Greely was increasingly depressed and apprehensive. He confided to his journal:

1 March 1883. The first day of Spring has come and with it a sense of relief that a second winter is over and none of the party in bad health or injured condition. The experience of other expeditions, the forebodings of my Doctor and the knowledge that no party had ever passed a second winter in such high latitudes, were all of such a character as to give me a deal of uneasiness. . . . It cannot be expected that I should feel ever again at ease until a ship is once more sighted. 15 March. If no vessel comes I will consider our chances desparate [sic]. If none come God help us.
26 March. Arrangements are being gradually made to retreat southward by boats early in August in case no vessel arrives.

Meanwhile, there was the business of science. Israel kept the magnetic work going. Pavy, supposedly doubling as surgeon and naturalist, had withdrawn from any service under Greely, while volunteering to continue his medical activities. Pavy was now only desultorily a naturalist.[44] Indeed, Pavy seemed increasingly to limit his activities to sleeping, grumbling, reading novels, and generally avoiding work. Greely complained that Pavy had made little or no effort to make collections "by dredging &c.," and decided to take on some of the naturalist's tasks himself. Marine invertebrates soon became his principal recreation: "23 April. Spent the forenoon at the tidal

[42] Charles A. Schott in ibid., p. 479.  [43] Greely Journal, 19 June and 19 Sept. 1882.
[44] Pavy notes that he submitted no official journal as a naturalist in part because "the scientific question was outside of my sphere" (NARS RG 27 Lady Franklin Bay Expedition Box 12).

hole. . . . I thought I would try and collect some medusae for the expedition. I was fortunate enough to catch after nearly half an hours patient waiting, an excellent specimen of *Ptychogastria polaris*," a small jellyfish of which only one specimen had previously been noted, by Feilden during Nares's expedition. Greely then caught three other kinds, and Lockwood made sketches of them.

Greely's pleasure in these creatures is evident in his descriptions, for example this one of 24 April:

I caught today a very large Medusa – 5″ long × 2½″ wide. . . . It was of the most delicate character . . . It had two spots of smoke color at the upper end which end was pointed like a melon. Indeed the shape was melon except that the lower end was as you may say, cut off. There were 8 ribs which were of smoke color and which as far as the lower end went were formed of a succession of annular formations which present a serrated appearance on either side. There appeared to be two large stomachs. Occasionally from the tentacles, iridescent colors with purple predominating were seen.

Medusae continued to fascinate him, and to give him respite from the pressures of command and the fear of a southward retreat without relief. He discovered here a new species, there a fine specimen of one already known. He made fascinated observations on their strange life cycles, and had Lockwood and others make fine drawings of these fragile creatures. He did assert himself so far as to order Pavy to submit a report on all the specimens that he had, and to prepare six sets of plants to carry on a retreat. On 22 May Pavy gave Greely his flora "in an unsatisfactory condition"; at the end of the month Pavy had still not submitted his report, and Greely finally and formally relieved him of the duties of naturalist, duties that he had in any case not been carrying out for some time. Lockwood, better inclined but worse prepared, took over, only to find that he had inherited chaos. There were, for example, twenty snowy owls killed, but none prepared as specimens. On 19 July, Greely informed Pavy that he was under arrest, and would be court-martialed – a doubtful prospect for a civilian.

On 9 August 1883, with no relief at hand, Greely decided to abandon the station, and to take to the boats. They got as far as Cape Sabine, where they passed a horrendous winter. There were deaths from famine and scurvy, suicide, and an execution of one man for stealing food; and there was cannibalism. Seven survivors, Greely among them, were rescued in June 1884; one of them died on the homeward journey. Greely's expedition had been the most tragic and disastrous since John Franklin's.

But unlike Franklin's, there were survivors, who brought with them the results of more than two years of observations in a broad range of sciences.[45] Ironically, the worst disaster in almost half a century was one of the more successful in science.[46] The most valuable material, occupying most of the second volume of the official report, was in astronomy and geophysics.[47]

The astronomical observations[48] were primarily for navigation and determination of time, latitude and longitude. Sergeant Israel, as the expedition's astronomer, was responsible for most of the observations, but a good deal of work was also done by Lockwood.

Meteorological observations were the most extensive of all those carried out during the expedition, and were the responsibility of three sergeants.[49]

Israel's magnetic observations[50] had been most thorough. His original records were in too heavy a volume for the expedition's retreat by boat, so Greely first reduced readings showing extreme oscillations of the needle to a mean point on the scale. The magnetometer and dip circle were also abandoned at Fort Conger, but Greely brought back "two magnets used in the magnetometer, also two of the dip needles," which enabled them to be checked back in Washington. Israel himself died in the expedition's final grim winter.

Tidal and pendulum observations completed the systematic and rigorous roster of geophysical work.

Because of Pavy's negligence, the incompetence of the other members of the expedition in natural history, and the loss of specimens during the retreat, the zoological results of the expedition were less impressive. Observations of mammals were the most reliable, and essentially coincided with those made on Nares's expedition.

No one except Pavy was able to identify more than a handful of species of arctic plants, but Greely worked hard at collecting specimens during the second summer, so that more than sixty species were brought back, several of them not found by Feilden and Hart. The collection of mosses was thorough; as a result of the exertions of Greely and two of his men, sergeants Brainard and Jewell, "fifty-eight species of mosses have been found north of latitude 81°30', of which seven species were found only by

---

[45]  A summary is given in Caswell, *Arctic Frontiers*, pp. 96–113.
[46]  These are principally given in vol. 2 of Greely's *International Polar Expedition*.
[47]  Ibid., vol. 2. See also W. H. Lamar and F. W. Ellis, *Physical Observations during the Lady Franklin Bay Expedition of 1883*, U.S. Army, Signal Corps, Signal Service Notes No. 14.
[48]  Greely, *International Polar Expedition*, vol. 2 pp. 59 et seq.
[49]  Ibid., vol. 2 pp. 91–453, followed (pp. 455–72) by a very thorough list of "Authorities on Arctic Meteorology."
[50]  Ibid., "Contributions to Terrestrial Magnetism," pp. 475–635.

Major Feilden, four by Mr. Hart, thirty-one by the Lady Franklin Bay Expedition . . . , and sixteen by two or more of these parties."[51]

The ornithological results were meager.[52] In 1883, Pavy had handed over his collection with a list of observations, and in spring and summer there were always hunters in the field, ensuring observation and collection of "all birds within reach." Greely found himself having to identify these birds from descriptions given in the appendixes of earlier arctic narratives, and his own journal was the main source of ornithological data, "a fact which is unfortunate, as neither personal taste nor scientific work has ever turned my attention to this or kindred subjects."[53] Greely was able to prepare notes on thirty-two species. Before the expedition, Sergeant Joseph Elison had gone to the Smithsonian Institution where Professor Baird had instructed him in the preparation of ornithological specimens, and Elison had prepared many specimens at Fort Conger, only to have to leave them there when they abandoned the fort. "The main point of ornithological interest," according to Greely, "rests in an identified egg of the knot . . . which was obtained from the bird itself. Unfortunately for ornithologists too much care was taken to ensure its safety, and the egg was packed with other specimens weeks before the retreat by boats and so remains at Conger."[54]

Greely's collection and observation of medusae was more productive, thanks largely to the careful drawings made by Gardiner and Lockwood. There were the expected species, as well as a probable new one, *Nauphanta polaris*. Greely's collecting through the hole in the ice for the tide gauge also resulted in a modest haul of other marine invertebrates.[55]

*Fort Rae, with a note on the aurora borealis.* The Lady Franklin Bay expedition had been to territory that now at least in theory was controlled by Canada. The Dominion had acquired an increased stake in the high Arctic in 1880, when the islands there had been transferred from Britain to Canada. The stake was vast, but also imprecise, since the extent of the archipelago was not known, and parts of it had first been mapped by citizens and ships of other nations.[56] The order in council, published in the *Canada Ga-*

[51]  Ibid., vol. 2 p. 18.
[52]  Greely, *International Polar Expedition,* vol. 1 appendix 91, "Inventory of Collections in Natural History"; vol. 2 appendix 131, "Ornithology." The latter appendix contains a useful list of arrival and departure dates for species observed at numerous locations by expeditions around the pole.
[53]  Ibid., vol. 2 p. 19.        [54]  Ibid.        [55]  Ibid., pp. 39–58.
[56]  M. Zaslow, ed., *A Century of Canada's Arctic Islands 1880–1980* (Ottawa, 1981); Zaslow, *The Opening of the Canadian North 1870–1914* (Toronto, 1971) pp. 251–55; G. W. Smith, "The Transfer of Arctic Territories from Great Britain to Canada in 1880,

*zette* on 9 October 1880, referred with deliberate vagueness to "all British Territories and Possessions in North America, not already included within the Dominion of Canada, and all Islands adjacent to any of such Territories or Possessions," with the exception of the crown colony of Newfoundland.

This transfer meant that Canada was truly an arctic nation, and so it was a proper recipient of invitations to take part in the IPY. The physicist H. Wild of St. Petersburg, Neumayer's successor as chairman of the Polar Commission, wrote to Charles Carpmael, the new director of the Canadian Meteorological Service, inviting Canada to contribute to the IPY by manning a station somewhere in the archipelago. Carpmael transmitted this request, with his endorsement, to the prime minister of Canada, Sir John A. Macdonald,[57] but there was in the event no British or Canadian station in the Canadian archipelago. Instead, there was Greely's expedition, and a German expedition[58] to Kingua Fiord on Baffin Island.

British and Canadian participation in the IPY could charitably be called modest. Canada, still a relatively new country, had no focus for lobbying government on behalf of science. The Royal Society of Canada was not founded until 1882, too late to have any voice in the IPY. The most successful and sustained scientific enterprise in Canada in the nineteenth century was its Geological Survey, and the government had an interest in any science that bore on settlement, development, and the exploitation of natural resources.[59] The IPY, on the other hand, concerned with the polar regions, was primarily devoted to the study of geophysical phenomena, and although meteorology was important for agriculture and settlement, and for the IPY, it was hard for politicians in the Canadian South to see how their interests could be served by meteorological and geophysical observations in the polar regions. The arctic archipelago was about as far from Ottawa as London was, and a great deal less accessible. Canada lacked the notions of glory associated with Royal Naval expeditions, and even the Hudson's Bay Company, although its activities ranged far into the North,

and some Related Matters, as seen in Official Correspondence," *Arctic* 26 (1973) 53–73. The immediate spur came in 1874 from two applications for grants of land in Baffin Island, one from an American for mining, the other from an Englishman to establish a whaling station. Inaction on the part of British officials encouraged the American to remove a cargo of mica and graphite in 1876. The Canadian government wanted legislation identifying the islands as part of Canada, and got it in 1880.

57  PAC MG 26A vol. 311 ff. 141300–141309.
58  See notes 70–87.
59  Zeller, *Inventing Canada* (Toronto, 1987) part 1, and Zaslow, *Reading the Rocks: The Story of the Geological Survey of Canada* (Ottawa, 1985) give complementary accounts of the history of the GSC.

had no interest in the pole and little in geophysics.[60] The British government's view of the IPY was apparent in its reluctant response to the first German initiatives, and in the hydrographer's dismissal of the proposals for Greely's expedition. The perceived failures of Nares's expedition were no encouragement, and the British government, unlike British scientists, was not particularly keen on international cooperation. Britain decided at the last minute to send a very modest expedition of four men as its contribution; Canada would merely help financially.

Robert Scott, Secretary of the Meteorological Council of the Royal Society, had been a signatory to Wild's letter in 1879 inviting national meteorological bodies to collaborate in the IPY. Perhaps because of the British government's clear lack of enthusiasm, Scott was not invited to join the International Polar Commission as the British representative until March 1882. Britain's commitment was made at the beginning of April, weeks before the start of the expeditions. The government, with advice from a committee of the Royal Society,[61] decided to establish a station at Fort Rae, at the northern exit of Great Slave Lake. Fort Rae was a Hudson's Bay Company post, the one nearest to the magnetic pole, and the most northerly one from which the party of observers could return at the end of the period of observation, before the rivers froze up and compelled them to spend a winter at the fort. The Royal Society of London asked for Canadian help in transporting one officer and three men to Great Slave Lake and back; Ottawa made available $4,000 for the purpose.[62]

Since this was to be a land-based operation, with no maritime component, the Royal Artillery was the obvious source of personnel, given its historic role in nineteenth-century geomagnetism. The leader appointed was Captain Henry Dawson R.A., assisted by an artillery gunner and two sergeants of the Royal Horse Artillery. They sailed from Liverpool to Quebec, thence via the Great Lakes to Winnipeg, and from there by steamer, canoe, and york boat to Fort Rae.

The time for the preparation of the expedition had been so brief – less than six weeks – that there was no possibility of having instruments spe-

---

60   The HBC did, here and there, make meteorological observations, and it was an important resource in the collection of natural history specimens for southern museums. See chap. 9.
61   The committee consisted of Dr. John Rae, Admiral Richards (hydrographer of the navy from 1863 until 1874), Robert Scott, and the officers of the Royal Society.
62   PRO Colonial Office MSS 42 v. 771; Barr, *Expeditions*, p. 142. H. P. Dawson, *Observations of the International Polar Expeditions, 1882–1883: Fort Rae* (London, 1886) p. vii. Lord Lorne to John A. Macdonald, n.d. [1882]: "What is the answer with reference to the President of the British R. Society's request for a small expedition in cooperation with other nations for 'Circum Polar' observation?"

cially made: "all that could be done was to select the most suitable of those that were in stock at Kew and at the Meteorological Office."[63] The instruments fared rather better than earlier ones had done in traveling with the company. The latter part of the journey

was not so trying to the instruments as might have been supposed, as at the portages . . . it was possible to see that cases containing fragile instruments were treated with care, but when travelling by rail they could not always be protected from rough usage at the hands of railway employés. Transport in bullock carts over exceedingly rough roads exposed the instruments to many unavoidable concussions.

The very last stage of the journey was rough on instruments and explorers alike. Stormy weather forced the party to take eight days to cross Great Slave Lake, during which "the boat was stove in, and sunk in a gale; . . . most of the cases of instruments were submerged."[64]

The principal instruments were barometers, anemometers for measuring wind velocity, a variety of thermometers and other meteorological devices; one unifilar and two bifilar magnetometers, two declinometers, a Lloyd's balance magnetometer, and a dip circle; navigating and surveying gear lent by the Royal Geographical Society, including a transit theodolite and chronometer watch; and a spectroscope with camera.[65]

They reached Fort Rae on 30 August. Meteorological observations began on 31 August. They were delighted that there was at the fort an unfinished and unoccupied log hut, readily converted into a magnetic observatory: "on the 3rd September the declinometer, on the 4th the bifilar, and on the 6th the balance magnetometer, were mounted in their places, and observations commenced therewith." The instruments worked well, except for the balance magnetometer. The magnetic observatory was finished by mid-September, and a new building for absolute magnetic observations was ready a month later; that building was also used for the transit instrument.[66]

The pattern of magnetic and meteorological observations was on the whole standard, including the measurement of temperature in the earth at different depths. They were too far south to contribute to an understanding of permafrost, but the program of observations was part of the first systematic study of that phenomenon. They had some problems with local animals. The earth thermometers were dug up and destroyed, probably by a wolverine, and they found it necessary to surround "the meteorological

[63] Dawson, *Observations*, p. ix.    [64] Ibid.    [65] Ibid, pp. vii–viii.
[66] Ibid., pp. xi, 119.

instruments with a fence, to prevent the attention of the observer on duty being distracted by the possible visit of a wolf. These animals, which are here large and formidable, often roamed at night amongst the buildings of the post."

The most remarkable feature of their observations was the frequency of auroral displays, which occurred every night throughout the winter, in striking contrast to the paucity of such displays in the highest latitudes.[67] Employees of the Hudson's Bay Company were familiar with this richness, but not with its cause.

Fort Rae was situated in the zone of maximum activity of the aurora borealis, a circular belt centered on the north geomagnetic pole and having a radius of about 1,600 miles. James Ross had located the dip pole, where the dip of a freely pivoted magnetic needle was 90°, and nineteenth-century observers regarded this as *the* north magnetic pole. But the dip pole owes its position not only to the magnetism of the earth's core, but also to local conditions of the earth's crust. When distortions arising from such local effects are eliminated, one has the most significant geomagnetic poles, which indicate the position of the earth's magnetic axis. The magnetic field of the earth about this axis is the main factor determining the position of the belts of active electrically charged gas particles. These Van Allen belts,[68] so named after their discoverer, are horned in shape, and they come closest to the earth in the belts of maximal auroral activity. Sunspots, associated with intense magnetic storms on the sun, send out storms of electrically charged particles that spiral along the Van Allen belts around the lines of the earth's magnetic dipole field, and collide in the upper atmosphere to produce auroras. The solar wind of ionized particles is also responsible for distorting the earth's magnetic field, so that the simple account of a magnetic dipole no longer applies; now, within what is called the earth's magnetosphere, the earth's magnetic field dominates, whereas outside that region, the solar wind rules. None of this could have been known by Captain Dawson in 1882–3, but his observations of auroras, together with those of the other expeditions of the IPY, and of former expeditions too, enabled Herman Fritz to produce the first maps showing the frequency of auroras, thus helping to shape questions that took the best part of a century to answer. The IPY indicated that the auroral belt of maximum intensity went from Fort Rae across Hudson Bay through Nain

---

[67] Ibid., pp. xiii–xiv.
[68] These belts were discovered by the use of satellites and rockets in the International Geophysical Year held seventy-five years after the first IPY.

in Labrador, south of Iceland, just missing Novaya Zemlya, along the Siberian coast, and over Cape Barrow in Alaska.[69] The party left Fort Rae at the beginning of September, and was back in England on 20 November. They had done strictly what they set out to do; their party was too small to have tackled more than meteorology and geomagnetism, and although they made incidental geological and zoological observations, their volume of observations was the shortest of any of the expeditions of the IPY.

*Kingua Fiord.*[70] As we have seen, Germany had been involved with proposals for an International Polar Year almost from the beginning. Neumayer had been the first president of the International Polar Commission, and its first conference was held at his observatory in Hamburg. In 1881 the Imperial government agreed to participate, and appointed Neumayer to take charge of the German contribution. There had at first been thoughts of manning two German stations in the Antarctic. The difficulties of establishing a high Antarctic station, and the needs of meteorological science, led to a change in plan. There would be one station in South Georgia, and another in Cumberland Sound, Baffin Island. The choice of site was determined partly by logistics, partly by a desire to be as close as possible to the magnetic pole, and partly by the need for a reasonably even circumpolar distribution of stations, several of which were already determined, and one already in place in Lady Franklin Bay. Germany's first choice, Jan Mayen, to the north of Iceland, had been preempted by the Austro-Hungarian expedition.[71]

---

[69] J. Tuzo Wilson, *I. G. Y.: The Year of the New Moons* (Toronto, 1961). G. D. Garland, "Another Centenary: Poles of the Unknown in Our Earth," *Transactions of the Royal Society of Canada*, Series 4, 20 (1982) 359–68; S. Chapman, *The Earth's Magnetism* 2nd. ed. (London, 1951); H. S. W. Masset and R. L. F. Boyd, *The Upper Atmosphere* (London, 1958); C. Störmer, *The Polar Aurora* (Oxford, 1955).

[70] Barr, *Expeditions*, pp. 46–59. Heathcote and Armitage, "First International Polar Year," pp. 53–9; Georg von Neumayer, ed., *Die internationale Polarforschung 1882–1883: Die deutschen Expeditionen und ihre Ergebnisse*, vol. 1, *Geschichtlicher Theil, und in einem Anghange mehrere einzelne Abhandlungen physikalischen und sonstigen Inhalts* (Berlin, 1890); vol. 2, *Beschreibende Naturwissenschaften in einzelnen Abhandlungen* (Hamburg, 1890). Neumayer and C. N. J. Börgen, eds., *Die Internationale Polarforschung 1882–1883: Die Beobachtungs Ergebnisse der Deutschen Stationen*, vol. 1, *Kingua-Fjord und die meteorologischen Stationen*, vol. 2 *Ordnung in Labrador, Hebron, Okak, Nain, Zoar, Hoffenthal, Rama, sowie die magnetischen Observatorien in Breslau und Göttingen* (Berlin, 1886).

[71] Neumayer and Börgen, eds., *Die internationale Polarforschung: Die Beobachtungs Ergebnisse*, vol. 1 pp. 1–15, I.

The expedition leader was the physicist Dr. W. Giese, who worked especially on terrestrial electricity and magnetism. The deputy leader and astronomer was Dr. Leopold Ambronn from Neumayer's naval observatory.
The other scientific staff were H. Abbes, a physicist and mathematician
who was officially an assistant, but who did first-rate ethnological work;[72]
Dr. W. Schliephake, the expedition's doctor and botanist;[73] and A. Müh-
leisen, a meteorologist, and a last-minute replacement for Dr. Ludwig
Rösch, who was injured on board the steamer at Hamburg during loading,
and died as a result. H. Seemann was the mechanic, and there were six support staff.[74]

They went by the steamer *Germania* to Cumberland Sound, encountering frustrating delays due to ice conditions and near-disaster from storms.
They managed to anchor successfully in Kingua Fiord on 21 August 1882,
where the ship stayed until they had built quarters on land, and then left
them for the year. Meanwhile, a supplementary German expedition had
gone to Labrador for meteorological work.[75]

Inuit families visited Kingua Fiord soon after the expedition arrived.
Some of the men helped to unload stores and erect the buildings,[76] enabling
the expedition to move into their new quarters on 4 September. Inuit visitors came in winter, spring, and summer, sometimes bringing caribou and
salmon to trade. Abbes carefully observed them, their snow houses, carving
tools, customs, and clothing.[77]

---

[72]  H. Abbes, "Die Eskimos des Cumberland-Sundes," *Globus* 46 (1884) 198–201, 213–18;
"Die deutsche Nordpolar-Expedition nach dem Cumberland-Sunde," *Globus* 46 (1884)
294–8, 312–15, 328–31, 343–5, 365–8; "Die Eskimos des Cumberlandgolfes," in Neumayer, *Die internationale Polarforschung*, vol. 2 pp. 1–61.

[73]  The natural history results from the expedition were slim, and the collections made were
scanty. See H. Ambronn, "Allgemeines über die Vegetation am Kingua-Fjord," together
with his list of cryptogamous and phaenogamous plants from the expedition, and his identification of plants collected around Cumberland Sound by Boas in 1883–4; these are all
published in Neumayer, ed., *Die internationale Polarforschung*, vol. 2 pp. 61–74, 75–96,
97–9. Boas's anthropological work is discussed in the next section. Collections of marine
invertebrates were also made: see Georg Johann Pfeffer, "Mollusken Krebse und Echin-
odermen von Cumberland-Sund nach der Ausbeute der deutschen Nordexpedition 1882
und 1883," *Jahrbuch der Hamburgischen Wissenschaftlichen Anstalten* Jahrg. 3 (1886)
23–50.

[74]  Neumayer and Börgen, *Die internationale Polarforschung: Die Beobachtungs Ergebnisse*,
vol. 1 p. 6.

[75]  The supplementary expedition under K. R. Koch is described in Neumayer, *Die internationale Polarforschung*, vol. 1 pp. 145–88. Besides his meteorological research, Koch did
valuable auroral work, undertook investigations of the plasticity of ice, and made valuable
observations on the Inuit.

[76]  Ibid., vol. 1 p. 58.       [77]  Abbes, "Die Eskimos des Cumberland-Sundes."

Figure 32. Abbes's drawings of Eskimo snowhouses, in Neumayer, ed., *Die Internationale Polarforschung 1882–83*. University of Toronto Library.

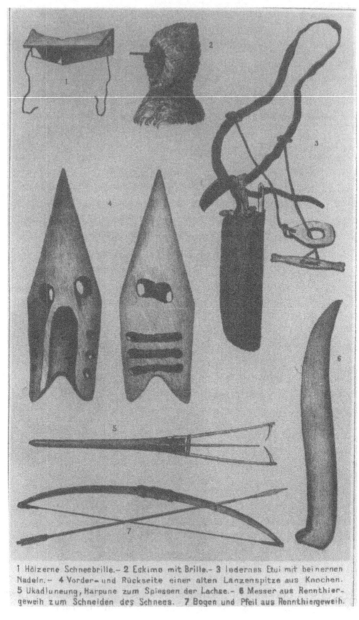

Figure 33. Abbes's drawings of Eskimo implements, in *Die Internationale Polarforschung*. University of Toronto Library.

Meteorology and geomagnetism[78] were the principal subjects of investigation, here as at the other IPY stations. But the first task was an accurate determination of position. This, and other astronomical observations for checking the chronometers and pendulum clock, were carried out using a transit instrument and other apparatus housed in the observatory, a stone building with a wooden roof; one of the expedition's prefabricated wooden observatories had been left in Hamburg. The astronomical observatory, like the headquarters building, had a raised wooden floor insulated by peat, to avoid subsidence into the permafrost.

Observations of the northern lights[79] were made using an auroral theodolite to mark their position, which was generally to the south. Here, although not in the Labrador station, they were outside the main auroral zone, and such observations were secondary. The scientists did seek to analyze auroral spectra, using a miniature spectroscope attached to the auroral theodolite, and although they always found the main auroral line,[80] other lines were generally too faint to measure.

The observatory for absolute magnetic observations was wooden, whereas the structure for relative observations was built of stone,[81] with brass, copper, or wooden fastenings, so as to keep the apparatus well away from iron. The instruments here, and in the Labrador station, included a Casella inclinometer (dip instrument), and, serving as the principal instrument for dip measurements, an earth inductor by Edelmann working on a quite different principle. The latter instrument had been devised by Wild of the Polar Commission as an alternative to the dip circle. It was based on the earth inductor invented by Weber in 1837, but that instrument had been inadequate for the precise measurement of current. Wild's inductor consisted of a coil of insulated wire that could be rotated about an axis, which was itself rotatable about two other perpendicular axes, and connected to a commutator and galvanometer. When the coil was rotated, the earth's magnetic field induced a current in the coil, which could be detected by the galvanometer, thus giving a sensitive indirect measurement of dip.[82] Each

---

[78]   Neumayer and Börgen *Die internationale Polarforschung: Die Beobachtungs Ergebnisse,* vol. 1 pp. 1 et seq. & pp. 185 et seq.

[79]   Ibid., pp. 465 et seq.          [80]   At ca. 5570 Ångstrom units.

[81]   Neumayer and Börgen, eds. *Die internationale Polarforschung: Die Beobachtungs Ergebnisse,* vol. 1 plate facing p. xx.

[82]   Ibid., vol. 1 p. 188; A. McConnell, *Geophysics & Geomagnetism: Catalogue of the Science Museum Collection* (London, 1986) p. 34; R. P. Multhauf and G. Good, *A Brief History of Geomagnetism and A Catalogue of the Collections of the National Museum of American History* (Washington, D.C., 1987) p. 37.

Figure 34. Magnetic and astronomical observatories at Kingua Fiord, in Neumayer and Börgen, eds., *Die Internationale Polarforschung 1882–83 . . . : Die Beobachtungs Ergebnisse,* University of Toronto Library.

station also had a magnetic theodolite, an instrument combining a suspended magnetic bar in a variation magnetometer with a theodolite for precise alignment and measurement of direction.[83]

The bulk of the meteorological apparatus was similar to that of the British and American expeditions, with some notable additions. There was a zinc aspirator for the measurement of absolute humidity, and an Osnaghi electrical self-registering apparatus for use with the anemometer. The development of self-registering apparatus in meteorology, leading to automatic continuous records, is of great importance;[84] its successful adoption in the particularly unfavorable and demanding environment of the polar regions represented a considerable technical achievement.

Most of the observations, however, were still dependent on observers, who made hourly observations throughout the year. On term-days, the first and fifteenth of each month, the magnetic variation instruments were read at five-minute intervals for twenty-three hours, and at twenty-second intervals for the final hour. Ambronn and Abbes had a very heavy schedule, as they were responsible for all the astronomical and absolute magnetic observations.

On top of their magnetic and meteorological work, the expedition undertook measurements of earth currents[85] on magnetic term-days, using a variety of galvanometers made by Siemens and Halske. One of these, the electrodynamic galvanometer, was unaffected by changes in the earth's magnetism. The observations showed rapidly fluctuating currents in the earth, often in excess of one volt, and once as high as 4.8 volts.

When the year's work was over, two boats came ashore, one with the commander of a whaling schooner who had visited them earlier in the season, the other apparently with three Inuit. One "Inuk" greeted them in German, and informed them that the steamer had come for them; he was Franz Boas, one of the leading arctic ethnographers of the late nineteenth century,[86] who had come on the steamer for geographical

[83] Multhauf and Good, *A Brief History,* p. 31 fig. 25.

[84] W. E. Knowles Middleton, *Catalog of Meteorological Instruments in the Museum of History and Technology* (Washington, D.C., 1969).

[85] Neumayer and Börgen, *Die internationale Polarforschung: Die Beobachtungs Ergebnisse,* vol. 1 pp. 313 et seq.

[86] Many would rank him as the leading ethnographer. But one could at least advance a claim for H. J. Rink; see for example Rink, *The Eskimo Tribes: Their Distribution and Characteristics, Especially as Regards Language: With a Comparative Vocabulary, and a Map* (London and Copenhagen, 1887), first published in *Meddelelser om Grønland* 11 (1887).

discovery,[87] and to spend a year with the Inuit of Cumberland Sound. His work was a progression from geography to ethnography.


## A POSTSCRIPT ON FRANZ BOAS AND ANTHROPOLOGY


Anthropology in the mid-nineteenth century had fallen somewhere between natural history and racial theory. German scientists and philosophers, notably Kant and Blumenbach, had in the late eighteenth century developed theories of race that became joined to theories of development or degeneration. Primitive peoples were then contrasted with civilized peoples, the former being either the underdeveloped remnant of the forebears of the latter, or their degenerate successors. Moral, social, biological and physical anthropology, and ethnology came together, for example in the mid-century work of James Cowles Prichard. Observations of native cultures around the world thus tended to incorporate the values of the Victorian age. Such values colored observations, so that, for example, Charles Darwin's account of the Indians of Tierra del Fuego evinced a disapproval and distortion far from his normal practice as the historian of living and mineral nature. Evolutionary theory was susceptible of being brought into line with racial theory, along the lines developed from Kant to Prichard. When, as often happened in spite of Darwin's protests, evolution was confused with progress, science could be used to reinforce the ideology of progress and the superiority of European civilization. That ideology predated evolutionary theory; for example, well before evolutionary theory had been distorted and co-opted to reinforce European supremacy, John Ross qualified his earliest observations of Inuit customs with remarks on their moral turpitude.

    Theories have a way of being applied well beyond the limits of their validity, in ways unacceptable to their original proponents. Anthropology was not the first or the only science to suffer in this way, but it had come increasingly to rest upon misapplied theories. The principal achievement of Franz Boas was the rejection of judgmental racial theories, and the recreation of anthropology and ethnography on the foundations of a purged natural history, as free from speculative theory and as systematic in its observations as possible. He was looking for empirically based laws gov-

---

[87]  Boas, "The Geography and Geology of Baffin Island," *Trans. Royal Society of Canada* 4 section 4 (1887) 75–8; *Baffin-Land: Geographische Ergebnisse einer in den Jahren 1883 und 1884 ausgeführten Forschungsreise*, Ergänzungsheft No. 80 zu *Petermanns Mittheilungen* (Gotha, 1885).

erning cultural phenomena, and rejected theories of cultural determinism because there were too many factors operating in unpredictable degrees.[88] The study of the Cumberland Sound Inuit was Boas's first major contribution to anthropology, and it reflects many of the approaches that he was to adhere to throughout his professional career.[89] First, in accordance with Darwinian theories of dynamic natural history, he saw anthropology as the study of the history of human society, using the widest range of evidence. Archeology, language, and geographical distribution were among the sources for historical understanding. His method was essentially a comparative one, looking at related forms to throw light on differentiation, for example in the development of languages. He was cautious, avoiding some of the simplistic intuitions that had characterized the heroic age of the study of etymology, before the rigorous incursion of German historical criticism.[90]

In distinguishing races, Boas was looking for anatomical characteristics distinguishing large groups from one another, and common to all members of the group. He emphasized skeletal measurements, not because bones were intrinsically more significant than other parts of the body, but because they were often all that was left of past generations. He proposed and undertook a comprehensive program of measurement. He noted, as evolution

---

[88]   Marvin Harris, *The Rise of Anthropological Theory: A History of Theories of Culture* (New York, 1968) p. 283. Harris emphasizes Boas's "inductive purity" and his use of "historical particularism" (p. 250).

[89]   The summary in this paragraph could have been derived inductively from Boas, *The Central Eskimo*, which is based on his Cumberland Sound sojourn in 1883–4, and is available in many different variants in print; I have used the reprint of his synthetic account, incorporating his own observations with those of previous explorers (Lincoln, Nebraska, 1964). For a brief account, see his "The Eskimo of Baffin Island," *Transactions of the Anthropological Society of Washington 3* (1884) 95–102. He published his linguistic material separately from his overall account of the Baffin Island Inuit: "Eskimo-Dialekt des Cumberland Sundes," *Anthropologische Gesellschaft in Wien: Mittheilungen 24* (1894) (neue Folge 14) 97–114. Explicit statements of method are also numerous, e.g., Boas, *The Ethnography of Franz Boas . . .* (Chicago, 1969), and G. W. Stocking, ed., *The Shaping of American Anthropology 1883–1911: A Franz Boas Reader* (New York, 1974). The most concise, and the one principally used in this and the following paragraph, is F. Boas, ed., *General Anthropology* (Boston and New York, 1938); Boas wrote the introduction (pp. 1–6), and chapters on race (pp. 95–123), language (pp. 124–45), invention (pp. 238–81), literature, music, and dance (pp. 589–608), mythology and folklore (pp. 603–26), and methods of research (pp. 666–86). Accounts of Boas include M. J. Herskovits, *Franz Boas: The Science of Man in the Making* (New York, 1953), and L. A. White, *The Ethnography and Ethnology of Franz Boas* (Austin, Texas, 1963). An admirable summary of knowledge about the Inuit of Canada is in *Handbook of North American Indians*, general ed. W. C. Sturtevant, vol. 5, *Arctic*, D. Damas, ed. (Smithsonian Institution, Washington, D.C., 1984) pp. 359–521; pp. 8–16 give a history of research before 1845.

[90]   E.g., John Horne Tooke, *Epea Pteroenta, or The Diversions of Purley* (London, 1798).

suggested, that gradual changes could be brought about by the interaction of hereditary factors and environment, and that isolation could lead to well-defined types. But he also stressed that there was currently no means to account for the varieties called races.[91] There was, moreover, no criterion by which races could be identified as inferior or superior:

The differences between the more fundamental racial types cannot be so interpreted that one of them would be higher than the other. . . . On the contrary, every human type [when compared with other primates] shows excessively human characteristics in certain directions. Thus the Eskimo have the largest brains and a slight amount of body hair; but proportions of limbs are not as excessively human as those of the Negroes who also excel in development of the lips.[92]

In writing of the Eskimo, he noted the crucial importance of caribou ("deer"): "That the mode of life of the Eskimo depends wholly upon the distribution of these animals will . . . be apparent, for . . . they regulate their dwelling places in accordance with the migration of the latter from place to place in search of food."[93] He carefully enumerated the tribes in the area, and considered the influence of geographical conditions upon the distribution of the settlements.

He gave detailed attention to the tools used in hunting and fishing, considering not only current usage, but also forms of harpoon and other items used elsewhere in the Arctic, and in former times. He adopted a similar approach, at once comparative and historical, in discussing Eskimo sledges, kayaks, snowshoes, and other aids to travel. Then he examined snow houses, tents, clothing, tattooing, and other forms of personal adornment. He thus moved from material to social, cultural, and religious aspects of Eskimo life. He noted that social order and laws were founded entirely upon the family, and on blood ties between individual families. He was constantly seeking to compare his own observations with those made by other observers, noting similarities and differences: for example, "although the principal religious ideas of the Central Eskimo and those of the Greenlanders are identical, their mythologies differ in many material points."[94] He collected tales, recorded songs, and altogether produced the most complete and the most valuable account of "the Central Eskimo" thus far compiled.

Boas's work was not part of the International Polar Year, but merely took advantage of the German expedition to Kingua Fiord in order to reach the Arctic. The IPY was essentially a geophysical project. Boas went for geography, and for anthropological studies at the interface between natural

[91] Boas, *General Anthropology*, p. 95, 116–17.    [92] Ibid., p. 115.
[93] Boas, *The Central Eskimo* (reprint 1964) p. 11.    [94] Ibid., p. 175.

history and the social sciences. His pioneering anthropological work provided models and methods for the next generation in his field. In his integration of data from other times and places, and in his development of standards for observation, he was far closer to Weyprecht's ideal of international cooperation in science than to Clements Markham's ideal of competitive national exploration.

### THE FIRST INTERNATIONAL POLAR YEAR: A PRECEDENT AND A MODEL

Alas, Weyprecht never saw the fruits of his labors; he died on 29 March 1881. His program was handsomely although posthumously realized. The data of the IPY are still important for geophysicists, and the meteorological and geomagnetic work in particular were remarkable achievements in the scientific exploration of the circumpolar physical world. Ten countries had committed themselves to establish fourteen major stations for a year's observations,[95] as well as several minor ones. They had followed guidelines laid down by Weyprecht and the International Polar Commission, so that comparisons and syntheses of their results were in order. This was by any standards a major achievement, and an unprecedented one. The IPY was of greater significance than its only real precursor, Gauss's and von Humbolt's magnetic crusade, and it was the inspiration for a second IPY fifty years later, and for the International Geophysical Year another twenty-five years later. Just occasionally, even the Circumlocution Office fails to stop things from being done.

[95]  Since the Netherlands didn't reach their intended station, only thirteen major stations were in fact manned.

# 9

## Science without Borders, or Scientific Territory? Imperialism and Emergent Nationalism before 1910

For most of the nineteenth century, the Royal Society of London acted as a kind of scientific civil service, advising the Admiralty and less frequently the government.[1] From John Ross's frustrated first voyage in 1818 to Britain's modest participation in the International Polar Year in 1882–3, the society provided scientific instruction for naval officers, and programs of observations for their expeditions.

There was a symbiotic relationship between explorers and the society. Naval officers who had contributed to geographical discovery were regularly elected as fellows.[2] They were then in a position to cooperate closely with more theoretical scientists, and with instrument makers, who in the early decades of the century still enjoyed considerable standing, not yet being regarded by scientific authorities as mere technicians and subordinates.[3] James Short in the eighteenth century had managed to acquire wealth, position, and standing within the Royal Society through his manufac-

---

[1]  M. B. Hall, *All Scientists Now* (Cambridge, 1984) pp. 162–81.

[2]  The point emerges strikingly from even this partial list of arctic officers who became F.R.S., with the year of their election: E. Sabine (1818 – he was R.A. rather than R.N., but it was the work on an arctic naval expedition for which he was elected), W. E. Parry (1821), F. W. Beechey and W. Scoresby (1824 – again, Scoresby was not R.N., but was elected for arctic work), J. Franklin (1823), J. Richardson (1825), J. C. Ross (1828), J. Barrow senior (1805) and junior (1844) (neither of them officers, but both were Admiralty officials who actively promoted arctic science; J. Barrow junior's election certificate leaves blank the section indicating his areas of scientific eminence!), G. Back (1847), F. L. McClintock (1865), E. Ommaney (1868), S. Osborn (1870), C. R. Markham (1873), and G. S. Nares (1875). E. Belcher was nominated, and supported by Barrow and Beaufort among others, but was not elected.

[3]  J. Bennett, communication to the Congress of the International Union of the History of Science, Hamburg, 1988.

338

ture of reflecting telescopes.[4] William Herschel built the biggest reflecting telescopes available, used them to excellent astronomical effect, and was one of Britain's scientific eminences.[5] The status of instrument makers within science declined as the nineteenth century progressed,[6] but exchanges like those between John Franklin and Robert Were Fox about the latter's dip instrument still show the importance for the navy and for science of a respectful exchange with instrument makers.

The emergence of a corps of experienced arctic officers led naturally to their functioning, first informally and then formally, as an advisory council for the Admiralty. We have already noted the prominence, indeed the dominance of Fellows of the Royal Society on the Arctic Council and the arctic committees. A further overlap with the Royal Geographical Society meant that there was a substantial scientific and geographical lobby to which arctic officers and administrators could appeal. Since the officers were themselves often FRS or FRGS, there was generally unsurprising harmony between the recommendations of the scientific bodies and the Admiralty arctic committees and council. That harmony enabled John Barrow to play the scientific card with frequent success; the extent as well as the nature of British naval involvement in arctic scientific exploration was much influenced by the close-knit nature of the key advisory groups.

These networks defined outsiders as well as insiders. John Ross found himself rejected by the Admiralty after his first arctic expedition. He also managed to confirm his position outside the scientific pale, first with an absurd theory that the aurora borealis was caused by reflection of the sun from icebergs and ice fields,[7] and later by an indignant assertion that he and not Michael Faraday was the discoverer of the effect of magnetism on light.[8] He also wrote fiercely criticizing Mr. C. R. Weld, Assistant Secretary and future historian of the RSL, for attempting in lectures to discredit him.[9] John Ross is an extreme example, but it remains true that it was hard to be out with the Admiralty and in with the RSL.

---

[4]  G. L'E. Turner, "James Short F.R.S., and His Contribution to the Construction of Reflecting Telescopes," *Notes and Records of the Royal Society of London* 24 (1969) 91–108.

[5]  M. Hoskins, *William Herschel and the Construction of the Heavens* (London, 1963).

[6]  J. A. Bennett, "Instrument Makers and the 'Decline of Science in England': The Effect of Institutional Change on the Elite Makers of the Early Nineteenth Century," in P. R. de Clerq, ed., *Nineteenth-Century Scientific Instruments and their Makers* (Leiden and Amsterdam, 1985) pp. 13–27.

[7]  John Ross, "Aurora Borealis," n.d. (watermark 1855), Museum of the History of Science, Oxford, MS Buxton 2 no. 84.

[8]  Ross to ?, 11 Dec. 1845, SPRI MS 486/5/21.

[9]  Ross to C. R. Weld, n.d., copy in Mus. Hist. Sci., Oxford, MS Buxton 2 no. 65.

The Admiralty had its own reward system, based on rank, money, and position; the Royal Society had another such system, operating through election, publication, and medals. Its presidents could wield considerable authority. Joseph Banks and Joseph Hooker were finely autocratic: Banks used his influence not only to promote the first group of nineteenth-century arctic expeditions, but also to provide an immunity that was not far removed from diplomatic immunity for men of science in time of war. This tradition was respected by his successors, so that Franklin was well protected during his second voyage by the passport from the Russian minister in London. The authority of the president was vigorously asserted by Joseph Hooker in the case of poor Feilden, who was caught between military duty and respect for science; and we have seen how George Nares, although not the Admiralty, was so overwhelmed as to yield in the affair. Dominance of a different kind is apparent in the extent to which Hooker's theoretical concerns in phytogeography dictated the nature of the naturalists' programs. The contrast between this theoretical shaping, and the relatively traditional natural history collections carried out by, for example, the Americans at about the same time, is striking.

The Royal Society referred officers to observatories for special training in the use of instruments. Thus, for example, Lefroy and others were sent to Lloyd in Dublin or to the magnetic observatory at Kew as part of their preparation for geomagnetic work. Sabine made his branch of the ordnance another center for geomagnetism. Similar support for the natural history sciences came from the Royal Botanic Gardens at Kew, for several decades the preserve of the Hookers, first William and then Joseph; from the Geological Society of London; and to a much lesser extent from the universities – Edinburgh had been an important center for geology and natural history.

Natural history specimens and collections found their way back to museums, especially the Royal Scottish Museum and the British Museum (Natural History), which provided expert assistance to Feilden, and to many later and some earlier arctic naturalists. Collectors often gave special collections or duplicates to local museums: after the Nares expedition, for example, Moss gave most of his specimens to the Dublin Museum, while some of Feilden's collection went to the museum in Norwich.

These remarks apply to the many British expeditions that explored the Arctic up to and including the British Arctic Expedition of 1875–6. Although individuals in the British government and learned societies regarded the Northwest Passage and the arctic archipelago as British, explicit claims of sovereignty waited until the 1870s. The Royal Society had expressed no more concern with questions of sovereignty in the Arctic than in the Antarctic. As long as there was free access for the purposes of scientific exploration, planting a flag was an irrelevance. This did not contradict the assumption that the Arctic between Alaska and Greenland was an extension of British North America: the United States' purchase of Alaska in 1867 was met as a countermove by the transfer to Canada of Rupert's Land and the remainder of the North-Western Territory. We have seen that Britain's claim to the arctic archipelago was vague; like its claim to much of the empire, it may be said to have had its origins in a fit of absence of mind. Until mid-century, the Royal Navy was the principal presence in the far North. Along the Smith Sound route, the third quarter of the century saw a series of American expeditions in territory that the British had previously come to regard as their own, and subsequently transferred to Canada in 1880. As an Admiralty communication noted in 1897, the Order in Council transferring jurisdiction over the islands to Canada "does not form a title of itself, & leaves the question of what is or what is not British undecided. No further action seems required."[10]

This left the problem squarely with Canada. In 1882, Canada made its position clear. Southern regions of settlement (together with the newly constituted District of Keewatin, from the Manitoba border to the Arctic Ocean) needed administering. For the rest of the North, continental and archipelagic alike, no steps should be taken "for the good government of the country until some influx of the population or other circumstance shall occur to make such provision more imperative than it would at present seem to be."[11] Native populations of Inuit and Indians clearly were assumed to need no southern provision for good government. Given this lack of aggressive territoriality on the part of Canada, on the mainland as well as in the archipelago, American collectors for museums tended to regard the entire continental arctic sweep as their preserve;[12] in matters of natural science, there are cogent arguments for ignoring Alaska's eastern border. But there was nonetheless a territorial question, not one between governments, but rather with the Hudson's Bay Company, which had long been all but sovereign in its territories, and oversaw any science prosecuted there.

---

[10] Wharton, Admiralty, 20 Oct. 1897, PRO Adm. 1 7341.
[11] Grant, *Sovereignty or Security?* (Vancouver, 1988) p. 5, quoting Dominion Order in Council, P.C. 1839, 23 Sept. 1882.
[12] See notes 81 and 82.

## THE ROLE OF THE HUDSON'S BAY COMPANY

We have seen that the Hudson's Bay Company sometimes extended hospitality to scientific observers, and sent them bills in businesslike fashion. Overland expeditions for science and exploration also depended on the company for logistical support.[13] Expeditions under the company's control sometimes had mandates that included scientific work; George Simpson's directions to John Rae are an extreme example. Thomas Simpson in his expedition with Peter Dease also incorporated science in a company venture.

Another category of scientific engagement by the company took the form of encouragement of systematic observations, for example in meteorology,[14] and the collection of natural history specimens for southern museums. When the Natural History Society of Montreal prepared its list of 253 questions for distribution far and near, even to the most northerly HBC posts, Simpson engaged to have meteorological journals kept at HBC posts in "Indian territories," as well as agreeing to help with the collection of natural history specimens. Later on, he sent rock specimens to William Logan for the Geological Survey of Canada.[15]

Simpson, who was no scientist, sometimes promoted the sciences for the company's reputation. Company employees were not selected with an eye to natural history,[16] and apart from logistical help, they could offer little more than the reading of meteorological instruments and the collection of animals and plants. Robert McVicar, first employed by the HBC as a clerk at York Factory, and a fierce fighter for the company in its rivalry with the North West Company, became a chief trader in 1821, on the amalgamation of the rivals. John Richardson wrote to him in 1824, seeking to

---

13    For example, see HBC Ref. E.15/1 (PAC microfilm HBC 4M21) for accounts and miscellaneous items concerning John Franklin's second arctic expedition. Similarly, see HBC Ref. E. 15/2 (PAC microfilm HBC 4M22), "Captain G. Back's first Arctic Expedition – Accounts Correspondence and Servants Contracts 1832–35," including (item 15) a printed notice about the search for John Ross: "Soon after the Committee had been appointed the Governor and Committee of the Hudson's Bay Company came forward, and in the handsomest manner undertook to furnish from their stores, for the use of the expedition, every thing that it might require during its absence, upon the understanding that all beyond their truly liberal subscription of one hundred and twenty bags of pemmican, two boats, and two canoes . . . should be regularly charged. The Company foreseeing also that Captain Back in passing through their territory would be materially benefited by being invested with a commission under their hands," provided one.

14    Peter Fidler (1796–1822), a surveyor on several HBC expeditions, frequently wrote asking for meteorological as well as surveying instruments.

15    See chap. 5, and Zeller, *Inventing Canada* (1987) pp. 5, 97, 123.

16    Cf. Simpson to Franklin, 27 Feb. 1824, copy in SPRI MS 248/281/1 pp. 162–3.

engage his whole-hearted support for Franklin's forthcoming expedition, and to ask him about the Indian names for the birds and beasts he hoped to see.[17] McVicar was on friendly terms with Richardson and Franklin; the latter, as justice of the peace, married McVicar to Christina McBeath at Fort Chipewyan in 1827.[18]

Among the minority of scientifically competent officers of the HBC was George Barnston, a naturalist and fur trader. Born in Edinburgh in 1800, he studied for the army, but instead joined the North West Company, leaving for Canada in 1820. After the amalgamation of the HBC with the North West Company, Barnston worked for them until his retirement in 1863. He met the Scottish botanist David Douglas at Fort Vancouver, corresponded with him after his departure from Canada, admired him greatly and with exaggeration as one who had "done more for the promoting of Botanical Science than any other of late times, with the exception of Sir Joseph Banks & the celebrated Humboldt," and was shocked to learn of his sudden death from being gored by a bull in the Sandwich Islands in 1834.[19] Encouraged by Douglas, Barnston became a keen botanist and entomologist. On leave in England in 1843–4, he presented a collection of insects to the British Museum. In 1849–50 he sent a collection of plants to Edinburgh, and also sent plants to the Smithsonian Institution and to McGill.[20] He corresponded with John William Dawson, geologist, naturalist, and Principal of McGill College, expressing his pleasure in 1879 that Dawson had found

so good a position for the Northern plants as alongside of the Lapland collection. Tis certainly the best place for them – especially if by comparison on a large scale of sub-Arctic Floras, light can be thrown on the various Problems & Theories, (not to call them Myths,) of the Glacial age & the great ocean currents of that remote Period.[21]

Barnston was sympathetic to Dawson, who interpreted geological change in accordance with the biblical narrative of creation.[22]

[17] Richardson to McVicar, 2 Jan. 1824, McCord Museum MS M2745.

[18] Franklin to McVicar, 28 Feb. [1824], McCord Museum MS M2746; E. Arthur, "Robert McVicar," *Dictionary of Canadian Biography* vol. 9, *1861–70* (Toronto, 1976) pp. 532–3.

[19] Zeller, "The Spirit of Bacon: Science and Self-Perception of the Hudson's Bay Company, 1830–1870," *Scientia Canadensis* 13 (1989), 79–101 at 97; Barnston's natural history is discussed on pp. 97–100.

[20] J. S. H. Brown and S. M. van Kirk, "George Barnston," *Dictionary of Canadian Biography* vol. 11, *1881–90* (Toronto, 1982) pp. 52–3, which lists some of Barnston's more significant publications; William Barnston to J. W. Dawson, 21 March 1883, McGill University Archives MG 1022 C9.

[21] G. Barnston to J. W. Dawson, 21 June 1879, McGill University Archives MG 1022 Container 7.

[22] C. F. O'Brien, *Sir William Dawson, a Life in Science and Religion* (Philadelphia, 1971).

THE SMITHSONIAN INSTITUTION[23] AND THE HUDSON'S BAY
COMPANY 1855–1869[24]

The Hudson's Bay Company had an uneven but generally supportive role
with regard to scientific collecting and observing. John Richardson, who in
his Franklin search expedition with John Rae had worked unusually closely
with an officer of the company, was able to obtain publications of the Geological Survey of Canada as well as geological specimens, through the good
offices of George Simpson. In general, however, British and Canadian museums and collections were less successful in obtaining help through the
company than was the Smithsonian Institution in Washington, D.C.[25]

The Smithsonian had been founded by a bill of Congress in 1846, using
an endowment willed by an Englishman, James Smithson, who had died in
1835, leaving instructions that the money was to be used for the establishment of an institution for "the increase and diffusion of knowledge among
men." It took years of debate before agreement was reached that knowledge
meant primarily scientific knowledge. Joseph Henry was elected as the first
secretary of the Smithsonian, and set about molding it in his own image,
with a major commitment to research, an emphasis on the physical sciences,
and a determination that no natural history collections be dumped on the
fledgling foundation. He showed a genius for the administration of science,
creating at very little expense a North American network for meteorological observations, to be reduced, interpreted, and published in Washington.
He was attentive to the quirks and needs of his far-flung observers, and they
responded with corresponding loyalty.

[23] Some Canadian museums will be noticed in this and the next chapter. Victorian museums
are discussed in S. Sheets Pyenson, *Cathedrals of Science: The Development of Colonial
Natural History Museums . . .* (Kingston, Ontario, 1988).

[24] Debra J. Lindsay, *Science in the Sub-Arctic: Traders, Trappers, and the Smithsonian Institution, 1859–1870* (Ph.D. thesis, University of Manitoba, 1990) examines the role of
northern collectors in testing new collecting procedures while adding to the Smithsonian's
collections. Her account is particularly valuable for its discussion of taxonomic issues.
Welcome, useful, and published while this book was in press is Lindsay's edition, *The
Modern Beginnings of Subarctic Ornithology: Northern Correspondence with the Smithsonian Institution, 1856–68* (The Manitoba Record Society, Winnipeg, 1991).

[25] For an account of the founding of the Smithsonian Institution, and for the tensions be-
tween physical science and natural history in its early years, see Robert V. Bruce, *The
Launching of American Science 1846–1876* (New York, 1987) pp. 187 et seq. See also
Thomas Coulson, *Joseph Henry: His Life and Work* (Princeton, N.J., 1950); William H.
Dall, *Spencer Fullerton Baird* (Philadelphia, 1915); George B. Goode, *The Smithsonian
Institution, 1846–1896* (Washington, D.C., 1897).

Henry may have wanted to stress the physical sciences, but was maneuvered in 1850 into accepting Spencer F. Baird as his assistant secretary. Baird's extensive personal collection became the nucleus of the institution's natural history collection. He was assigned not only to take charge of the collections, but also to request army and navy officers and others to help in collecting specimens; in this capacity, he assembled collecting gear for Kane's expedition in 1853. His enterprise succeeded so well that by 1858 Henry found himself with what he had sought to avoid, an official natural history museum within the Smithsonian.[26] Baird was inclined to splitting rather than lumping when it came to classification; his publications, based on his own collections and on those made for the Smithsonian under his guidance, added 210 new bird species and 70 new mammals. More significant, he added one new genus and seven new subgenera to mammals and nineteen genera and two subgenera to Audubon's classification.[27] The growth in northern collections was especially impressive, and occurred largely because the courtesy and personal skills that Henry had used in setting up the meteorological network were applied first by Baird, and then by Henry and Baird, to the organization of collecting in natural history.

The result was a willingness on the part of some collectors to oblige the Smithsonian where they were less willing to oblige British and, later, Canadian institutions. Nowhere is this more apparent than in the case of the Hudson's Bay Company. A typical illustration is offered by Baird's correspondence with John William Dawson, who was then representing the Natural History Society of Montreal, and would later be Canada's most prominent scientist, offering the Smithsonian's services in obtaining for McGill specimens collected through the HBC. Baird stressed that in the science of natural history, they were citizens of a common country; and certainly Smithsonian collectors managed to cross, ignore, or transcend national boundaries with ease. But at this early stage, Smithsonian collectors were not strong in the field, and employees of the Hudson's Bay Company, traveling and sojourning in the far North, were to Baird

---

[26] The natural history collections of the Smithsonian Institution "grew under the studied inadvertence of its secretary until the 1870s" [S. G. Kohlstedt, "International Exchange and National Style: A View of Natural History Museums in the United States, 1850–1900," in N. Reingold and M. Rothenberg, eds., *Scientific Colonialism: A Cross-Cultural Comparison* (Washington, D.C., and London, 1987) pp. 167–90 at 171].

[27] Lindsay, *Science in the Sub-Arctic*, pp. 133–4; W. H. Dall, "Professor Baird in Science," *Smithsonian Institution Annual Report* (1888) p. 732; Baird, J. Cassin, and G. Lawrence, *The Birds of North America . . .* (Philadelphia, 1860; reprinted New York, 1974).

obviously a resource to be cultivated. He found himself urging Dawson to start a museum in Montreal by asking George Simpson to secure materials through the company.[28]

Henry acted on Baird's advice, made the necessary approaches to Simpson, and then worked with Baird to secure the cooperation of company employees in the field, in making collections for the Smithsonian, and in assisting the Smithsonian's own collectors. Among the earliest of the company's men to assist the Smithsonian was Bernard Rogan Ross,[29] who came from Londonderry to Canada at the suggestion of George Simpson, and who from 1858 to 1862 was chief trader at Fort Simpson, in charge of the Mackenzie River district.[30] Ross was ambitious, seeing science as offering a route to personal advancement, and anxious to secure Henry's and Baird's patronage. He was, as George Barnston observed, busy "digging away into the mines of scientific enquiry, opening up to him in the north," in a region still little trodden by other naturalists.[31] He collected and had others collect for the Smithsonian, and from the beginning of his tenure as chief trader was vigorous on behalf of that institution; a list of specimens that he was attempting to procure in the fall of 1858 includes mountain goat, musk-ox, reindeer, and whistling marmot. He sent birds, fossils, and plants too, sometimes at his own expense, and even sought to use Henry and Baird as avenues to publication in natural history.[32] "I will spare neither cost nor labour," he wrote, "to render the collection sent out by myself & brother officers as one of the largest & most valuable contributions to Arctic science ever made."[33] He provided not only specimens, but also data about the distribution of fur-bearing animals based upon HBC trapping records over the previous decade. Here was information that was commercially sensitive, and Ross was nervous about a conflict of interest. The *"strictly confidential"* information

---

[28]  S. F. Baird to Dawson, 7 Nov. 1855, Smithsonian Institution Archives [hereafter SIA] Record Unit [hereafter RU] 53 vol. 13 p. 110.

[29]  Lindsay, *Science in the Sub-Arctic,* p. 171 identifies Bernard Rogan Ross, Roderick MacFarlane, James Lockhart, and Strachan Jones as the main HBC collectors for the Smithsonian.

[30]  Hartwell Bowsfield, "Ross, Bernard Rogan," *Dictionary of Canadian Biography* [hereafter *DCB*] vol. 10 (Toronto, 1972) p. 629.

[31]  Barnston to Baird, 8 June and 1 April 1861, SIA Hudson's Bay Company Correspondence Collection (hereafter HBC collection) folder 2.

[32]  B. R. Ross to J. Henry/S. F. Baird (the recipient is not always identified), 28 Nov. 1858, 25 July and 26 Nov. 1859, 10 July 1861, 1 June 1862, in SIA HBC collection 1858–69 and undated, folder 36.

[33]  Ibid., 18 March 1861. Baird named Ross's goose after him.

is what I would not communicate to any but a thorough man of Science – the publishing of the details would seriously comprimise [*sic*] me: but the inferences of a Scientific nature drawn from it cannot be objected to. Might I turn your attention to the remarkable circle of increase and decrease that each decades [*sic*] exhibits. In nearly all the Fur-bearing animals this is observable, but particularly so in the Martens.[34]

His eagerness to help the Smithsonian led to problems; in his view, the company unwarrantably resisted efforts at serving two masters.[35] Lawrence Clarke, first a clerk but later factor and office-holder in the company, saw things differently, explaining to Robert Kennicott, the Smithsonian's charismatic field naturalist: "The truth is . . . Barney has done incalculable damage by his dishonest dealings . . . it was brot [*sic*] home to him of having misappropriated much of the companys property in obtaining his collections, and has been fined heavily by minutes of council this year, the result is, that people on this side feel an a[n]tipathy to meddling with collections of any sort."[36] Here was a matter of too much zeal. The damage turned out to be reparable. Collecting for the Smithsonian prospered, fostered in part by the example and inspiration provided by Kennicott, with whom company employees cooperated and corresponded while he was in the North, and after his departure for the United States.[37]

Kennicott began collecting in the Northwest in 1859. In March of that year, Joseph Henry had written to George Simpson introducing him, and explaining that the naturalist proposed to visit the HBC's territory to make collections for the Smithsonian Institution, and requesting Simpson's and the company's assistance. Simpson replied that "The Hudson's Bay Company are always happy to be instrumental in promoting the interests of science," and the company would accordingly offer hospitality to Kennicott. "I shall not fail to see that everything in our power is done to facilitate his operations and to promote his comfort and convenience."[38] He issued Kennicott with a virtual passport to the company's territory, informing all its officers that Kennicott came recommended by Lord Napier, the British

---

[34] Ibid., 26 Nov. 1859 (typescript copy).
[35] Lindsay, *Sciences in the Sub-Arctic*, p. 230 remarks that "The officers who became prodigious collectors for the Smithsonian shared a certain amount of disaffection with regards to their employer."
[36] L. R. Clarke to Kennicott, 16 Jan. 1865, SIA HBC collection folder 9. "Barney" appears to refer to Bernard Ross.
[37] Lindsay, *Science in the Sub-Arctic*, p. 161 notes that under Kennicott, specimens sent out from the Mackenzie River District filled at least twenty cases per year.
[38] Simpson to Henry, Hudson Bay House, Lachine, 28 March 1859 (copy), SIA HBC collection folder 38.

minister[39] in the United States, as well as by the Smithsonian, and that he was to be hospitably received and offered every facility.[40]

Kennicott, sponsored by the Audubon Club of Chicago as well as by the Smithsonian, headed north and west for the Mackenzie River and the Yukon.[41] His journey from Lake Superior was by canoe with a company brigade. He was at Fort Simpson by mid-August. At the end of August, George Barnston reported to Baird that he had heard that Kennicott "was advancing like a Hero" – and added thanks for books and a keg of alcohol.[42]

In the fall, and then again in the new year, Kennicott visited Fort Liard. In the summer of 1860, he traveled down the Mackenzie[43] and across to Fort Yukon, where he wintered. It had become clear that he could profitably extend his collecting for at least another year, and Henry had accordingly written to Simpson, sweetening the request with a gift of two boxes of well-chosen books, many of them bearing on Simpson's collection of works on the discovery, exploration, and history of North America, and on its aboriginal inhabitants. Simpson was more than happy to oblige Henry about extending Kennicott's trip, and also about the development of a comp? ny network of northern meteorological stations.[44] In return, he sought Henry's help in adding specimens of southern birds to his private collection; "I have," he concluded, "a tolerably complete collection of those found in its northern portions."[45]

Only once did Kennicott offend the company, when he reportedly sought a supply of strychnine to poison foxes. Simpson wrote to Henry pointing to the company's prohibition of this practice in most of its territory.[46] In all

[39]  I.e., ambassador.
[40]  Simpson to all HBC officers, c/o Kennicott, 28 March 1859, SIA HBC collection. folder 38.
[41]  Cooke and Holland, *The Exploration of Northern Canada*, p. 216.
[42]  George Barnston to S. F. Baird, Michipicoten, 26 Aug. 1859, SIA HBC collection. folder 2. Alcohol is important in this part of the story: see notes 51–5.
[43]  Kennicott had been advised by Lawrence Clarke to spend the summer at Fort Rae. In 1865, Clarke wrote to him: "I am as sorry as yourself that you did not take my advice . . . for I believe with you that with the exception of, perhaps, the Anderson, it is the best place in the district for both mammals and birds and just conclusion could be formed from the specimens found there of the Arctic fauna than any other place in the district" (16 Jan. 1865, SIA HBC collection folder 9).
[44]  It was not until 1883 that Charles Carpmael, superintendent of the Meteorological Service in Toronto, prepared a circular for circulation to HBC officers to set up a Canadian network of observers. See Carpmael to J. W. Dawson, 22 March 1883 and Francis Walter de Winton to J. W. Dawson, 17 March 1833, McGill University Archives, Dawson papers.
[45]  Simpson to Henry, 11 Feb. 1860 SIA HBC collection folder 38.
[46]  Simpson to Henry, 7 May 1860, SIA HBC collection folder 38.

other respects, Kennicott was to prove not only an assiduous and efficient collector, but a perfect northern ambassador for the Smithsonian Institution and for natural history.

His work in the field was ably supported by Henry and Baird. We have seen that Bernard Ross was almost overzealous on the Smithsonian's behalf. The intensity of his cooperation, like that of many others in the North, was strengthened because the secretary and assistant secretary took it upon themselves to write supportively to him, to acknowledge his help, to supply his needs, and generally to flatter him with their attention. British institutions, in contrast, were often less anxious to secure the goodwill of mere traders, and were accordingly the losers. Ross complained that the museum in Edinburgh had failed to acknowledge receipt of bird specimens, which "will certainly be the best way to prevent me from taking the trouble of sending . . . any more."[47] Here is a recurrent theme in the correspondence between officers of the company and scientific institutions. Encouragement, whether from collectors like Douglas and Kennicott in the field, or from high officials like Baird and Henry at their desks in Washington, secured and maintained the goodwill of company collectors in natural history; a lack of courteous encouragement, typified by Ross's experience with Edinburgh, naturally had the reverse effect.

It was not just in Britain that lack of tact led to lack of cooperation. William Hardisty,[48] son of a chief trader for the company, and like Ross promoted to that rank in 1858, succeeded Ross in administering the Mackenzie River district. In 1869 he wrote from Montreal to the Smithsonian:

Since I came down, Barnston and others have given me some under hand hits about sending all our collections to Washington and none to Canada. I told him that if they would ask us, *as friends, in a private way,* to collect for them, we would be most happy to give them a share of what we might collect, but if the demand was made *through, or by permission of the H B Co,* thereby implying that even our leisure moments were at the disposal of the Co. they certainly would not get anything from me.[49]

There was quite a contrast between the considerate personal tone adopted by Baird and Henry in their dealings with the traders, and the impersonal and even peremptory tone of requests through the company.[50]

[47]  Ross to Henry/Baird, 26 Nov. 1859.
[48]  J. S. H. Brown, "WILLIAM LUCAS HARDISTY," *DCB* vol. 11 (Toronto, 1982) pp. 384–5.
[49]  SIA HBC collection folder 22, Hardisty to [Baird], Montreal, 15 Jan. 1869; cf. Strachan Jones to [Baird] n.d., SIA HBC collection folder 24: "I have got one or two raps over the knuckles from Canadians for sending you my collections & notes they say I ought to send everything to the Institutes of my own country."
[50]  SIA HBC collection folder 38, Simpson to C. Drexler, 2 May 1860.

Baird and especially Henry could be brusque in giving orders to their own subordinates, but they knew how to win loyalty from the HBC men by courtesy and by supplying their intellectual and material needs. Books were crucial to instruct the men in the rudiments of the theory and practice of natural history. Tools were necessary for taxidermy, for mounting specimens, and, not least, for securing those specimens; the Smithsonian sent excellent hunting rifles to those company forts that sent good specimens. Then there was the ticklish problem of alcohol, which was necessary for the preservation of specimens, but much given to mysteriously leaking away so that more was needed. Alcohol had been used in trade with the Indians until 1843,[51] having done enormous damage to native cultures. Its continued consumption by company men led in the 1860s to a ban on the import of alcohol into the company's territory, but the import of alcohol for "scientific purposes" continued quietly and unofficially.[52] Ross, writing to Baird in 1860, indicated that there was a desperate need for preservatives, poisonous and nonpoisonous alcohol, and books. He shouted his message loudly in the postscript to his letter:

P.S. Preservatives!-
Alcohol (poisonous)!!-
Alcohol (not poisonous)!!!-
Books!!!![53]

In like fashion, Lawrence Clarke[54] wrote from Fort Rae that he needed cotton, paper, forceps, books, and alcohol for preserving specimens; and, if the Smithsonian would "remit a small quantity of real *good whisky* to cheer the heart of us bipeds I shall not fail to drink success to your museum."[55] Baird and Henry generally managed to supply what was wanted.

They determined in 1860, while Kennicott was in the field, that his northwestern collecting should be complemented by a more easterly effort. Simpson gave permission for a collector to go to James Bay.[56] In April 1860, C. Drexler, taxidermist at the Smithsonian, received his orders. He was to visit James Bay to make a collection of all the animals there, while

51 P. C. Newman, *Caesars of the Wilderness* (Markham, Ontario 1987) p. 227.
52 James Lockhart wrote to Kennicott from Fort Resolution (5 Dec. 1864, SIA HBC collection. folder 26) that one of his colleagues "thinks it is a damned shame that permission to send us a 'horse' to keep the cold out of our toes, has been again refused. He thinks however, that you might 'unbeknownst' like, ship in a little in case without saying anything about it."
53 B. R. Ross to S. F. Baird, 2 July 1860, SIA HBC collection. folder 36.
54 S. Gordon, "Lawrence Clarke," *DCB* vol. 11 (Toronto, 1982) pp. 194–5.
55 L. R. Clarke to [Baird], 21 June 1861, SIA HBC collection. folder 9.
56 Simpson to Henry, 11 Feb. 1860.

making the collection of eggs his first business. Officers of the company would help him choose the best locations for egging.[57] Drexler duly reported to Simpson in Montreal,[58] offered to procure birds for his personal collection,[59] and on 26 May reported to the Smithsonian from Moose Factory that there was

no prospect in sight in regard to egging at this point, i ashure you, that more of wariety of eggs can be found and in larger numbers in Washington then ther is her. . . . – this is the wors place for egging i have seen yet, everthing is said to breed further north. . . . i will not stay at this inf[ornel] post if otherwise can be helpt, as it is shure, wher ther ar no birds, ther con be no Eggs.

He would therefore head for Fort George at the first opportunity, where everyone assured him that "there should scarsely be a Rock without som eggs to be found on it."[60] On 4 June he set out with Indians by canoe along the shore. He left Fort George on 12 July, having found that birds and eggs were numerous enough, late though it was in the season, but species were few – a familiar pattern in the Arctic.[61] Drexler was dissatisfied with the results, and was happy to head home that same season.[62]

While Drexler was completing his journey from James Bay to Canada, George Simpson died following an attack of apoplexy, leaving uncertain the future of the company's support for science. An essential part of that support depended on men at the forts and in the field, who would continue to help. George Barnston wrote reassuringly to Baird: "The Company have always countenanced scientific pursuits, and all of my Brother Officers to who I have written, requesting subjects of Natural History, have showed willingness wherever there was the ability."[63] Henry and Baird were nonetheless concerned that Simpson's supportive policy be continued under his successor. Henry wrote to London requesting that the company's officers in Hudson's Bay and the rest of their northern territory be asked to maintain their scientific cooperation with the Smithsonian; his request

---

[57]  Draft instructions to Drexler, 24 April 1860 SIA HBC collection folder 16. See also Henry, 26 April 1860, introducing Drexler (folder 16), and Simpson to the officers of the HBC's service, Southern Dept., 30 April 1860 (folder 38).

[58]  Drexler to Baird, 1 May 1860 (SIA HBC collection folder 16).

[59]  Simpson to Henry, 7 May 1860 (SIA HBC collection folder 38). Simpson to Drexler, 2 May 1860, folder 38: "Sir George Simpson will feel obliged if Mr. Drexler will collect for him in the course of his travels some specimens of grouse &c. for his private collection."

[60]  Drexler to [Baird], 26 May 1860 (SIA HBC collection folder 16).

[61]  L. L. Snyder, *Arctic Birds of Canada* (Toronto, 1957) p. 10.

[62]  Drexler to [Baird], 26 Aug. and 21 Oct. 1860 (SIA HBC collection folder 16).

[63]  George Barnston to Baird, 19 Dec. 1860 SIA HBC collection folder 2.

352    Science and the Canadian Arctic

was granted.[64] Similar although less authoritative reassurance was sent to Baird by James Clouston from Hudson's Bay House at Lachine: "I do not think that the death of Sir George Simpson is likely to occasion any interruption in your Arctic explorations, as whatever was undertaken with his sanction is likely to be carried out by whoever succeeds him."[65]

So it proved. Simpson was succeeded by the man he had recommended, Alexander Grant Dallas, a member of the committee of the Hudson's Bay Company, and their representative for the Western Department following the creation of the crown colony of British Columbia in 1858. His commission as "President of Council and Governor in Chief in our Territory of Ruperts Land" was dated 3 February 1862.[66] His appointment did no harm to the Smithsonian's program of northern collecting. As Edward Hopkins wrote to Joseph Henry from the company's house at Lachine, Dallas "takes a lively interest in all kinds of scientific research."[67]

It was just as well because Kennicott was still up north, making his own collections. The friendships he forged with the company's men engendered in them a wholly unaccustomed interest in natural history, reinforced by hopes of meeting their need for alcohol to preserve specimens. Lawrence Clarke, asking for this alcohol, as well as for good whisky, explained that Kennicott had inspired him "with a more lively interest in gathering for your Institution. I must acknowledge, however, that I find it sad up hill work, without preservatives for [the] skins." He had bagged a pair of trumpeter swans, one of which, unpreserved, had subsequently been destroyed by maggots – a real loss, "as this species are very seldom shot." Clarke had tried to enlist the Indians in the business of collecting, but they "have an insurmountable antipathy to this pursuit, superstitiously believing that, were they so guilty as to cure a specimen to be sent out of the country, the Deer would all leave their land!"[68]

Another company man, James Lockhart, wrote from Fort Yukon that he could not skin specimens properly until Kennicott taught him, and now he would gladly collect for the Smithsonian, "with a view, in the event of my going down to Canada to settle, a few years hence, of making a request to the Institution, for a small collection for myself. Should this not be practicable, n'importe – they are welcome to all that I can do, to further the cause

64  Thomas Fraser to Joseph Henry, Hudson's Bay House, London, 13 Feb. 1861, SIA Baird correspondence with HBC box 1.
65  J. S. Clouston to [Baird], 12 Oct. 1860, SIA HBC collection folder 10.
66  W. Kaye Lamb, "Dallas, Alexander Grant," DCB vol. 11 pp. 230–1; B. A. McKelvie, "Successor to Simpson," Beaver outfit 282 (1955) 41–5.
67  E. M. Hopkins to Joseph Henry, Lachine, 23 April 1862, SIA HBC collection folder 23.
68  L. R. Clarke to [Baird], 21 June 1861, SIA HBC collection folder 9.

of science."[69] He asked for zoological and other books from the Smithsonian to make his collecting profitable.

Robert Campbell had been collecting for the Smithsonian from Athabasca, had made good use of the books and alcohol that Baird had sent him, and looked forward eagerly to assisting Kennicott at Fort Chipewyan, where the American naturalist would be a great acquisition to their fireside circles.[70] When it turned out that Kennicott would be wintering elsewhere in the territory, Lockhart wrote to Baird that "collecting will be rather uninteresting in his absence." As a parting gift, Kennicott had presented him with a Maynard rifle, "a perfect gun: out of all comparison the most commanding and efficient arm that I have ever seen, or even heard of." Lockhart would use it to continue to collect, and his fellows were similarly inspired, one with "the Oölogical fever," another with the determination "to 'go in' to the Fishes of the Youcon Valley." Preparing specimens of fishes without preserving alcohol was a particularly unpleasant task. In 1865, Lockhart wrote to Baird that Kennicott had told him

that the greatest want felt in your scientific world now, is the Northern Fishes. I must confess, that we have all shown extreme reluctance to undertake that branch. . . . The difficulties attending their collection and preservation have hitherto deterred many of us from entering on it. The alcohol sent us some years ago, you are no doubt aware, all *'leaked out'* long ago, and 'the powers that *were*' put their veto on the supply being renewed.[71]

Lockhart and his fellows sent Baird not only specimens and requests for alcohol, but also a stream of information about the occurrence and distribution of species.[72] From Fort Rae, Lawrence Clarke wrote that "all the credit of our present dabbling in Natural History, belongs to the exertions and persuasions of my amiable friend Mr. Kennycott."[73]

In the spring of 1862, Kennicott walked from the Yukon to Fort Simpson, whence, hearing of his father's illness, he set out for Chicago, intending to return to Ruperts Land. He had served science and the Smithsonian well,

[69]  J. Lockhart to [Baird], 24 June 1861, SIA HBC collection folder 26.
[70]  Robert Campbell to Baird, Norway House, 6 July 1861, SIA HBC collection folder 6.
[71]  Lockhart to Baird, Fort Resolution, 4 July 1865, SIA HBC collection folder 26. On 26 June 1865, Lockhart had written to Baird (folder 26): "For those who like fish, I am certain that no branch of Natural History would be so full of interest as the Fishes of this district. For at every different Fort, the same kinds vary so much in size and shape, that one wants almost to swear they were of different species – and doubtless many of them are too."
[72]  J. Lockhart to Baird, Fort Norman, 4 Sept. 1861; Fort Resolution, Great Slave Lake, 7 Dec. 1862, SIA HBC collection folder 26.
[73]  Clarke to [Baird], 1 Dec. 1861, SIA HBC collection folder 9.

and his satisfaction with the company's reception of him in the North was happily noted in London.[74]

Meanwhile, there was the question of publishing Kennicott's findings. Dallas wrote to Henry to set out the ground rules: Kennicott could publish everything, except for "tariff prices and similar matters of our private business, which, to the uninitiated, do not fairly represent our dealings with the natives." Parenthetically Dallas, who in the following year retired to his estates in Scotland, asked for the Smithsonian's assistance in sending natural history specimens to the museum at Inverness. The Smithsonian was still better able than the HBC to get specimens from the company's own territory.[75]

Kennicott kept up a running correspondence with friends in the North. He planned to develop a "suckling Smithsonian" in Chicago. The idea was discussed in council at Fort Simpson in the fall of 1864: "The general wish expressed was, that you deserved all the aid we could give; that we want to continue to send any collections we might make, to the Smithsonian as usual, with the understanding, that in the distribution of duplicates, your Institute should have the preference over all others, home or foreign."[76]

Kennicott planned to return to the Mackenzie district and the Yukon as a naturalist and collector. But his last visit to the North was in a different capacity. In 1865 he was joint leader of an expedition for the Western Union Telegraph Company to survey a route for an overland telegraph line to Europe, by way of Alaska and Siberia. He died suddenly[77] in the Yukon in 1866, at Nulato, the trading post built by the Russian explorer Lieutenant Zagoskin during his exploration of the Yukon River in 1842–3; the Western Union expedition was abandoned. Kennicott had been so strikingly the first and most popular of naturalists collecting in the North in the 1860s that news of his death came equally as a shock to his company friends and a blow to the natural history of the Yukon. William McLean wrote from Fort Liard that there were "several persons in this dist[rict] contributing many valuable specimens to the Institution, merely for Mr.

---

[74]  Thomas Fraser to Joseph Henry, Hudson's Bay House, London, 24 Feb. 1863, SIA Baird correspondence with HBC, box 1.

[75]  Dallas to Henry, Fort Garry, 18 May 1863, SIA HBC collection. folder 14.

[76]  James Lockhart to Kennicott, Fort Resolution, 21 Nov. 1864, SIA HBC collection folder 26. Lockhart wrote to Baird (28 Nov. 1864) requesting that this policy for distributing duplicates be followed (folder 26).

[77]  Possibly from a heart attack.

Kennicott's sake; though they had no idea whatever of the objects for which they were labouring, but because it was his wish, and wishing to please him."[78] Kennicott's scientific monument is in the cases of specimens in Washington,[79] but more remarkable was his achievement in winning "the affection and respect of every one of us in the North both officers and men."[80]

His legacy encouraged modest further cooperation between the company and the Smithsonian following the Alaska purchase. The Smithsonian offered to protect the company's interests in Alaska, where the HBC had functioned as lessee, "besides having certain rights of navigation &c. secured by the Convention of 1818." In return, the company provided the Smithsonian with extensive reports on the natural history of the territory.[81] As the Smithsonian came to terms with Alaska's American status, it had its own territory for arctic investigation. Migrating birds and animals, however, took no heed of political boundaries. The Smithsonian, in its dealings with the HBC, recognized the company's dominion in the North, but in pursuing arctic and subarctic specimens and knowledge, sought to transcend borders. In this instance, it succeeded remarkably well. Henry and Baird deserve much of the credit, but Kennicott, their man in the field, deserves even more.[82]

---

[78] McLean to Baird, Fort Liard, 17 Nov. 1867, SIA HBC collection folder 32.

[79] SIA RU 7215 includes an edited manuscript on arctic birds, apparently compiled from the notes of Kennicott, as well as those of Bernard Rogan Ross, Roderick Ross McFarlane, and other HBC men.

[80] Strachan Jones to Baird, 1 Dec. 1866, SIA HBC collection folder 24.

[81] See, e.g., E. M. Hopkins to Baird, Montreal, 18 April 1867, SIA HBC collection folder 23. For a fuller account, see James Alton James, *The First Scientific Exploration of Russian America and the Purchase of Alaska* (Evanston and Chicago, 1942).

[82] The Smithsonian continued to send collectors, either on its own, or jointly with other American organizations. For example, in one of the auxiliary northern expeditions of the International Polar Year, Lucien Turner went to Ungava Bay, where he made meteorological observations for the United States Signal Office, and collected natural history and ethnological specimens for the Smithsonian: L. M. Turner, "List of the Birds of Labrador," *U.S. National Museum, Proceedings* 8 (1885; pub. 1886) 233–54; "Physical and Zoological Character of the Ungava District, Labrador," *Trans. Roy. Soc. Canada* 5 (1887; pub. 1888) 79–83: "Ethnology of the Ungava District, Hudson Bay Territory," *Annual Report, Bureau of American Ethnology* 11 (1889–90; pub. 1894) 159–350. He went with permission and support from the Hudson's Bay Company, to whom the Smithsonian promised a series of his collections. Baird wrote to J. W. Dawson, now Principal of McGill, that if he thought "proper to obtain the permission of the Company to take charge of them for the Museum of McGill College, such an arrangement would be perfectly agreeable to us" (Baird to Dawson, 25 May 1882, SIA RU 33 vol. 24 p. 90). It was ironic that the Smithsonian was ahead of Canadian institutions in obtaining natural history specimens from the HBC.

GEOLOGY, EXPLORATION, AND TERRITORY: ROBERT BELL

The best collections in natural history in Canada in the mid-nineteenth century were those of the Geological Survey of Canada.[83] The Survey had collected geological specimens, but also birds, animals, and flowers that were long inadequately and improperly housed, accumulating in boxes and cases with no room for display. Founded in 1842, the Survey had at first examined the rocks of what was then called the Province of Canada, a union made in 1840 of the provinces of Upper and Lower Canada, approximately corresponding to those parts of today's Ontario and Quebec lying within the drainage basin of the St. Lawrence and the Great Lakes. Today's Atlantic maritime provinces were separate British colonies, and the rest of modern Canada was controlled, here nominally, there rigorously, by the Hudson's Bay Company. As the century advanced, Canada came to embrace more and more of British North America. In 1867 the Province of Canada was united with Nova Scotia and New Brunswick in the federated Dominion of Canada; three years later, the Hudson's Bay Company's territory was acquired by Canada and a southern portion of it was turned into a fifth province. British Columbia joined the Dominion in 1871, followed by Prince Edward Island in 1873. The arctic archipelago was transferred to Canada in 1880. The Geological Survey of Canada thus enjoyed an expanding field for its work, with an enormous leap in 1870. Over the next two decades, the Survey tackled the coasts of Hudson Bay, Labrador, and northern Quebec. Robert Bell was a geologist who vigorously espoused the Survey's northern role.

Bell, the son of a Presbyterian minister and amateur geologist, became something of a protégé of William Logan. In 1857, at the age of fifteen, he worked as a summer assistant for the Geological Survey of Canada. In 1861 he graduated from McGill in civil engineering, and from 1863 to 1869 taught chemistry and natural history at Queen's University in Kingston, where his connections were evidently more important than formal qualifications in those areas. He studied in Edinburgh in 1864, he continued to do summer survey work for Logan, and became a well-seasoned field geologist and naturalist. In 1869, Logan retired and Bell joined the Survey's staff as a full-time geologist. In 1877, he was one of four appointed as assistant directors of the Survey, and in the following year earned his M.D. from McGill.

[83]  Morris Zaslow, *Reading the Rocks: The Story of the Geological Survey of Canada 1842–1972* (Ottawa, 1975) pp. 3–4.

His fieldwork was by then extensive; he was to become the most traveled Canadian geologist of his generation. As Canada expanded its boundaries, officers of the Geological Survey were often advance agents for the Dominion government. Bell was regularly in the vanguard. His work was characterized by breadth of observation, and by the union of topographical and geological surveying. He believed that mapping the new country was an essential prelude to investigating its geology, and that geologists were the best topographers. Indeed, he wrote that "It seems to me almost impossible to separate the work and do the geology without the topography."[84] More generally, his aims were closer to those imposed by Simpson on John Rae than those of his predecessors and successors in the Geological Survey of Canada:

While working in new territory we also take advantage of the opportunity to obtain heights of banks or cliffs, hills and mountains and comparative levels of water, grades and depths of streams and lakes, records of the temperature of the air and water and of other meteorological observations as indications of climate, notes as to the kinds and characters of the forest trees and on the flora generally; also as to the fauna, the collection of zoological and botanical specimens, making notes on the nature of the surface of the country, whether hilly or level, rocky, swampy or covered with soil, the character of the land, and on various other matters. We also enquire from the natives as to the topography, etc., of regions beyond our own explorations. Photographs are taken to illustrate the geology, scenery, the character of streams, etc.[85]

In 1877, Bell headed for Moose Factory, and from there sailed along the east coast of James Bay and Hudson Bay, making a track survey of the coast and studying the geology of the coastline and islands as far as Cape Dufferin. He also made extensive notes on natural history, topography, and meteorology.[86] Two years later, he made a similar track survey of the country west of Hudson Bay, between Norway House and Crosse Lake.[87] In 1880 Alfred Selwyn, Logan's successor as director of the GSC, told Bell to go to York Factory, and from there travel by ship to England, so as to acquire a cursory knowledge of the geology of Hudson Bay and Hudson Strait. This was frustrating – geology is not the easiest science to practice at sea – but Bell gleaned some useful information.

[84] Bell to R. W. Brock, 3 July 1901, Geological Survey of Canada [GSC], Director's Letterbook no. 29 p. 486, quoted in Zaslow, *Reading the Rocks*, p. 215.
[85] Bell, GSC *Annual Report 14* (1901) 14a, quoted in Zaslow, *Reading the Rocks*, p. 153.
[86] Zaslow, *Reading the Rocks*, pp. 162, 171.
[87] Bell, "Report on an Exploration of the East Coast of Hudson's Bay 1877," *GSC Report of Progress for 1877–78* (1879) part C.

This was a time of renewed interest in northern lands and waters. The International Polar Year had helped to concentrate minds. The whale fishery in the eastern Canadian Arctic, although declining,[88] was still commercially significant, and coastal and inland mapping, as well as the sounding of coastal waters and the observation of ice conditions, were important for navigation. Bell's experience, unlike that of earlier navigators, suggested that the ice-free season in Hudson Bay and Strait was a long one. With a railway from the prairies to Hudson Bay, and given a long ice-free and therefore navigable season, it would be possible to ship grain to the Canadian maritime provinces and to Europe. Expanded markets for western agriculture would aid prairie settlement and development, and the advice of the GSC was important for the government.[89]

In Britain, Dr. John Rae, growing more impatient with the years, and having had very different experience of northern ice conditions, saw the proposed Hudson Bay railway route as utterly misguided. He was indignant at the errors, arising from "carelessness or something worse," in the advice being given to the government, especially by the GSC; and the GSC's errors and misinformation were for Rae personified in Robert Bell. Rae accused Bell of having resorted to falsehoods against him, and of having dismissed him as a foolish jealous old man. Rae's own experience contradicted Bell's topographical assertions, as well as his notions about the northern climates (for example, at Moose Factory) being suitable for growing cereals. Bell, in short, "must either be a little mad, or else he does not quite know how to tell the truth." Given northern ice conditions, surely the St. Lawrence and the Great Lakes would offer a better route for the railway and the shipment of grain.[90]

Grain and immigration were not the only issues. Mining was always an important concern for the Geological Survey, and Bell was optimistic that minerals might "become in the future the greatest of the resources of the Hudson's Bay." Little direct search had been made for them, but he had found iron, lead, and other ores around the bay. There were also lead ores on the northern side of Hudson Strait, and "some capitalists have applied to the Canadian Government for mining rights" there. Bell, however much

[88]   The last whaling ship left Hudson Bay in 1915.
[89]   "Speeches Delivered by Messrs. Dawson, Royal and Macdonald on the Hudson's Bay. Ottawa, 21st. February, 1883," *House of Commons Debates: First Session, Fifth Parliament* (Ottawa, 1883); the same session included "Return Compiled by Robert Bell, M.D., &c., of the Geological Survey, under Instructions from the Director, Dr. Alfred R. C. Selwyn. 7th. May 1883" [copy in PAC MG 29 B 15 vol. 30].
[90]   Rae to J. W. Dawson, 26 June 1883, 18 Dec. 1884, 26 April 1885, McGill University Archives MG 1022 C9.

Rae criticized him, was well regarded as a topographical and geological surveyor, and was eager for an opportunity to extend those activities into the arctic archipelago.[91]

Sir John A. Macdonald rose in the House to sum up the debate:

It has not escaped the attention of the Government that there is, in the future, a great prospect of wealth and prosperity being created in connection with the fisheries and mineral wealth of Hudson's Bay. I do not know that there is any precise information to be found in our Archives as yet in this regard. . . . There are three railways now procuring Acts of incorporation for the purpose of connecting older Canada with Hudson's Bay: two to Hudson's Bay proper, and one from a point on Lake Superior to Hudson's Bay. These projects are in the hands of gentlemen whose names are guarantees of respectability and wealth, and of enterprise; and the question – and of course the great question of all – at issue . . . is the navigation of Hudson's Bay and Straits. I may say, that, at this moment, there are unofficial communications passing between Sir Alexander Galt and the Admiralty, for the purpose of ascertaining whether the Admiralty will be willing to enter into some joint arrangement with Canada for the survey of Hudson's Bay and Straits, [by] putting on this work a vessel fitted out for navigation in the Arctic seas, and arranged to undergo all the casualties to which arctic voyagers and ships are liable. . . . [92]

Completion of the Hudson Bay Railway was well in the future, in spite of the eagerness of gentlemen of respectability and wealth. Over the ensuing decades, the northward extension of the track reflected the northward expansion of the frontier of the province of Manitoba.[93] But the anticipation of wealth, the assertion of sovereignty, and the dynamism of science all reinforced continued geological and other exploration of Canada's arctic coasts and islands. Bell, who had been frustrated previously by attempting to geologize while at sea, came closer to geological fieldwork along Hudson Strait in 1884, on an expedition in the *Neptune*, commanded by Andrew Gordon for the Canadian Department of Marine and Fisheries, with Bell as official naturalist and geologist for the GSC.[94]

[91]  Bell to A. R. C. Selwyn, Director GSC, 4 June 1884, in Bell letterbooks, PAC MG 29 B15 vol. 22. *House of Commons Debates*, 21 Feb. 1883; Bell, "Return," 7 May 1883.

[92]  *House of Commons Debates*, 21 Feb. 1883.

[93]  For a brief account, see Zaslow, *The Opening of the Canadian North 1870–1914* (Toronto, 1971) pp. 218–19, 221, 222, 226, 244, 256. See also Howard A. Fleming, *Canada's Arctic Outlet* (Berkeley, Calif., 1957).

[94]  A. R. Gordon, "Report of the Hudson's Bay Expedition under the Command of Lieut. A. R. Gordon, R.N. 1884," *Canada: Department of Marine and Fisheries: Annual Report 17* (1884 [pub. 1885]) appendix 30, 189–228; R. Bell, *Observations on the Geology, Minerals, Zoology and Botany of the Labrador Coast, Hudson's Strait and Bay* (Montreal, 1884) [Geological and Natural History Survey of Canada, Report of Progress, 1882–83–84, Section DD]. See also Bell, "Report to Lieut. A. Gordon, R.N., Commanding Hudson's Bay Expedition," PAC MG 29 B15 vol. 22.

In spite of Macdonald's optimistic notions that the British Admiralty might contribute to the survey, Gordon's northern voyage was a Canadian one, the first sponsored by Canadian government agencies. The ship was the strongest of the Newfoundland northern fishing fleet, a tough steamer eight feet thick in the bow. Their primary task was to establish six stations with wintering parties at each, to observe ice conditions, tides, and surface temperatures in Hudson Strait; they were also to make studies of fisheries, and gather a variety of zoological, botanical, and geological specimens and information. Strikingly lacking was the range of geophysical studies that had been at the core of the Polar Year's program of researches, and that had characterized many earlier voyages. Utility, commerce, national interest, and natural history now formed an alliance justifying northern science.[95]

In the following year, Gordon and Bell returned, this time in the whaler *Alert,* to relieve the men along Hudson Strait. Ice conditions prevented them from reaching one station, but they were able to make contact with the staff. At another station, they found that the observers had run out of food, and had left for Fort Chimo; at a third, on Nottingham Island, they found that one of the men had died. They returned to Churchill, and, before relieving the stations with fresh crews, surveyed one of the Ottawa Islands, in Hudson Bay west of the Ungava Peninsula. This was one of the rare occasions during these voyages when Bell was able to do fieldwork ashore.[96] As he complained, geological work had been very limited, because "most of our time was spent either at sea or in the ice, or in releiving [sic] the stations, which I had already visited on the 'Neptune' expedition of the

---

[95] A similar shift, sharing some causes, was apparent in Australia at the same date: "it was not until the last two decades of the nineteenth century that Australian science began clearly to respond to local imperatives rather than to 'imperial' demands. Until then, Australian scientific development was intermittent, heroic, dependent and pragmatic. . . . Only when, from the 1880s, the demand for the application of science became acute, did the character of scientific investigation alter. It was . . . the need for resource-based efficiency improvements . . . that eventually transformed the nature of things scientific." [I. Inkster and J. Todd, "Support for the Scientific Enterprise, 1850–1900," in R. W. Home, ed., *Australian Science in the Making* (Cambridge, 1988) p. 125.]

[96] Bell, "Observations on the Geology, Zoology and Botany of Hudson's Strait and Bay, made in 1885," *Geological and Natural History Survey of Canada: Annual Report* (1885 [pub. 1886]) Part DD; F. W. Payne, "The Mammals and Birds of Prince of Wales Sound, Hudson's Strait," *Proc. Royal Canadian Institute* Series 3, 5(1887 [pub. 1888]) 182–9; Canada, Dept. of Marine and Fisheries, "Report of the Second Hudson's Bay Expedition, under the Command of Lieut. A. R. Gordon, R.N., 1885," *Canada: Parliamentary Sessional Papers* 9 (1886).

previous year and had done as much geological work as possible in their neighborhood."[97]

In 1886, Gordon returned, this time without Bell, to take off the observers and take down the observing stations. That year, gold was discovered in the Yukon, and the GSC's attention shifted there. George Mercer Dawson, Bell's rival on the Survey, was sent to survey the region; others were sent out to do similar work in 1887.[98] The GSC and the government, however, had recognized the economic and political importance of Hudson Bay and its seas and islands, and so sent a succession of expeditions to the eastern Arctic. Bell continued to geologize and map in the Arctic. He wrote of himself in 1896:

Previous to the present expedition [a GSC expedition on the steamer *Diana*] he had had more opportunities . . . of examining our northern regions than had fallen to the lot of any other man, so that from personal knowledge he may be regarded as the best . . . authority on matters connected with this part of the Dominion. Since 1875 he had made various explorations on both sides of Hudson Bay and in the interior of the surrounding country and he has now passed through Hudson Strait nine times either way, including a voyage in one of the Hudson's Bay Company's ships in 1880, the voyages of the government expeditions by the *Neptune, Alert, Diana,* and the yacht [now] under his own command. . . . [99]

Another frequent geological explorer was Albert Peter Low, who went to James Bay in 1887–8, and to the east coast of Hudson Bay in 1898–9.[100] Low was commander of several expeditions, but his most important, and the first Canadian expedition officially concerned with sovereignty and science, was the far-ranging cruise of the *Neptune* in 1903–4. That cruise, which was vigorously advocated by Clifford Sifton, Canada's minister of the interior and self-appointed protector of northern sovereignty, was meant to show "quasi-occupation" by showing the presence of official authority.[101] Not coincidentally, *Neptune* sailed soon after the turn of the century's geographically most fruitful expedition to the northern and still imperfectly known islands of the Canadian archipelago, the Norwegian expedition led by Otto Sverdrup on the *Fram* (*Forward*). Sverdrup discovered new land: but the archipelago was not his intended destination.

[97]  Bell, "Report as Geologist and Naturalist on the Second Expedition to Hudson's Strait and Bay (*Alert* 1885)," PAC MG 29 B 15 vol. 22 f. 68.ᵛ

[98]  Cooke and Holland, *The Exploration of Northern Canada* (Toronto, 1978) pp. 256–60.

[99]  Bell, PAC MG 29 B 15 vol. 24, f. 212.ᵛ

[100]  Low, *Report on Explorations in James Bay and Country East of Hudson Bay, Drained by the Big, Great Whale and Clearwater Rivers* (Montreal, 1888) [GSC *Annual Report 3* (1887) Part J]; *Report on an Exploration of Part of the South Shore of Hudson Strait and of Ungava Bay* (Ottawa, 1899) [GSC *Annual Report* new series 2 (1898) Section L].

[101]  S. Grant, *Sovereignty or Security?* pp. 9–10.

OTTO SVERDRUP, ZOOLOGY, AND BOTANY

The years around 1900 were those of Scandinavia's arctic ascendancy.[102] Nordenskiöld, Sweden's foremost explorer-scientist of his day, had achieved the Northeast Passage, and his other polar work had included important geological and zoological studies in Greenland.[103] The Norwegian Fridtjof Nansen was the next great figure, a zoologist who became a major authority in oceanography. His first experience of arctic zoological fieldwork came in 1882 when he joined a sealer working in the seas to the east of Greenland. In 1884, he read about Nordenskiöld's penetration of the Greenland ice sheet. Inspired by that example, he planned to cross Greenland on skis. He and his party, including Otto Sverdrup, did so, wintering in 1888–9 among the Inuit.

Then came Nansen's plans for a drift in the ice across the polar basin. In February 1890 he explained his ideas to the Norwegian Geographical Society. Previous explorers, including Nares and his men, had been frustrated and exhausted by trying to advance against the drift of the ice; he would go with that drift. "If we pay attention to the actual forces of nature as they exist here, and try to work with them and not against them, we shall find the safest and easiest way of reaching the Pole. It is useless to work . . . against the current."[104] Evidence of the drift was there in plenty, from the wreckage of the *Jeannette,* which had sunk off the New Siberian Islands and turned up off the southwest coast of Greenland, to Siberian driftwood found along Greenland's east coast.[105]

Nansen's project was controversial, but it received governmental and private support. The *Fram,* specially designed, set off for the Arctic in 1893, commanded by Sverdrup, with an elaborate scientific program. They entered the pack ice to the north of the New Siberian Islands in September, and were soon frozen in. After two years, it was clear that the drift would

[102] Kirwan, *The White Road* (London, 1959) pp. 190–213. A survey is given by Tore Frängsmyr, "Swedish Polar Exploration," in Frängsmyr, ed., *Science in Sweden: The Royal Swedish Academy of Sciences 1739–1989* (Canton, Mass., 1989) pp. 177–98. See also Gunnar Eriksson, *Kartläggarna: Naturvetenskapens Tillväxt och Tillämpningar i det Industriella Genombrottets Sverige 1870–1914* (Umea, 1978) numerous refs. but especially pp. 140–43.

[103] *Ymer* 22 (1902) was dedicated to his memory, and includes A. G. Nathorst, "Nordenskiöld som Geolog," 207–24. See also G. Kish, *North-East Passage: Adolf Erik Nordenskiöld: His Life and Times* (Amsterdam, 1973). Nathorst himself did important geological work in Svalbard and Greenland (see, e.g., his *Polarforskningen: Föreningen Heimdals Folkskrifter* 74 (Stockholm, 1902).

[104] Quoted in Kirwan, *The White Road,* p. 198.      [105] Ibid., pp. 198–9.

not take them over the pole; Nansen, accompanied by H. Johansen, left the ship with dogs, sledges, kayaks, and food, and tried for the pole. They attained 86°14' N, 160 miles further north than any of their predecessors, before the ice ridges, cold, and drift persuaded them to turn south. In an astonishing journey, they reached Franz Josef Land, where they wintered. Heading south again in the spring of 1896 they met Frederick Jackson of the Jackson-Harmsworth expedition,[106] and returned to Norway on his ship. Meanwhile the *Fram* under Sverdrup completed its drift, and reached Norway in August, having accomplished outstanding scientific work, most notably in oceanography and hydrography.[107]

Sverdrup lacked Nordenskiöld's scientific formation, but he had become highly skilled in polar exploration and ice navigation. As soon as the *Fram* was back in Norway, there was a renewal of Norwegian arctic initiatives, led by Sverdrup and Roald Amundsen;[108] "in the nine years 1898 to 1906 their voyages and discoveries added almost as much new land and sea to Canada's future Northwest Territories as all the ships of the thirty-year Franklin search."[109] Sverdrup intended to sail on the *Fram* up Smith Sound to the north of Greenland, where the American Robert E. Peary had done dramatic work, including the first European crossing of the northern Greenland ice sheet. His plan was to go "as far [as possible] along the north coast of Greenland before wintering. From there we were to make sledge expeditions to the northernmost point of Greenland, and as far down the east coast as we could attain."[110] The expedition was to resolve

---

[106] Alfred Harmsworth (later Lord Northcliffe) sponsored Jackson's attempt to head for the pole via what he hoped would be a land bridge from Franz Josef Land. Jackson's published account of his expedition is *A Thousand Days in the Arctic* (1889): his expedition diaries are in the SPRI.

[107] F. Nansen, *Farthest North: Being the Record of a Voyage of Exploration on the Ship Fram 1893–96 and of a Fifteen Months' Sleigh Journey by Dr. Nansen and Lieut. Johansen with an Appendix by Otto Sverdrup, Captain of the Fram*, 2 vols. (London, 1897); *The Norwegian North Polar Expedition 1893–1896: Scientific Results*, F. Nansen, ed., 6 vols. (London, 1901–5; reprinted New York, 1969); vol. 3 (1902) pp. 1– 427 presents Nansen's account of "The Oceanography of the North Polar Basin."

[108] Amundsen was the first to navigate the Northwest Passage, the first to reach the South Pole, a navigator of the Northeast Passage, the first to fly to latitude 88° N, and the first to fly across the Arctic Ocean (Harald U. Sverdrup, "Roald Amundsen," *Arctic* 12 221– 36 at 221). In the course of his Northwest Passage expedition on *Gjoa* in 1903–6, his expedition made extensive magnetic observations, especially around southeast King William Island, and they discovered the Nordenskiöld Islands in Queen Maud Gulf (Roald Amundsen, "To the North Magnetic Pole and through the Northwest Passage," *Geographical Journal* 29 (1907) 485–518; *The North West Passage: Being the Record of a Voyage of Exploration of the Ship "Gjoa" 1903–1907 . . .* 2 vols. [London, 1908]).

[109] Kirwan, *The White Road*, p. 206.

[110] Sverdrup, quoted in Kirwan, *The White Road*, p. 206.

questions about Greenland – the pole was not at issue – although Peary was immediately convinced that Sverdrup was trying to beat him to it. Peary promptly tried to get there first, but both he and Sverdrup were prevented by heavy ice from advancing far up Smith Sound.

Caught in the pack ice near Cape Sabine, the *Fram* was forced in August to find winter quarters, retreating to Hayes Fiord on Ellesmere Island. In the spring of 1899, Sverdrup and one of his men[111] went west by dog sledge across the island, to the head of Bay Fiord on the west coast. From there, they were the first Europeans to see Axel Heiberg Island.[112] Gunerius Ingvald Isachsen, Sverdrup's second in command, led another sledge party to the west of the island.[113] In summer, Sverdrup tried again to sail north, and was once more frustrated by ice. Instead, the expedition headed into Jones Sound, where they passed a second winter. In the spring of 1900, Isachsen began mapping the western coast of Ellesmere Island with exemplary accuracy, crossed with Sverdrup to Axel Heiberg Island, saw new land, Amund Rignes Island, to the west, and landed on it before heading back to the ship.[114] The new lands were claimed for Norway and King Oscar. Another sledge party, led by the young geologist Per Schei,[115] explored the islands in and around Norwegian Bay, off the southwest of Ellesmere Island.[116] A third winter was then spent at Goose Fiord, on the southern shore of Ellesmere Island at the western end of Jones Sound. Spring sledging parties extended the discoveries of the previous year; they were iced in at Goose Fiord for another winter, did more exploration in the region in the following spring, and were able to leave their winter quarters in early August. They were back at Stavanger on 9 September 1902, after they passed four winters safely in the ice – a feat matched only by John Ross on his second arctic expedition. The channels and coasts mapped by Sverdrup's expedition are complicated and rugged, the currents violent and conflicting, the ice threatening, and navigation a nightmare; their achievements rank with the greatest in arctic exploration.[117] Besides extensive geographical

---

111  Edvard Bay.
112  Sverdrup, *New Land: Four Years in the Arctic Regions*, 2 vols. (London and New York, 1904) vol. 1 pp. 120–41.
113  Ibid., pp. 170–82.                    114  Ibid., pp. 330–461, vol. 2 pp. 1–8.
115  Peter R. Dawes and Robert L. Christie, "Per Schei," *Arctic* 39 (1986) 106–7.
116  Sverdrup, *New Land*, vol. 1 pp. 330–59, 466–83.
117  Another feature shared by Ross and Sverdrup was their sponsorship. The philanthropic distiller Felix Booth had been Ross's main supporter; Ross named Boothia Isthmus and the Boothia Peninsula after him. In the group that now bears Sverdrup's name, four islands, Axel Heiberg, Isachsen, Amund Ringnes, and Ellef Ringnes, were similarly named after supportive brewers.

discoveries, there were wide-ranging scientific observations, notably in natural history and meteorology.[118] Greely ungraciously noted that "unfortunately, the meteorological observations were not supplemented by magnetic work."[119]

Botany on Sverdrup's expedition was particularly well served by Herman G. Simmons. The primary field for the expedition's botanical researches was Ellesmere Island, but in August 1898 they first visited Foulke Fiord in northwest Greenland, returning there a year later. This was territory that had previously been visited by the expeditions of Kane, Hayes, Hall, and Nares. On Nares's expedition, Hart brought back a "tolerably large" collection. Working through the botanical specimens that he and his colleagues had collected on Sverdrup's expedition, Simmons found

at least thirty-five of the forty-four phanerogams which make up Hart's list, and possibly also a few more, which, however, in such a case I classify differently. I am able to augment the list of Foulke Fjord higher plants with thirty-three species, among which the following are new for the whole of North-west Greenland: *Arabis Hookeri, Eutrema Edwardsii, Ranunculus affinis, Carex glareosa, C. incurva, Woodsia glabella, Equisetum arvense,* and probably a few more which I have not had an opportunity of examining since my return. About seventy species should thus be known from Foulke Fjord, a number of which not inconsiderably exceeds the previous tale from any part of North-west Greenland. . . . [120]

As this account of the flora indicates, Foulke Fiord was remarkably fertile, far more so than any other place that Simmons knew to the north of Danish Greenland, with vegetation rather than rock determining the color of the landscape. He attributed this fertility to "the manure of the millions of little auks which breed here."[121]

Simmons's chief work was on Ellesmere Island, from August 1898 until August 1902. Snow in the first season restricted botanizing to a few days in August, and it was not until June 1889 that he was able to work extensively. He was able to correct Hart's list for some locations, "partly because in his list of species found he includes forms that are now united with other species also in his list, and also because (partly through intermixture with plants collected in Greenland?) he includes species which, doubtless, do not

---

[118] Preliminary scientific reports are in Sverdrup's *New Land*. The main scientific report is Norske Videnskaps Akademi, *Report of the Second Norwegian Arctic Expedition on the "Fram" 1898–1902*, 4 vols. (Kristiana, 1907–19).

[119] A. W. Greely, *The Polar Regions in the Twentieth Century: Their Discovery and Industrial Evolution* (London, 1929) p. 81. Vol. II of the *Report* does include observations on terrestrial magnetism, but these were not a major part of the expedition's work.

[120] H. G. Simmons, Appendix II, "Summary of the Botanical Work of the Expedition, and its Results," in Sverdrup, *New Land* vol. 2 p. 467.

[121] Simmons in Sverdrup, *New Land*, p. 468.

appear here at all."[122] The perennial debate between lumpers and splitters was at work here, and so, in the revealed inadequacies of Hart's list, was a lack of system like that which Taverner later found in John Rae's collection of birds.[123] Simmons, in contrast, was always precise about the location of his finds, and aware of the importance of such information:

Among the new contributions to the list of the Ellesmere Land phanerogams met with this summer, *Chrysosplenium tetrandrum* is entitled to special mention, as it has never been found anywhere in Greenland. *Saxifraga Hirculus*, first met with in Framfjord, was quite common farther west, along Jones Sound. This species, in Greenland, is confined to the northern parts of the east coast.[124]

Joseph Hooker had been right in anticipating that discoveries of many new species were unlikely,[125] whereas major contributions to phytogeography could still be made in the high Arctic.[126] He had characterized the Greenland flora as essentially a remnant of the European flora, distinct from that of the archipelago to the west. He attributed the difference, including the relative paucity of the Greenland flora, to the possibility enjoyed by the archipelago's plants of moving south during ice ages and then returning north, while Greenland's plants had no southern avenue of retreat and subsequent recolonization open to them. Simmons's fieldwork, reinforced by his study of Scandinavian and British herbaria, confirmed the American character of the flora of Ellesmere Island, but also noted in northwestern Greenland the presence of species included in the American flora of the archipelago. This appeared to contradict Hooker's European characterization of the Greenland flora. "But still," Simmons noted,

the opinion of HOOKER holds true if it is only altered in so far as the comparison is not made with the flora of Greenland as a whole, but with that of the northern parts and especially of the region which lies nearest to Ellesmereland – north-western Greenland. . . . [T]he Greenland flora is no unity, there are great differences in the communities of species belonging to the different parts, which clearly show that an immigration from different quarters must have taken place in post-glacial times, and the region north of Melville Bay especially, has a number of American immigrants large enough to show that here the influence of the near neighborhood to the American flora has been considerable, i.e. the invasion of American species – from Ellesmereland – has put a conspicuous mark upon the flora of that region.[127]

---

[122] Ibid., p. 470.     [123] See chap. 5.     [124] Simmons in Sverdrup, *New Land*, p. 471.
[125] See chap. 5. Simmons p. 472 indicated that as far as he could then judge, the collections included "a number of cryptogams which have not been found before."
[126] See chaps. 5 and 7.
[127] Simmons, "The Vascular Plants in the Flora of Ellesmereland," *Report of the Second Norwegian Arctic Expedition in the "Fram" 1898–1902* vol. I no. 2 (Kristiana, 1906)

Simmons's interpretation of the complex Greenland flora is close to modern ones. Greenland shares some plants, for example the arctic buttercup (*Ranunculus sulphureus*), with most of the circumpolar Arctic; others, like the pink lousewort (*Pedicularis hirsuta*), are found in Greenland, the central and eastern arctic archipelago, and the western Russian Arctic, whereas the arctic avens (*Dryas integrifolia*) is one of those plants endemic to Greenland and the North American Arctic.[128] Simmons and his colleagues collected some 50,000 botanical specimens in Greenland, Ellesmere Island, North Devon Island, and other parts of the archipelago. There were 115 species of vascular plants on Ellesmere Island, excluding 10 listed by Hart but not found on Sverdrup's expedition.[129] Simmons estimated that the total number of plant species, including the "lower plants," was at least 400.[130]

Animal life in the newly discovered lands was locally plentiful. Walruses, often seen lying on drifting ice floes in "most imposing heaps – perfect 'meatbergs' – . . . are mines of wealth as food for the dogs during the winter." The Greenland whale (*Balaena mysticetus*) had long been exterminated in the vicinity of Jones Sound: "but numerous bones near the old Eskimo houses tell that it flourished here before the whalers found their way through the ice and brought death and destruction with them." Now only belugas and narwhals were to be seen. Polar bears seemed to have learned about long-range rifles, and kept their distance by day, but "at night it is sometimes quite another animal, and one which it is well to beware of." The musk-ox was the most interesting to the expedition, as big game, and for its excellent meat. Caribou were present, but in much smaller numbers than the musk-ox; their relative scarcity was explained by predation by the wolf, the "most noxious animal of these regions." Arctic hares were found sometimes "in quite large numbers; indeed, I may say, in absolute flocks."[131]

The expedition's observations of land birds were few, although they did kill several gyr falcons. Seabirds were more numerous, "both as regards species and individuals. It is they which bring life and turmoil to these barren coasts." Among waders, the knot was singled out as most noteworthy; they found its young, but no eggs.

pp. 1–197 at p. 11. Vol. I no. 1, pp. 1–22, is A. G. Nathorst, "Die oberdevonische Flora des Ellesmere-Landes."
[128] E. Haber in B. Sage, *The Arctic and its Wildlife* (New York and Oxford, 1986) pp. 67–70.
[129] Ibid., p. 9.      [130] Simmons in Sverdrup, *New Land*, pp. 475–6.
[131] E. Bay, Appendix III, "Animal life in King Oscar Land, and the Neighbouring Tracts," in Sverdrup, *New Land* vol. 2 pp. 477–9.

Noteworthy also was the wealth of marine invertebrate life brought up by dredging in the fiords and sounds, sometimes, for example at Hell Gate at the western end of Jones Sound, in spite of violent currents.[132]

GEOLOGY AND PER SCHEI[133]

Greely later singled out the expedition's geological work as, after geographical discovery, its most important: "The geological collections of Schei – Cambrian, Silurian, and Devonian formations in South Ellesmere Island; Mesozoic formations and Tertiary deposits on Axel Heiberg Island and the shores of Greely Fiord – were extensive and illuminating."[134]

Per Schei, the expedition's geologist, was born in Norway in 1875, the year in which Nares's ships first sailed and steamed north. A farmer's son, he did well at school in Denmark, and returned to Norway to study geology, graduating in 1898 from the University of Kristiania (now Oslo) in mineralogy and geology. In the same year, he embarked with Sverdrup on the *Fram*. Schei was vigorous in fieldwork throughout their four years in the Arctic, and he came back with valuable notes and an extensive collection of rocks and fossils. His geological sketch map[135] shows geological mapping not only of western and southern Ellesmere Island, but also of north Devon Island, and the east coast of newly discovered Axel Heiberg Island – all in all a very substantial achievement. He was appointed chief scientific editor for the official reports, but died before they began to appear.

Schei's preliminary report, published in Sverdrup's *New Land,* began by noting that previous geological knowledge of the archipelago had come mostly from "occasional observations" made during expeditions with some other object, notably in the course of searching for Franklin. If the opportunities for research had been limited, those for transporting collections home from the Arctic had been even worse. "In more than one place are still lying whole collections got together by interested and energetic discoverers, who in the end were obliged to abandon them because it was impossible to transport them on their already heavily-laden sledges, manned often

[132]  Ibid., pp. 479–83.        [133]  Dawes and Christie, "Per Schei," *Arctic* (1986) 106–7.
[134]  Greely, *The Polar Regions in the Twentieth Century,* p. 81.
[135]  Schei, "Preliminary Account of the Geological Investigations Made during the Second Norwegian Polar Expedition in the 'Fram,' " in Sverdrup, *New Land* vol. 2 pp. 455–66, map facing p. 466.

by crews devastated by sickness."[136] Schei confirmed the formations identified by his predecessors, while introducing some modifications to their geological maps. "From the Silurian in the Hayes Sound tracts Cambrian deposits have been separated, while those identified as Cambrian" by G. M. Dawson in the Geological and Natural History Survey of Canada's *Annual Report* (1886) were, in Schei's opinion, possibly much more recent.

The oldest or Archaean rocks occurred on the south and east coasts of Ellesmere Island, and where examined were found to consist of granite and gneiss-granite. The Devonian formation was the chief one along Jones Sound and the southwest of Ellesmere Island, but there were extensive Silurian rocks on the north shore of the western end of Jones Sound. Mesozoic formations (principally Triassic) predominated on the west coast of Ellesmere Island and the opposing east coast of Axel Heiberg Island. Schei wrote not only of the principal geological formations in their succession, but also of the sandstone, limestone, and other strata he observed, and of the fossils they contained. Besides the sedimentary deposits, he found a variety of eruptive rocks, such as porphyry and basalt, on Ellesmere and Axel Heiberg Islands.

Schei was struck by the way in which the various strata had been subjected to radical disturbance, folding, intrusion, fissures, elevation, and depression. "Near the great Archaean plateau in Ellesmere Land proper, . . . the small areas have been violently dislocated." The rising of some beds and the sinking of others, coupled with the planing off of younger strata overtopping older ones, could bring older and younger rocks to the same level in a plateau. Schei was clearly thinking not only in terms of the classification of rocks and formations, but also of the earth's dynamic history. He envisioned a process from the deposition of the earliest rocks to the recent action of glaciers, and the still more recent evidences of subsidence and elevation. He speculated about the implications of the folding and inclination of strata: "Can it be the axes of the folds from the north side of Greely Fjord which appear at Black Cape, Cape Rawson, and Cape Cresswell? And could Feilden's Cape Rawson beds, within whose horizon probably Mesozoic as well as Tertiary deposits (Cape Murchison) are known to occur, possibly be the Mesozoic shales and sandstones of Heureka Sound?"[137]

Schei's work was promptly recognized. In 1903, he and Sverdrup were honored by the Royal Geographical Society, which heard their accounts.[138]

---

[136] Ibid., p. 455.    [137] Ibid., pp. 462–3.
[138] Schei, "Summary of Geological Results," *Geographical Journal* 22 (1903) 56–65.

The Secretary of the Society, J. S. Keltie, sent Feilden proofs of Schei's paper. Feilden's response could have been more generous. Schei, he responded to Keltie,

is a good geologist and has done excellent work. But he tells us nothing new, and only confirms the exactness of the observations of his predecessors. Indeed from our knowledge already obtained from the British and American expeditions to Ellesmere Land, & Hayes Sound <and the Franklin search expeditions> it was easy to predict what Sverdrup's expeditions would meet with.
    It is gratifying to find that where Schei crosses the path of the B. Pol. Ex. 75–76, [*viz*] at Cape Camperdown, Cape Hilgard, Norman Lockyer I. his observations are absolutely in accord with mine, and he and that distinguished Palaeontologist Kjaer, have from the examination of the faunal remains, come to the same opinion of the age of these rocks, as pronounced by Etheridge as absolutely certain nearly 30 years ago, after examining the large series of fossils brought back by us. It is therefore as reasonable to suppose that the stratigraphical horizons of the further 600 miles that we laid down from Norman Lockyer I. to Aldrich's farthest, are approximately correct.[139]

Fair enough, as far as it went; Feilden and the British Arctic Expedition had done a splendid job in the Smith Sound region, and where Schei and the others overlapped, it was unsurprising that a man of Schei's ability should have confirmed and extended their prior work. But Feilden should have recognized the extent of the new discoveries and observations that Schei reported in regions that Sverdrup's expedition was the first to explore.
    What rankled Feilden most was Schei's relatively brief acknowledgment of the work of his predecessors, with McClintock and Aldrich being credited with carrying back from their sledging expeditions only the few specimens that they could fit into their pockets. Schei simply ignored the contributions of the earliest nineteenth-century explorers from the era of Ross and Parry. Feilden complained that Schei

shows the most appalling ignorance and presumption, in the opening point of his paper, from that we might be led to believe that nothing had been done in the North American archipelago save by himself and his companions. He has skipped or perhaps overlooked the geological work of Parry, and Sabine, of the Ross[e]s<,> of the men of the Franklin search expeditions, of Sherard Osborne<,> of Mecham<,> of McClintock<,> of the two Markhams, of Belcher<,> of Kane of Peter Sutherland, Walker and many others. Why McClintock has done more by himself than this expedition, and he is dismissed with a notice that he left his collections behind. . . . we remember that the observations of these explorers, have been illuminated by the pens of such great geologists as Jameson, Murchison, Salter, [John Whitaker] Hulke,

[139]  Feilden to J. S. Keltie, 26 April 1903, RGS MS, Feilden Correspondence.

[Andrew] Leith Adams, the Woodwards, Etheridge, and Heer, and last but not least by that distinguished man Samuel Haughton, whose monumental map of the geology of the N. A. archipelago will ever remain the basis of further discoveries.[140]

Feilden had identified much significant work. But Schei's contributions to geology were and remain among the most significant in arctic exploration. Schei "carried out an extensive reconnaissance of the sedimentary rocks of parts of what are now named the Arctic Platform, Franklinian Geosyncline, and Sverdrup Basin. . . . [most of his units in the stratigraphic column] have subsequently been retained with new formational names."[141]

## LOW, BELL, AND THE CRUISE OF THE *NEPTUNE*

Sverdrup had claimed his new lands for Norway. Canada regarded all lands within the archipelago as Canadian, whether or not they had been discovered by 1880. It was far from clear that Britain had the legal authority to transfer to Canada lands of which it had no cognizance. Besides, Ottawa was not in effective possession of the outer islands of the archipelago, and effective possession was beginning to emerge implicitly as a requirement for sovereignty.[142] Sverdrup's claims were thus a problem for the Dominion government, and were to be resolved in the long run by diplomacy, coupled in 1925 with the payment to Sverdrup of $67,000.

The years immediately before Sverdrup's expedition had been eventful ones for the Geological Survey of Canada. In 1895, George Mercer Dawson had succeeded Selwyn as director of the Survey, to Bell's infinite and jealous disgust. The appointment was nonetheless a reasonable one: Dawson was as good as Bell in making a reconnaissance survey, better educated,[143] and

---

140   Ibid.
141   R. L. Christie and J. W. Kerr, "Geological Exploration of the Canadian Arctic Islands," in M. Zaslow, ed., *A Century of Canada's Arctic Islands* (Ottawa, 1981) pp. 187–202 at 191.
142   S. Grant, *Sovereignty or Security?* (1988) p. 5. G. S. Schatz, *Science, Technology and Sovereignty in the Polar Regions* (Lexington, Mass., Toronto, and London, 1974) deals mainly with the Antarctic, but shows (p. 74) that effective possession remains important: "Sovereign claim to *terra nullius* [i.e., territory for which not even an implicit national claim exists] is extremely difficult to establish, but the International Court of Justice has shown willingness to accept the criterion of effective authority in 'Arctic and inaccessible latitudes,' " Whitman, ed., *Digest of International Law* vol. 2 (Washington, D.C., 1963) pp. 1029–31
143   He had studied at the Royal School of Mines, where he finished first in his class, and was made an associate of the school. T. H. Huxley was one of his teachers there, and

Bell's superior in synthesizing a mass of geological and topographical data.[144] Dawson's experience as a field surveyor for the GSC in the Yukon, coupled with the Klondike gold rush in 1897, and the lack of movement on plans for the Hudson Bay Railway, combined to make arctic work less urgent for him than exploring the mineral resources of the Northwest. Hudson Bay expeditions continued, with Bell continuing to advocate geological expeditions around the shores of what he saw as Canada's Mediterranean. Dawson made his very different priorities clear to Bell, who was a resentful and difficult subordinate. It is easy to imagine how he must have felt on receiving this letter from his exasperated superior in 1897:

> Low has been to Halifax and ordered two boats for your Hudson Bay expedition one of which I suppose is for you, so you can have a splendid time cruising about if you like that sort of thing, which I for one would not in that area. . . . I understand your expedition to the Bay will leave Halifax about 15th May so you will not have a very long time here if you decide to go but I should think you would find more profitable fields in the mining areas nearer home. There is someone wanted for the north of Lake Superior as well as all over the country and I can't see that there is much practical outcome from the Hudson Bay expedition. I think the country could get better value for the money in the mining districts. However that is a matter of opinion I suppose.[145]

The sniping between Bell and Dawson was ended by the latter's death: Bell thereupon became acting director of the GSC, a very half-hearted recognition. But he was now in a position to encourage arctic geology, including work on the first Dominion Government expedition to Hudson Bay and the arctic islands, the cruise of the D. G. S. *Neptune* in 1903–4.[146] Low, commanding the expedition, also served as its geologist. Scientific activity was being used to reinforce claims to sovereignty, with scientists as advance guards of government. The expedition also carried a naturalist, a botanist, and a topographer and meteorologist.[147]

recommended him for the Edward Forbes Medal and Prize for coming first in natural history and paleontology; he also received the Director's (Murchison) Medal and Prize: D. Cole and B. Lockner, eds., *The Journals of George M. Dawson: British Columbia, 1875–1878* vol. 1, *1875–1876* (Vancouver, 1989) pp. 6–7.

144  See, e.g., his splendid work in the North American Boundary Commission, *Report on the Geology and Resources of the Region in the Vicinity of the Forty-Ninth Parallel* (Montreal, 1875). That work put him at once above Bell and others on the GSC staff (*The Journals of George M. Dawson* vol. 1 pp. 7–8), a professional advantage that he maintained.

145  G. M. Dawson to R. Bell, 5 April 1897, PAC MG 29 B 15 vol. 8.

146  A. P. Low, *Report on the Dominion Government Expedition to Hudson Bay and the Arctic Islands on board the D. G. S. Neptune 1903–1904* (Ottawa, 1906).

147  L. E. Borden [see *Arctic* 16 (1963) 279] was the expedition's surgeon and botanist; A. Halkett was the naturalist, and C. F. King was the topographer and meteorologist.

John Macoun[148] and members of the staff of the Geological Survey assisted in the preparation of the expedition's scientific reports.

The *Neptune* left Halifax on 23 August, heading via Hudson Strait for Hudson Bay "to patrol the waters of Hudson bay and those adjacent to the eastern Arctic islands; also to aid in the establishment, on the adjoining shores, of permanent stations for the collection of customs, the administration of justice and the enforcement of the law as in other parts of the Dominion."[149]

They managed some geological work on the way to winter quarters at Fullerton Harbour, on the west coast of Hudson Bay facing Southampton Island. They avoided winter boredom with recipes developed by Parry and his successors: games, cards, a piano and other musical instruments, weekly lectures, dances, and a newspaper; but no outdoor scientific work was done until April 1903, apart from regular observations of weather and ice conditions. Then they mapped the neighboring coast, made surveys of exposed rocks, collected natural history specimens, and chafed to be on their way. They raised the anchor in mid-July, returned through Hudson Strait to Baffin Bay, and in August crossed Melville Bay into Smith Sound, noting ice conditions as they went, until they were stopped by heavy sheets of ice in the vicinity of the Littleton islands. "Into this neighbourhood," Low observed,

it would be dangerous and foolish to force the ship for no definite purpose. A crossing was therefore made to Cape Sabine, and considerable anxiety for the ship's safety was felt passing between the great pans of thick solid ice. . . . Some very hard knocks were given to the ship as she was forced through the heavy ice from one lead of water to the next. . . . [150]

They landed, and visited Peary's last winter quarters. Low was a scientist with little time for racing to the North Pole: "The pluck and daring of such men are to be admired, but the waste of energy, life and money in a useless and probably unsuccessful attempt to reach the pole can only be deplored, as no additional scientific knowledge is likely to be gained by this achievement."[151]

Turning south, they headed for Lancaster Sound and Beechey Island, at the southwest point of Devon Island, site of the Franklin expedition graves. There they found a record left by Amundsen's expedition, describing their attempt to get as near as possible to the magnetic pole. *Neptune* returned to

---

[148]  W. A. Waiser, *The Field Naturalist: John Macoun, the Geological Survey, and Natural Science* (Toronto, 1989).

[149]  Low, *Report on the Dominion Government Expedition*, p. 3.

[150]  Ibid., p. 45.                              [151]  Ibid., p. 46.

Davis Strait, and thence, via Hudson Strait, to Hudson Bay, then back to Port Burwell at the northeast point of Ungava Bay. There they met another Canadian patrol on C. G. S. *Arctic*, and learned of their recall to Halifax.

Along their extensive route, they had taken possession of islands not previously explicitly claimed for Canada, and they made observations and collected specimens along the shore. As Low remarked, the coastline was generally well mapped, but the interior of even the more southerly islands in the archipelago remained mostly unexplored, "with only a few isolated lines run across" them.[152] Using the information gleaned by his predecessors, as well as his own observations, he and the GSC produced the most comprehensive geological map of the archipelago and neighboring continental shores to date. The principal remaining lacunas (apart from islands still undiscovered) were along the shores of Melville Bay, and those parts of Ellesmere Island explored neither by Feilden nor Schei.

Low followed the narrative of his voyage with an account of the Inuit, drawing heavily on the work of Boas, as well as his own observations. He discussed the Eskimo population of the eastern half of North America, tribe by tribe and overall (between 3,400 and 3,700). He discussed the organization of the tribes, their subsistence, the dependence of their annual routine on the demands of the hunt, their houses, sledges, dog teams, hunting and trapping techniques and tools, social organization, religion, taboos, amusements, clothing, and more besides. He was clearly taking Boas's instructions and example seriously.[153]

Then came an account of the geology, based on his own observations as far as the southern and eastern parts of the Canadian Arctic were concerned, and supplemented by the reports of previous explorations. He was generally unimpressed by those reports: "The geological work of the Arctic explorers until recent years was necessarily poor and disconnected owing to the absence of trained men, and to such work being of secondary importance among the objects of the expeditions."[154]

Beyond the GSC's general compilations for northern Canada, including the arctic islands, Low's principal sources were Schei's recent work on and near Ellesmere Island, and Bell's observations of southern Baffin Island.[155] Most of Low's additions to arctic geology consisted in characterizing the formations in the southeast of the archipelago as Archaean, with lesser but significant Silurian tracts of coast.

---

[152] Ibid., p. 113.        [153] Ibid., pp. 131–82.        [154] Ibid., pp. 183–4.
[155] Ibid., p. 184; see Appendix V pp. 337–42 for a list of Low's sources.

The arctic islands and neighboring continental shores "present an almost continuous ascending series from the Archaean to the Tertiary, while the upper loose material represents . . . the Glacial age and . . . the subsequent Post-Glacial deposits." The oldest rocks were found in the north of Ellesmere Island; younger Mesozoic rocks were found "on the northern Parry islands, on the Sverdrup group and on the western and northern sides of Ellesmere island." The still younger Tertiary formations "occur on the northwestern islands, on the northern part of Ellesmere, as well as on the northern and eastern parts of Baffin island." Low gave an elegant dynamic account of the changes that had produced the present arrangement, followed by an account of island group after island group, and formation after formation.[156]

Then he discussed economic minerals, although there had been no systematic prospecting for minerals. There was a mica mine on the north shore of Hudson Strait, and previous expeditions wintering on Melville and Ellesmere islands had quarried coal for their ships. In general, Low argued, the occurrence of Laurentian and Huronian rocks over much of the eastern Arctic indicated that major mineral finds could be expected. Gold, silver, copper, iron, mica, lignite, and coal deposits were all known there.

Then there was whaling.[157] The first permanent stations in the eastern Arctic had been American, and since the mid-1860s, whaling in Hudson Bay had been almost wholly in the hands of the Americans (although Swedish whaling there had not been negligible). Over-fishing by the Americans had led to a steadily decreasing annual catch, and the future of the industry looked gloomy without regulation and control.[158]

The unstated implication was clear: in mining and in whaling there were economic reasons for asserting and maintaining Canadian sovereignty in the far North. If one added to these reasons the growing importance of northwest grain and proposals to build a railway to Hudson Bay, then shipping through Hudson Bay and Hudson Strait became all the more attractive, if only ice conditions would allow a long enough season. That was why Gordon's expedition, and subsequent ones, had been instructed to record

---

[156] Ibid., pp. 185–236.
[157] Walrus as well as whales had been over-hunted, and "it is only a question of a few years, if the present methods of killing are continued, before the walrus will become as rare as the Right whale in the waters of Hudson bay. . . . Taking into consideration the value of the animal to the native, the great waste of life in the killing, and the comparative small value to civilization, it might be as well to pass regulations reserving this animal wholly for the use of the Eskimos." Ibid., pp. 281–2.
[158] Ibid., pp. 250–72.

ice conditions and currents, with a view to assessing the commercial navigability of the Hudson Strait route. Low concluded from their experience and his own that

Hudson strait and Hudson bay do not freeze solid, but are so covered with masses of floating ice as to be practically unnavigable for at least seven months in the year. The ice does not begin to melt until well into the month of June, and is not sufficiently melted for safe navigation with ordinary steamers until the middle of July. No ice is formed in the strait and bay sufficiently heavy to obstruct ordinary navigation until the latter part of November, but towards the close of this period there is danger from the early passage of the northern pack across the mouth of the strait, and also, to a much less degree, from the ice from Fox channel partly closing the western entrance to the strait.[159]

He concluded that, in spite of the drawback, expanding prairie produce could all be shipped to Europe in the season following the harvest, and that, besides offering a short route, there was the advantage that grain would be less inclined to spoil when stored in a cool climate.[160]

The Arctic could become a major highway for the expanding Dominion and its increasing harvests; it could prove important for its mineral wealth; a regulated whale fishery might yet return to health; and government posts that monitored ice conditions and weather, as well as extended geological surveying, would increase safe access and profitability in the North. Scandinavian claims to land, and American control of mines and fishing, could best be countered by the assertion of sovereignty and the pursuit of science. Here was a typically although not uniquely Canadian formulation. The history of science in Canada has been a striking story of government sponsorship and the concomitant development of maps, tabulations, and inventories of natural resources. The reciprocal role of science in nation building, and in the style of Canadian nationalism, has likewise been striking.[161]

It was in this context – Scandinavian assertions, doubtful sovereignty in the limits of the archipelago, prospects for trade and transportation, and resistance to American whaling and mining – that scientific exploration of the Arctic became, in the early years of this century, important for the Dominion government in Ottawa. No longer would government responses be as minor and reluctant as Canada's in the International Polar Year. Instead, the coast guard,[162] the Department of Mines and the GSC, and other gov-

---

159   Ibid., p. 294.                          160   Ibid., pp. 295–8.
161   These themes are explored at length in Zeller, *Inventing Canada* (Toronto, 1987).
162   The expeditions of Joseph-Elzéar Bernier on the Canadian Arctic Patrol in 1904–5, 1906–7, 1908–9, and 1910–11 were the most efficiently continuous patrols for the

ernment agencies institutionalized scientific exploration. When Vilhjalmur Stefansson developed plans for a major scientific expedition that promised the discovery of extensive new lands in the Arctic Ocean, he found a receptive government in Ottawa.

assertion and maintenance of national sovereignty; as mentioned in the introduction, Bernier, leaving nothing to chance, in 1908–9 laid claim to the entire archipelago from the mainland to the North Pole. The contributions of his expeditions to science were, however, relatively slight: see Bernier, *Master Mariner and Arctic Explorer: A Narrative of Sixty Years at Sea from the Logs and Yarns of Captain J. E. Bernier* (Ottawa, 1939). See also Bernier, *Report on the Dominion Government Expedition to the Arctic Islands and the Hudson Strait . . . 1906–1907* (Ottawa, 1909), and *Report on the Dominion of Canada Government Expedition . . . on Board the D. G. S. 'Arctic'* (Ottawa, 1910).

# 10

‖—‖—‖—‖—‖—‖—‖—‖—‖—‖—‖—‖—‖—‖—‖—‖—‖—‖—‖—‖—‖—‖—‖—‖—‖—‖—‖—‖—‖—‖—

## Vilhjalmur Stefansson: Science, Territory, and Politics

Vilhjalmur Stefansson[1] was born in 1879 into the Icelandic community on the shores of Lake Winnipeg. Two years later his immediate family moved with him to North Dakota. Icelandic and Canadian by birth, American by residence, Stefansson in his career was to make good use of his multivalent national status. He entered the University of Dakota, which suspended him, and graduated instead from the University of Iowa in 1903. The Harvard Divinity School then offered him a scholarship, which he accepted on the understanding that he would treat religion as a branch of anthropology. He moved to the department of anthropology, and was encouraged by its head, Frederick Ward Putnam, who appointed him a teaching fellow without requiring him to satisfy tiresome degree requirements. Notes by Stefansson from this period, on the native cultures of the Americas, anticipate later emphases in his work: "Technical side of cult. controlled by gen[eral] env[ironment]. . . . Dwellings permanent or imp[ermanent] acc[ording] to migratory needs . . . "[2]

While Stefansson was at Harvard, Robert Peary set out from New York, sponsored by the Peary Arctic Club, in his attempt to reach the North Pole by way of Smith Sound. He failed to reach the pole on this occasion; it is generally believed that he succeeded on a subsequent trip in 1909. He wintered in 1905–6 at Cape Sheridan on Ellesmere Island, and in the fol-

---

[1] There are three main published sources for Stefansson's life: William R. Hunt, *Stef: A Biography of Vilhjalmur Stefansson, Canadian Arctic Explorer* (Vancouver, 1986); VS, *Discovery: The Autobiography of Vilhjalmur Stefansson* (New York, Toronto, and London, 1964); Richard J. Diubaldo, *Stefansson and the Canadian Arctic* (Montreal, 1978). Useful also is Robert W. Mattila, *A Chronological Bibliography of the Published Works of Vilhjalmur Stefansson (1879–1962)* (Hanover, New Hampshire, 1978).

[2] Stefansson [hereafter VS], Anthropology notes, Harvard University, in Dartmouth College Library [hereafter DCL] VS MSS 98, Folders Ia-11, p. 147.

378

lowing summer reached the northernmost point of Axel Heiberg Island, the next island to the west, discovered by Sverdrup's expedition. From that point, Peary looked north and thought he saw land, which he named Crocker Land, after his patron George Crocker. A good deal of controversy followed.

In the 1870s Sherard Osborn had asserted that the huge ice fields enclosed by the mainland and the arctic archipelago must also be enclosed to the north by a major landmass or masses; otherwise the ice fields would be dispersed to the north by southerly storms.[3] There had recently been a revival of this hypothesis, supporting Peary's suppositious Crocker Land, and based on the arguments of R. A. Harris of the U. S. Coast and Geodetic Survey.[4] Harris used the evidence of currents, as evinced by the drifts of the *Jeannette* and the *Fram*,[5] as well as that of the age of ice in the Beaufort Sea, to infer the existence of land toward the pole, where

the sea seems to have no broad outlet through which the ice can escape, as it does north of Siberia. . . . It seems probable that land, continuous or nearly so, must extend far westward from off Banks Land, for this supposed land and the eastward currents might well explain why it is that the ice never recedes far northward from the northern coast of Alaska nor westward from Banks Land.[6]

The evidence from the tides reinforced his argument, albeit more tentatively; tides along the shores of the Beaufort Sea differed in timing and extent from those along the other side of Bering Strait, and yet the flood came from the west, suggesting the disturbing influence of a major landmass of about half a million square miles to the north of the Beaufort Sea.

Such arguments had been welcome to Alfred Harrison, who in 1905 had set off looking for the new polar continent, using Herschel Island as a base for exploration in the Beaufort Sea.[7]

[3] S. Osborn, "On the Probable Existence of Unknown Lands within the Arctic Circle," *Proc. RGS* 17 (1873) 172–81, with discussion on 181–3.
[4] R. A. Harris, "Evidence of Land Near the North Pole," *Report of the Eighth International Geographic Congress 1904* (Washington, 1905) pp. 397–406. The greater portion of the paper was published in the *National Geographic Magazine* for June 1904, pp. 255–61.
[5] F. Nansen, *Farthest North* (1897); "The Oceanography of the North Polar Basin," in *The Norwegian North Polar Expedition 1893–1896: Scientific Results*, F. Nansen, ed., 6 vols. (London 1901–5; reprinted New York, 1969), vol. 3 (1902) pp. 1–427 with 33 plates. G. W. De Long, *The Voyage of the Jeannette*, 2 vols. (Boston, 1884). Nansen's oceanographic evidence about ocean depths and the continental shelf could be and was used to argue against new land in the Arctic ocean: see J. W. Spencer, "On the Physiographic Improbability of Land at the North Pole," *American Journal of Science* Series 4 19 (1905) 333–40.
[6] Harris, "Evidence of Land," p. 399.
[7] Alfred H. Harrison, *In Search of a Polar Continent 1905–1907* (London and Toronto, 1908).

One other expedition had similar goals. The American geologist Ernest deKoven Leffingwell and the Dane Ejnar Mikkelsen, having been part of a recent unsuccessful North Pole expedition, were raising funds and organizing a further expedition to look for new land in the high Arctic. On Putnam's advice, they asked Stefansson to accompany them as an ethnologist. The *Boston Sunday Globe* gave them an enthusiastic and wildly exaggerated puff: "This polar expedition, of which Mr. Stefansson is one, promises to prove the most prolific in scientific investigation ever sent into the north. It is composed of scientists of all nationalities, representing nearly every class of research. / Their main object is to discover a large tract of land, as yet unknown. . . . "[8]

Stefansson, perhaps representing three of "all nationalities," traveled overland to the Arctic, arrived well before the expedition's ship, did his own exploring, and was captivated by the North and its people. As he went, he filled his notebooks[9] with geological observations of an elementary and nontechnical nature: "June 20 [1906] Sand along east bank is turning to soft sandstone; all sorts of petrified objects along shore – shells, wood . . . " He also began to compile an Inuktitut vocabulary, made notes on Eskimo houses, graves, food caches, boats, and tools, and made measurements of individuals, following Boas's prescription and Putnam's instruction. His response to the individuals he studied was not entirely scientific:

1 Sept. At about 3 p.m. another Omiak arrived with Nunas bound for the delta. I got measurements of six. One of these was not a Nuna – a woman from the West somewhere (I got her people's name as Akwamiuk). She had olive complexion, and is the prettiest Husky I have seen – an Italian face of the best oval type. One man refused to be measured, and all but one child ran off, and he began crying.

Stefansson spent part of the winter living with the Inuit: "One could live on a more agreeable diet, conceivably, but one could not dress better or travel in greater comfort than they do, considering the circumstances. . . . I have enjoyed the winter and suffered no hardships." Leffingwell and Mikkelsen's ship had been crushed in the ice, but "the trip does not seem a failure. One of the main purposes of the expedition was to determine the width of the continental shelf. . . . The theory is that if the shelf extended northward from Alaska . . . to any considerable distance, there was likely to rise from it to the surface more or less land. . . . "[10]

8   *Boston Sunday Globe* cutting 1906 n.d., DCL VS MSS 98(2) Folder II-13.
9   VS, Diary 18 May–9 Sept. 1906, DC VS MSS 98(2) Folder II-1; 10 Sept.–19 Dec. 1906 Folder II-2.
10  VS to "Charlie," 21 May 1907, DCL VS MSS 98 Folder II-14.

## THE STEFANSSON–ANDERSON ARCTIC EXPEDITION
## 1908–1912

The results of the 1906–07 expedition had not resolved the issue of new land, so that it was still very much alive for Stefansson and for the geographical community. Setbacks merely whetted his appetite. Besides, he had discovered the satisfaction of independent exploration, and the fascination of learning about the Inuit. He promptly set about planning his own expedition, and by February 1908 was writing about it to Rudolph Martin Anderson, whom he had met at Iowa. Anderson was a zoologist with a doctorate in ornithology, and a veteran of the Spanish-American War. Serving as assistant commandant at a military academy in Missouri, he was frustrated by the total incompatibility of his occupation with zoological fieldwork. He was looking for a post with the American Museum of Natural History, and he expected to get an offer as a field agent for them, with prospects of a first trip to South America.[11]

Stefansson, however, had other ideas. He told Anderson that Bumpus, the director of the American Museum of Natural History, had intimated to him that he would like to see Anderson go north with Stefansson, in which case "he would certainly pay all expenses and perhaps even salary." As an added incentive, Stefansson wrote that "you would have to go many times to South America before your work would command the public attention that the North trip would." The museum was now "making a specialty of arctic collections. They are now the first museum of the world in the field and want to keep their pre-eminence."[12] On 20 March Stefansson wrote to Anderson that he had just received "a tentative offer from the Carnegie Institution of Washington" to provide navigational instruments and an honorarium in exchange for magnetic observations down the Mackenzie and along the arctic coast. Anderson was persuaded to go with Stefansson, particularly by the hope that, if successful, he might become "superintendent or director of explorations, or head of a sub-department" at the museum.[13]

---

[11]  R. M. Anderson [hereafter RMA] to [? Mae Bell Allstrand, whom he married in 1913; I shall refer to her in these notes as MBA], 3 March 1908, MS National Museum of Natural Science, Ottawa [NMNS]; RMA to Frank M. Chapman of the American Museum of Natural History, 2 March 1908, PAC MG30 B40 vol. 1; RMA to Dr. J. A. Allen, curator, birds and mammals, American Museum of Natural History, 2 March 1908, PAC MG30 B40 vol. 1.

[12]  VS to RMA, 8 Feb., 28 Feb., 10 March, 11 March 1908, PAC MG B40 vol. 1.

[13]  RMA to [? MBA], 21 and 25 March and 5 April 1908, MS NMNS.

Stefansson was also in correspondence with R. F. Stupart, head of the Meteorological Service of Canada, who wanted to establish observation stations along the Mackenzie; Dr. Bauer of the Carnegie Institution wanted ten such stations.[14] Anderson headed for Washington, D.C. to learn how to make magnetic and meteorological observations, as well as how to determine latitude and longitude. He had lunch with Bauer, visited the Biological Survey of the U. S. Department of Agriculture, where he met E. A. Preble, who had collected in the Northwest Territories, and who gave him valuable information and a tour of northern collections in the National Museum.[15] Anderson was busy in Washington, but the work for the Carnegie Institution fell through: "The only instruments they had belonged to the Mass. Institute of Technology, and their permission had to be secured before alterations could be made; then the fine adjustments would have taken two or three weeks, which we simply could not spare."[16] He had a more productive week in Toronto, where he spent a week at the Dominion Meteorological Bureau, "learning the technique, as we are to take down six sets of instruments . . . and instruct observers at various Hudson's Bay posts . . . as well as repair instruments at Fort Chipewyan." Stupart had been north, and so had other scientists whom Anderson met in Toronto, where "the people are more interested in the North than on the U. S. side of the line."[17] The Geological Survey of Canada was also interested: it invited Stefansson to Ottawa, offered cosponsorship of the expedition, and indicated that it wanted both geological specimens and a share of the ethnological results.[18] The American Museum of Natural History was the chief sponsor, but paid expenses not salary. Bumpus laid down the aims of the expedition; Stefansson and Anderson would go down the Mackenzie to the arctic coast, and then explore west or east along the coast, remaining in the Arctic over the winter, and returning in 1909. The chief object of the expedition was "the scientific study of the Eskimo," followed by the securing of collections to

---

[14]  Stupart to VS, 10 March 1908; see also VS to Stupart, 28 Feb. and 27 March 1908, DCL VS correspondence Box 2. There were potential problems in an American institution seeking to carry out a regular survey on Canadian territory: see G. A. Good, "Scientific Sovereignty: Canada, the Carnegie Institution and the Earth's Magnetism in the North," *Scientia Canadensis* 14 (1990) 3–37.

[15]  RMA, Field Notes, 1908, Book I, entries for 8–11 April, PAC MG30 B40 vol. 12. Preble did important work on the zoology of the continental Arctic west of Hudson Bay, as well as around Great Slave Lake. He could reasonably have featured in the discussion of the Smithsonian Institution (see chap. 9), but Kennicott's interaction with the men of the Hudson's Bay Company made him a more rewarding subject for this book.

[16]  RMA to [? MBA], 5 and 19 April 1908, MS NMNS.        [17]  Ibid., 5 May 1908.

[18]  Diubaldo, *Stefansson and the Canadian Arctic*, p. 37; R. W. Brock, Director, Geological Survey of Canada, to VS, 16 May 1908, DCL VS MSS 98 Folder III-5.

illustrate their material cultures, the collection of zoological specimens, and lastly taking geological and paleontological photographs, notes, and if possible collecting specimens.

At the beginning of May Stefansson learned that some of their goods had miscarried at San Francisco, which increased the chances of their needing an extra year in the Arctic; Stefansson was to show a real genius for prolonging his expeditions. The boat they had hoped for was not ready when they arrived at Athabaska Landing, and rather than wait around for a month or more, they took passage with the Hudson's Bay Company. At the delta of Slave River, they bought a whaleboat for $175. "Dr. Anderson and I disagreed as to its being sound and seaworthy and I took it in spite of his most energetic protest." Anderson set up the meteorological posts along the way. Stefansson went ahead and secured the use of another whaleboat from a missionary leaving the country, "so I now have two boats – enabling Dr. A. and me to separate if necessary, and giving added carrying power both in food and specimens."[19]

They were at Herschel Island, west of the Mackenzie delta, by early August, but were disappointed in their hopes of meeting American whalers there: the ice conditions, it seemed, were the worst in years, and the industry was running down.[20] There was little work that they could do there: Stefansson had obtained cranial measurements of most of the Herschel Islanders on his previous visit, and the Inuit were opposed to archeological excavation of their grave sites. Stefansson and Anderson separated at Herschel Island, and spent the winter traveling with Inuit along the coast of Alaska. Anderson had collected specimens until the freeze up, and put up "about a hundred" bird and mammal skins along the route. His field notes consist of lists of specimens killed, as well as descriptions of bird behavior:

28 August 1908 [Barter Island, Alaska] Hundreds of Red Phalaropes are seen everywhere in large and small flocks, swimming far out on the lagoon, wading in shallow water, or whirling about in the water as if on a pivot.

He described the flamboyant display flight of the pectoral sandpiper:

[June 1909] Saw a Sandpiper flying about in an irregular zig-zag course, keeping within 10 or 20 ft. of the ground, with neck inflated or conspicuously puffed out (neck looked larger than body) and continually uttering a deep, hollow, muffled

---

19  VS to [Bumpus], 6 May, 24 June, 9, 13, and 15 July 1908, DCL MSS VS 98 Folder III-5.
20  Ibid., VS to Bumpus, Herschel Island, 9 Aug. The first whalers had reached Herschel Island in 1888, and had fished out the region within twenty-five years.

*toot, toot.* The bird finally lighted on the ground and I killed it – a Pectoral Sand-piper (*Tringa maculata*).²¹

Most of his winter collecting was for the cooking pot rather than for science, but he did collect some fine Dall sheep. He and Stefansson rejoined one another at Flaxman Island off the Alaska coast in the spring of 1909. That summer, they explored the Colville River in Alaska. The river had been partially mapped by the U. S. Geological Survey. Anderson reported to Mae Belle Anderson, whom he married in 1913: "We can add a few discoveries to their map, if you can call it a *discovery* to look at something the natives have always known. . . . " Then he described the inevitable mismatch between expectations and achievement, given the conflicts between the demands of survival and exploration with the needs of zoological and geological science.

I suppose nobody ever got out a scientific report that pleased everybody. I tremble to think of what the critics will do to me for "neglecting my opportunities" in the North. . . . The ornithologists and mammalogists will wonder why I did not get more specimens, the "bug men" will surely be sore, for I have caught only a few vials full of insects, the botanists will expect hay-bales of plants, and the geologists think that our boats and sleds should be ballasted with "rocks" at all times. But what is a man to do! We haven't a ship, and a good portion of our time has to be devoted to the prosaic labor of knocking a living out of a stern and rock-bound coast. It's much like touring the world on a wager, without money – scientific tramps. All summer, working like a deck hand, or long shoreman – dragging heavy boats over mud flats, and unloading them for every gale; and in winter, as a dog teamster on a continual chase for something to eat, no matter what.²²

Anderson's priorities were to be shaped by this hard-won experience, and his next expedition with Stefansson would set him, as the spokesman of disciplined science, solidly against his leader. So far, however, he was only moderately exasperated. He and Stefansson had both done useful work; he had made "the finest collection extant of *Ursus richardsoni,* a rare species in collections, seventeen mountable skins of Barren Ground Bear, all ages, sexes and seasons, besides odd skulls. This is my most important contribution for the year and there may be a new species among them"; but there wasn't.²³ Stefansson meanwhile had spent the winter "completing [an over-

²¹  RMA Field Notes 1908, Book I (4 April – 12 Aug. 1908), Book III, Dec. 1908 – June 1909, PAC MG30 B40 vol. 17. Is this the first description of the pectoral sandpiper's display flight?

²²  RMA to MBA, Herschel Island, Northwest Territories, 29 Aug. 1908, MS PAC MG30 B40 vol. 12 file 12.

²³  There was no new species. A. W. F. Banfield, *The Mammals of Canada* (Toronto and Buffalo, 1974) pp. 310–11 notes that a "distinct population known as the barren-ground griz-

sanguine term] a grammar and vocabulary of the Mackenzie River Eskimos . . . also writing down folk lore stories, etc."[24]

They were back at Herschel Island in August, and from there they went east by whaler to Cape Parry, traveled westward along the shores of Franklin Bay, and discovered the mouth of the Horton River, where they wintered. Stefansson hoped eventually to cover "the entire Eskimo area." His interest was in going further east from Cape Parry, toward the mouth of the Coppermine River in Richardson Bay, at the western end of Coronation Gulf and south of Victoria Island.[25] The museum, however, was unwilling to countenance further extensions of the expedition. Bumpus wrote, in a letter crossing in the mail with Stefansson's statement of his ethnological ambitions, that the museum would pay for his and Anderson's return with their specimens to New York. "If, however, you or Dr. Anderson desire to continue longer in the Arctic at your own expense and quite independent of the Museum, I see no reason why you should not do so. I am of the opinion, however, that Dr. Anderson desires to return to the Museum."[26] The whole correspondence was enormously cumbersome, depending on whalers in the western Arctic, Inuit who might serve as messengers, and lengthy delays and uncertainties; the interval between dispatch and receipt of letters in 1908–9 ran from three to eight months. In August 1910 Anderson received Bumpus's letter with its message about termination of arctic work, but felt unable to leave without ascertaining Stefansson's whereabouts.[27] He would otherwise have been happy to come home, for he had had enough of giving almost all his energies to garnering a living from the land, while continually traveling: "collecting work on the move is difficult and usually unfruitful, and, at the best, only skims the ground superficially. While this method of exploration may demonstrate that a white man can live where an Eskimo or an Indian can . . . , the corollary seems to be that the aborigines are not able to accomplish much more than a bare living and a white man cannot be expected to do much scientific work under similar conditions."[28]

Suffering not at all from such frustrations, Stefansson, well before the arrival of Bumpus's recall, was traveling east, looking for Inuit tribes

zly occurs on the arctic tundra of northern Yukon Territory and of northern and eastern Mackenzie District. . . . Because of the extinction of grizzly bears over much of their range, it may never be possible to assign reasonable sub-specific names to different populations."

24  RMA to MBA, 12 Aug. 1908, MS PAC MG30 B40 vol. 12 file 12.
25  VS to Bumpus, Cape Parry, 13 March 1910, DCL VS MSS 98 Folder III-5.
26  Bumpus to VS, 12 May 1910, DCL VS MSS 98 Folder III-5.
27  Anderson to Bumpus, Baillie Island, 13 Aug. 1910, DCL VS MSS 98 Folder III-5.
28  RMA to Bumpus, Baillie Island, 15 Nov. 1909, PAC MG30 B40 vol. 12 p. 17.

unknown to science. He encountered one on the ice of Dolphin and Union Strait on 13 May 1910. This

was the day of all my life to which I had looked forward with the most vivid anticipations, . . . for it introduced me, a student of mankind and of primitive men especially, to a people of a bygone age. . . . I had nothing to imagine; I had merely to look and listen; for here were not remains of the Stone Age, but the Stone Age itself, men and women, very human, entirely friendly, who welcomed us to their homes and bade me stay.[29]

Two days later, he crossed the Strait to Victoria Island, and there met the people whom readers of newspapers and magazines would, with his encouragement, come to know as the "Blond Eskimos."[30] They

differ strikingly from those of the mainland except from the Akuliakattagruit, who are much intermarried with the people north. They have a definitely European appearance, especially in the matter of beards, which are abundant and uniformly blond some even red. I have seen none with blond hair, but . . . others . . . report hair dark brown and blue eyes. . . . This European appearance . . . may doubtless have a climatic or other physical cause than admixture of white blood, but two known historical facts really come to mind, that Franklin's numerous expeditions may have had survivors, and that 3000 people of Icelandic (Scandinavian) parentage disappeared from Greenland in the 15th or 16th centuries.[31]

One of Stefansson's historical facts, concerning Franklin survivors, was decidedly factitious. His characterization of the Victoria Island tribes[32] was not accepted by subsequent ethnologists, who either rejected his story, or ignored it, allowing the "blond Eskimo" excitement to fade.[33] But it did

---

[29]  VS, *My Life With The Eskimo* (New York, 1922; first ed. 1912) p. 174.

[30]  VS, "The 'Blond' Eskimos," *Harper's Monthly Magazine 156* (1928) 191–8.

[31]  VS to Bumpus, 12 Aug. 1910, DCL VS MSS 98 Folder III-5. See also VS to Clark Wissler, Curator, Dept. of Anthropology, AMNH, 8 Dec. 1910, DCL VS MSS 98 Folder III-5.

[32]  *Anthropological Papers of the American Museum of Natural History* vol. XIV pt. I, *The Stefansson-Anderson Arctic Expedition of the American Museum: Preliminary Ethnological Report by Vilhjalmur Stefansson* (New York, 1914).

[33]  VS to H. F. Osborn, 27 Dec. 1923, DCL VS MSS Correspondence Box 10 1923A: "Mr. Diamond Jenness . . . maintains that the lightness of eye is due to snowblindness and to certain diseases, and claims further that my reports exaggerate the 'blond' characteristics." D. Jenness, "The Copper Eskimos," in V. Stefansson, ed., *Encyclopedia Arctica* (compiled 1946–8: see VS, *Discovery, The Autobiography of Vilhjalmur Stefansson*, pp. 359–67 for VS's account of the rise and fall of the *Encyclopedia Arctica*), typescript in DCL VS collection: "Another name [for the Copper Eskimos], 'Blond' Eskimos, given to them occasionally in former years, is now generally discarded. . . . [VS's theory of Viking admixture] has not yet been substantiated, and Stefansson himself in his writings has preferred the more usual term 'Copper Eskimos.'" RMA to Roald Amundsen 4 March 1926, PAC MG 30 B40 vol. 10: "The famous 'blonde Eskimo' hoax was not original with him [VS]. Richardson started it, and old Captain Charles Klengenberg [Klinkenberg] seems to have brought the idea to Stefansson. . . . " Klinkenberg wintered near Bell Island on S. W.

give him a splendid lever to pry further support from the museum. In December 1910, Stefansson wrote to Bumpus hoping that the museum would reconsider his recall; if not, he suggested, the Geological Survey of Canada might take up the slack and ensure continued support for the expedition.[34] Stefansson now and later sought with fair success to make reluctant sponsors vie for the uncertain privilege of supporting his explorations. He and Anderson stayed in the Arctic for yet another winter, with Stefansson working especially on Inuktitut.[35] At last, in the spring of 1912, he received the answer he wanted from the museum: congratulations for accomplishing the study of the Eskimo of Coronation Gulf, and a vote of an additional $1,000 toward the cost of the expedition.[36] He relayed the good news to Anderson, in palpable glee, and followed up with advice. He was rightly nervous about his "blond Eskimo" claim, and coached Anderson in handling the press:

> The Museum is thoroly [sic] satisfied with our work and foots all the bills – has footed all of them. . . . No doubt the newspapermen will ask you if you saw any "blond Eskimo" – the phrase has a national currency just now. By sticking to the facts we can, of course, have no fear of contradicting each other; the only thing is that you did not see the two blondest groups – those of Prince Albert Sound and near Pt. Williams. It will be well for you to point out that fact to prevent misunderstandings; of course you can in connection with them repeat – if you like – what you must have heard about them from [our Eskimo companion] Natkusiak.[37]

The "blond Eskimo" provided the most popular publicity for Stefansson and his expedition. As his reports circulated, the American press came out in a rash of headlines: "A NEW RACE, OR DESCENDANTS OF FRANKLIN'S PARTY?" "BLONDE ESKIMO STORY CONFIRMED," "NATIVES OF THE FAR NORTH HAD CAUCASIAN CHARACTERISTICS," "AMERICAN EXPLORER DISCOVERS LOST TRIBE OF WHITES, DESCENDANTS OF LIEF ERIKSEN," and "FINDER OF NEW RACE RETURNS. Dr. Vilhjalmar [sic] Stefansson Ends Four-Year Expedition. TELLS OF BLOND ESKIMOS. Believes They Descend from Scandinavian Glory."[38] The *New York Times* first announced Stefansson's

Victoria Island in 1905–6, where he was visited by the Inuit (VS, *The Friendly Arctic: The Story of Five Years in Polar Regions* [New York, 1922] p. 425).

34  VS to Bumpus, 6 Dec. 1910, DCL VS MSS 98 Folder III-5.
35  RMA to [MBA], 18 Jan. 1912, MS NMNS.
36  VS to RMA, 16 June 1912, PAC MG30 B40 vol. 1.
37  VS to RMA, 13 Oct. 1912, PAC MG30 B40 vol. 1.
38  *Arkansas Gazette*, 15 Oct. 1911; unidentified press cutting; *The Daily Nome Industrial Worker*, 20 Aug. 1912; *The Seattle Daily Times*, 9 Sept. 1912; *The Evening Sun*, New York, 16 Sept. 1912 (press cuttings in DCL VS MSS 98 Folder III-25).

discovery on 10 September 1912; the next day it announced "NEW RACE SOLVES MYSTERY OF AGES" and the day after recanted with "DISCREDIT STORY OF WHITE ESKIMOS. AUTHORITIES IN EUROPE THINK PROF. STEFANSSON'S REPORTED DISCOVERY IS IMPROBABLE."[39] For good and ill, Stefansson was a celebrity, and the focus of controversy. Distressed at imputations of quackery, he still knew how to take advantage of the publicity, and used it to reinforce his plans for another and more ambitious expedition.

NEW LANDS VERSUS SCIENCE: THE CANADIAN ARCTIC
EXPEDITION 1913–1918

*The argument: A new arctic continent.* His starting point, here as in his first arctic travels, was the hypothesis that there was extensive undiscovered land in the Arctic. In 1911, R. Harris published an expanded argument for the existence of new land, drawing on detailed hydrographic evidence from a variety of expeditions and voyages, including those of Peary and Mikkelsen and Leffingwell.[40] There were, asserted Harris, some "well-established facts which show at once the necessity for land or shoals":

> The range of the semidaily tides at Bennett Island is 2.5 feet, while it is only 0.4 feet at Point Barrow and 0.5 feet at Flaxman Island. . . . The observed tidal hours and ranges of tide show that the semidaily tide is not propagated to the Alaskan coast directly across a deep and uninterrupted polar basin. . . .

Bennett Island is north of the New Siberian Islands, off the Russian arctic coast, whereas Point Barrow and Flaxman Island are along Alaska's northern shore on the Beaufort Sea. The implication of the differences in tides was that some landmass intervened to moderate the tides at the latter locations. Harris continued to reinforce the point by looking at the discrepancy between observed tidal ranges from Teplitz Bay in Franz Josef Land in the Kara Sea east to Flaxman Island, and tidal ranges calculated upon the assumption of an uninterrupted and deep polar basin. Then he started to make remarkably precise predictions:

> From various indications it will be assumed that the land in question is trapezoidal in form and that it contains nearly half a million square miles. One corner has

---

[39]  See W. R. Hunt, *Stef*, pp. 57 et seq. gives an account of the newspaper response and that of scientists interviewed.

[40]  R. A. Harris, *Arctic Tides* (Coast and Geodetic Survey, Washington, D.C.: Government Printing Office, 1911) pp. 7 et seq., 30 et seq.

been placed to the northward of Bennett Island [using tidal evidence and the drift of the *Jeannette*]. Another corner has been placed to the northward of Point Barrow. . . . From this corner the coast line is assumed to trend in an easterly direction nearly to Banks Land, thus forming the north boundary of the Beaufort Sea. . . . Another corner has been placed to the north-west of Banks Land. . . . Another corner has been placed north of Grant Land [Ellesmere Island]. This is indicated by the discovery of Crocker Land by Peary. . . .

After this wonderful construction, breathtaking in its confidence, it is bathetic to find Harris allowing that "The coast line next the Pole is somewhat uncertain in position."[41]

Stefansson could not have asked for better timing. On his return from the Arctic, he found Harris's theory being vigorously debated. Peary, with his sighting of Crocker Land at stake, was in eager agreement with Harris, and soon found himself corresponding with Stefansson about the next arctic expedition.[42] Greely, who unlike Franklin had survived disaster and was now a general and major authority on arctic affairs, also took Harris's side. On 16 November 1912, the *New York Times* proclaimed: "SCIENTISTS BELIEVE IN ARCTIC CONTINENT. Gen. Greely and Dr. Harris Convinced There is New Land for Stefansson to Discover." There was, of course, a slight problem; the refutation of Harris's theory implicit in Nansen's epic drift in *Fram,* from the vicinity of the New Siberian Islands across the arctic basin, with no evidence of any new continent. But Harris argued that tides proved the existence of a continent. His model was relatively simple; the reality was quite otherwise, arising from enormously complicated fluid dynamics, land contours, ice distribution, and wind patterns. In effect, his model was a better candidate for chaos theory than for definitive prediction. When it came to a contest in North America, Nansen's experience counted for less than Harris's model. Popular and official opinion in Washington and Ottawa coalesced around Harris's views, and Stefansson knew how to benefit. He was in good odor with the American Museum of Natural History, thanks to his ethnological work and Anderson's zoological collections. Peary and, less consistently, Greely encouraged the geographical establishment to support Stefansson. The initial sponsors for his new expedition were to have been the American Museum of Natural History and the National Geographic Society, but the support they offered still fell short of what he needed for a ship for oceanic exploration away from the coast.

[41] Ibid., pp. 90–2.
[42] VS to R. E. Peary, 19 and 29 Nov. 1912, National Archives, Washington, D.C. [NARS] RG 401/1, Admiral R. E. Peary, letters received 1912.

*Enter Canada.* Stefansson widened his search for funds, and once more approached Reginald Brock, director of the Geological Survey of Canada.[43] Brock arranged an interview with Sir Robert Borden, the prime minister of Canada.[44] He also made Stefansson's case for him, in a letter to W. J. Roche, minister of mines, telling him that Stefansson had made good scientific collections on his previous expedition, and "the work was done remarkably cheaply." The new expedition would be almost entirely in Canada, so the government ought at least to be involved. The best thing would be a purely Canadian expedition, but with a total estimated cost of $78,000, this might be too expensive. "Participation to some extent is advisable in case any new lands should be discovered, a contingency that is not beyond the range of possibility. It is practically the one remaining place in the world where great geographical discovery is possible."[45] Brock's advice was transmitted to Borden;[46] Stefansson then talked with the prime minister and submitted a proposal for the expedition.[47]

The main object was "To discover new land, if any exists, in the million or so square miles of unknown area north of the continent of North America and West of the Parry Islands." Secondary objects were "To gather scientific information and collections . . . " Stefansson observed that, "As the larger part of the labours of the proposed expedition will be in Canadian territory, and as the Dominion of Canada may desire to lay claim to any lands that may be discovered, the Government is respectfully asked to support the expedition by a grant of Twenty-five thousand dollars, the remaining third of the money needed."

A subcommittee of the Privy Council then met for further discussion with Stefansson:

"We decided that if it could be arranged we thought it advisable for the Dominion to pay the whole cost of the proposed expedition,[48] on condition that Mr. Stefansson would become a naturalized British subject before leaving and that the expedition would fly the British flag. In this way we would get the entire benefit of the expedition and Canada would have any land that might be discovered.[49]

43   The GSC had supported the Stefansson–Anderson expedition 1908–12.
44   VS, *The Friendly Arctic*, p. viii.
45   Brock to Roche, 4 Feb. 1913, Borden Papers Series 3 File 2117 (RLB) microfilm FB674 reel 90, Robarts Library, University of Toronto.
46   Borden's office also had a copy of Spencer's paper arguing against land at the pole (*American Journal of Science* [1905] 333–40): see Borden Papers Series 3 File 2117.
47   VS to Borden, 4 Feb. 1913, PAC MG 26H vol. 234 file RLB 2117 ff. 130238–41.
48   Net expenditures by the Department of the Naval Service and the GSC were $516,332.17 and $43,630.13 respectively: PAC MG30 B40 vol. 10, summary of amounts expended for CAE 1913–18 (compiled from the Reports of the Auditor General, Sessional Papers).
49   7 February 1913, PAC MG26H vol. 234 file RLB 2117 f. 130245.

Stefansson indicated that it would be very satisfactory to have the whole cost of the expedition paid from one source. He had been born in Canada, and simply stated that he resumed his citizenship in 1913.[50] Funds were committed, and the venture became exclusively the Canadian Arctic Expedition. Borden was unwilling to extend partial aid, but he was willing to take over the entire operation, and to fund it to the tune of $75,000.[51] Henry Fairfield Osborn,[52] President of the American Museum of Natural History, and Gilbert H. Grosvenor, Director and Editor of the National Geographic Society in Washington, D. C., gracefully withdrew on being informed of the Canadian government's sponsorship; because Stefansson had earlier finessed the museum into offering support, withdrawal would have come easily, even gratefully.[53]

The government was right to be concerned over the territorial issue. Sverdrup's discoveries had opened the way for Sweden and then Norway to lay claim to part of the archipelago. Then there was the Alaska Boundary dispute,[54] which had made Canadians nervous about American aims in the North. Russia had not been a perceived contender since the time of Confederation.[55] Still, the observation made in 1905 by W. F. King, chief astronomer of Canada, was still true in 1913: "Canada's title to some at least of the northern islands is imperfect."[56]

Borden's government was concerned, and sought guidance from the British government. The response was not reassuring.[57] The full extent of the islands annexed to Canada in 1880 "has nowhere been formally

[50]  VS, *Discovery*, p. 246. Stefansson had become a naturalized American subject when his family moved to the U.S.A. Both countries have claimed or ignored him as it suited them.
[51]  MS note n.d., PAC MG30 B44 vol. 10.
[52]  Osborn wrote to Borden on 27 Feb. 1913: "While it is a great disappointment for us to forego this expedition and to relinquish the services of an explorer who has been carrying on scientific work in the North for so many years, I feel that it is eminently fitting that this work should be undertaken by the Canadian Government, to which the scientific and all other practical results naturally accrue." (Borden Papers Series 3 [RLB] file 2117).
[53]  Osborn to VS, 12 Feb. 1913, PAC MG30 B40 vol. 1. Diubaldo, *Stefansson and the Canadian Arctic*, p. 65.
[54]  *Alaskan Boundary Tribunal, Cases, Counter-cases, Arguments, Atlases of United States and Great Britain* (Washington, D.C., 1903, et. seq.); Canada had made significant territorial concessions in the negotiations. S. R. Tompkins, "Drawing the Alaska boundary," *Canadian Historical Review*, 26 (1945) pp. 1–24.
[55]  T. E. Armstrong, *Russian Settlement in the North* (Cambridge, 1965) and F. A. Golder, *Russian Expansion on the Pacific 1641–1850* (Cleveland, 1914) give a general account of Russia's northern American ambitions.
[56]  W. F. King, *Report Upon the Title of Canada to the Islands North of the Mainland of Canada* (Ottawa, 1905) p. 8, cited in Diubaldo, *Stefansson and the Canadian Arctic*, p. 5.
[57]  L. Harcourt, Downing Street, 10 May 1913, to the Officer Administering the Government of Canada, PAC MG26H vol. 234 file RLB 2117 f. 130279–81.

defined . . . " His Majesty's Government gave Canada authority to annex
any lands to the north that had not been acquired by any foreign power,
but that might not yet be British. However, as "it is not desirable that
stress should be laid on the fact that a portion of the territory may not
already be British, I do not consider it advisable that this despatch should
be published . . . "[58]

Given these concerns, Stefansson's proposal to Borden was shrewdly calculated to win approval. Instructions from Desbarats, deputy minister[59] of
the Department of Naval Service, conformed closely to Stefansson's wishes.
The expedition was to have two principal objectives: "First the exploration
of unknown lands and seas, and second the gathering of scientific information with respect to these areas, and also the partly unknown lands and seas
in the vicinity of Coronation Gulf." To achieve these aims, the expedition,
under Stefansson's overall command, was to be divided into two, a northern party led by Stefansson to explore the Beaufort Sea, and a southern
party led by Anderson devoted to scientific work.[60]

Stefansson was delighted with the Canadian government's acceptance of
his proposals, and its generosity in facilitating their realization. He cabled
to Peary, with whom he was in regular touch at this time, asking him "to
give newspapers interview commending government's action."[61] That action, as he had told Peary, evinced

a broad minded scientific attitude quite beyond my ideas of politicians in general,
but this is due largely, no doubt, to the Prime Minister, . . . who is beyond the gen-
erality of politicians. He takes the position that governments should be especially
interested in the increase of knowledge and that Canada should be especially interested in the exploration of her own dominions by land and sea.[62]

New land and sovereignty were Borden's concerns; Stefansson was especially excited by the prospect of new land. "I have always thought," he
wrote in his account of the expedition, "that the discovery of land which
human eyes have never seen is about the most dramatic of possible experiences. I don't pretend to be used to it or past the thrills that go with it."[63]
The prospect of new land was "the main reason for there being an expedition at all."[64] Now Stefansson, explorer and adventurer, needed scientists
for his expedition. The Geological Survey provided geologists, cartogra-

[58]   Ibid., ff. 130280–81.
[59]   Equivalent to Permanent Secretary in the British Civil Service.
[60]   G. J. Desbarats to VS, Ottawa 29 May 1913, PAC MG30 B40 vol. 1 file 13.
[61]   VS to Peary, 25 Feb. 1913, NARS RG 401/1, Adm. R. E. Peary Correspondence received
       1913.
[62]   Ibid., 16 Feb. 1913.          [63]   VS, *The Friendly Arctic*, p. 517.          [64]   Ibid., p. 97.

phers, and meteorologists. O. E. LeRoy, Brock's lieutenant, advised him that the geological and topographical surveying parties "should be complete and distinct units of the main expedition. . . . [They] should be enabled to leave the field on the completion of their work without regard to the main expedition."[65] Anderson was the senior zoologist; Stefansson had to select the other scientific personnel, including oceanographers and anthropologists, and recommend them to Brock, who, as director of the Geological Survey of Canada, could then approve the appointments. Here was a division of responsibility with the Department of Naval Service that caused inevitable jealousies.

Anderson received his official appointment as mammalogist in the Geological Survey Branch of the Department of Mines,[66] and took on most of the desk work for outfitting and personnel, which Stefansson happily delegated to him. Thus when Stefansson received a letter from a Norwegian applicant, Bjarne Mamen, seeking to join the expedition, it was Anderson who replied to Mamen urging him to approach Brock, and on the same day Anderson wrote to Brock recommending Mamen, who thus became the expedition's assistant topographer.[67] Meteorological instruments would be needed: Stefansson asked Anderson to write to Stupart about them, and also to see if the Dominion Meteorological Office could send someone to make "complete meteorological observations." Anderson also asked Stupart about tying in magnetic with meteorological observations; in the event, William Laird McKinlay, a teacher of mathematics and science in Glasgow, was the expedition's magnetician and meteorologist, who was chosen by Stefansson on his way through Edinburgh.[68]

Stefansson was there in search of advice about oceanography, sounding apparatus, and related hydrographic and oceanographic equipment. There was little time if the expedition was to leave that summer. Rapid consultation, acquisition, hiring, and travel were needed – and Stefansson was best at travel.

Roald Amundsen had advised Stefansson and Anderson that the only place to order deep-sea sounding apparatus was from Richter of Berlin – but it was too late in the day to order from him. "Amundsen said that the best chance would be to try to [get] that apparatus from the Prince of Monaco

---

[65] PAC MG30 B40 vol. 10, n.d.     [66] Letter 3 May 1913, PAC MG30 B40 vol. 10.
[67] Mamen to VS, 25 Feb. 1913; RMA to Mamen, 21 March 1913; RMA to Brock, 21 March 1913: PAC MG30 B40 vol. 1.
[68] RMA to Stupart, n.d. [March 1913], and Stupart to RMA, 14 March 1913, PAC MG30 B40 vol. 1. McKinlay's account of the expedition, a powerful indictment of VS, is *Karluk: The Great Untold Story of Arctic Exploration* (London, 1976). VS cable to RMA, 23 April 1913, McKinlay cable to RMA 26 April 1913, PAC MG30 B40 vol. 1.

as a favor. The Prince is said to keep a good supply of such things on hand. Mr. Stefansson went to Europe prepared to try Richter as well as the Prince of Monaco."[69] Stefansson would try to get all the hydrographic apparatus, including sounding machines, while he was in Europe.[70]

Meanwhile Fritz Johansen, a Danish oceanographer then working for the U. S. Department of Agriculture, was supplying Anderson with his want lists. Anderson and Stefansson wanted to secure Johansen for the expedition; there were precious few experienced marine biologists and oceanographers anywhere, and even fewer who wanted to go to the Arctic. After some wrangling about salary, Brock approved the appointment.[71] Johansen turned out to be an enthusiastic collector, but lazy or inefficient or both at everything else.[72] His list for zoology included trawls, dredges, and seines. For hydrography, he requested salinometers from Christiania, reversing thermometers to measure the temperature at determined depths from Negretti and Zambra in London or Richter in Berlin, reversing water bottles to collect samples at determined depths, 6,000 fathoms of thin wire for soundings on sledge trips, and a steam-sounding machine with snappers.[73]

Sounding machines were the most important pieces of scientific equipment where Stefansson was concerned, because of their utility in mapping the continental shelf and thus finding new land. They had undergone rapid development in the second half of the nineteenth century, thanks to the development of telegraphy and the corresponding need to find the best routes for submarine cables.[74] William Thomson, Lord Kelvin, had worked on telegraph cables, and in 1872 was led by that work to design the first sounding machine. Piano wire was wound on a wheel, and a weight was attached to the free end of the wire. As the weight ran into the sea, "a resistance exceeding the weight of the wire [and its attached weight] actually submerged at each instant was applied tangentially to the circumference of the wheel, by the friction of a cord wound round a groove in the circumference, and kept suitably tightened by a weight."[75]

[69] RMA to F. Johansen, 7 March 1913, PAC MG30 B40 vol. 1.
[70] RMA to Johansen, 15 March 1913, PAC MG30 B40 vol. 1.
[71] RMA to Brock, 7 March 1913; VS to RMA, 24 March and 5 April; Johansen to RMA, 14 April; RMA to G. J. Desbarats, 14 April; Johansen to RMA, 30 April; VS to Johansen, 3 May: PAC MG 30 B40 vol. 1.
[72] O'Neill to Le Roy, 13 Oct. 1918, PAC MG30 B66 vol. 1 file 2.
[73] Johansen to RMA, 12 March 1913, PAC MG30 B66 file 9.
[74] See, e.g., Great Britain, Privy Council, *Report of the Joint Committee Appointed by the Lords of the Committee of the Privy Council for Trade and the Atlantic Telegraph Company to Enquire into the Construction of Submarine Telegraph Cables* (London, 1861).
[75] W. Thomson, "On Deep-Sea Sounding by Pianoforte Wire," *Proceedings of the Philosoph-*

In the 1880s, Francis Lucas, a telegraph engineer, developed a particularly compact machine with a spring-loaded paying out wheel and an easily regulated friction strap; by the end of the century, his machine had become standard on Royal Naval vessels. Because of its compactness, it was useful for sounding from small vessels, and since it carried, attached to the sounding weight, a mechanical scoop or grab (a "snapper"), it was ideal for sampling gravel bottoms. It remained in use until recently superseded by echosounders.[76] It was natural that Johansen would want a Lucas machine, and that Stefansson would concur.

Stefansson hoped to get the Lucas sounding machine from the Prince of Monaco,[77] but first consulted Sir John Murray of *Challenger* fame.[78] He was treated by Murray to "a technical discussion of two or three hours as to various forms of sounding machines, dredges, nets and other paraphernalia for ocean investigation." He also had the advice of W. S. Bruce, director of the Scottish Oceanographical Laboratory. Bruce recommended an oceanographer named James Murray, who had served with Shackleton and worked with John Murray on the Scottish Lochs Survey.[79] There could thus be two oceanographers on the expedition, one with the northern exploring party, and one with the southern party, which was to have fixed bases and an elaborate scientific agenda.

Stefansson cabled wildly from Europe. On 24 March he announced: "EXPECT SAILING WEDNESDAY OCEANIC CAMPGEAR, CLOTHING, INSTRUMENTS PURCHASED. SHACKLETON'S MURRAY SELECTED OCEANOGRAPHER JOHANSEN STILL DESIRABLE." Then on 26 March came: "BUY IMMEDIATELY AMERICAN THINGS JOHANSEN REQUISITIONS INCLUDING 6000 FATHOMS SOUNDING WIRE. MURRAY WILL PROVIDE EUROPEAN APPARATUS." On 11 April: "STAYING LONDON NEGOTIATING LARGE MONEY GRANTS FOR EXPEDITION AND OCEANOGRAPHIC OUTFIT."[80] He later explained to Desbarats of the Naval Service that he had planned to go from London to Berlin and to Monaco to secure instruments and seek

*ical Society of Glasgow* 9 (1875) 488–93; quoted in McConnell, *No Sea Too Deep* (1982) p. 60.

[76] McConnell, *No Sea Too Deep*, pp. 70–1 and figs. 73, 74, and 75.
[77] RMA to Johansen, 15 March 1913, PAC MG30 B66 vol. 9.
[78] VS to RMA, London, 21 March 1913, PAC MG30 B66 vol. 9.
[79] VS, *The Friendly Arctic*, p. 30. Anderson commented (PAC MG30 B40 vol. 10 file 4) that Murray "was rather old for strenuous Arctic work, being the oldest member of the expedition, and exceedingly over-confident, largely as a result of his participation in one of the Shackleton Antarctic Expeditions."
[80] PAC MG30 B40.

Figure 35. Lucas deep-sea sounding machine. Science Museum London 1975/390, neg. no. 960/76.

advice, but found instead that it would be quicker, cheaper, and just as efficient to go to Rome instead.[81] Anderson became indignant at all these changes, and began to think that Stefansson was enjoying European irresponsibility, leaving him behind to do the work.

---

[81]   VS to Desbarats, 1 June 1913, PAC RG42 vol. 475 file 84-2-29.

Somehow, the apparatus was assembled. The Prince of Monaco supplied Légér sounders and Richard water bottles.[82] Also acquired for the expedition were piano wire for deep-sea sounding, as requested by Johansen and recommended by the cable companies;[83] braided copper wire, as recommended by Bruce; and Kelvin and Lucas sounding machines.[84] Jack O'Neill, a geologist appointed to the expedition by the GSC, wrote back to Ottawa from Nome, Alaska, that they were pretty thoroughly equipped, and that Stefansson had paid more attention to detail than he had first given him credit for.[85]

Stefansson also rushed around looking vainly for somebody to do magnetic work for the Carnegie Institution, wondering about a possible entomologist, and scooping up Henri Beuchat, an anthropologist from Paris: "He . . . seems to be admirably qualified, (considering the fact that we have no choice) for anthropologists seem to be exceedingly rare."[86] Once again, Stefansson was willing to take a scientist more because he was willing than because he was ideally or even appropriately qualified. Anderson later described Beuchat as "a scholarly man with a deep knowledge of American anthropology and archaeology derived mostly from museum and library study, a man of agreeable personality and great ambition and willingness but of no experience at all in field or outdoor work . . . "[87] Finding qualified explorer-scientists proved remarkably difficult. The expedition was Canadian, and so, as Stefansson noted,

we preferred Canadians in our choice, yet we were able to get in Canada only five out of a staff of thirteen. We turned next to other parts of the Empire, and secured three men from Scotland, one from Australia, and one from New Zealand. Even so, we had to look further and take one man from France, one from Denmark, one from Norway, and two from the United States.[88]

[82] See McConnell, *No Sea Too Deep*, p. 133–4.
[83] McConnell, *No Sea Too Deep*, p. 60 discusses different kinds of sounding wire.
[84] PAC MG30 B40 vol. 10, accounts for CAE include: To US. Steel Products Co. N.Y. – 3600 ft. steel hoisting rope, plus 3600 feet no. 11 music wire for deep sea sounding; to [Ernest] Gichner, Washington, D.C. – Frames for beam trawl, rectangular and triangular dredges & c; to Sorenson Bros., Washington, sieves and plankton buckets; to Gichner, Washington, 12 Agassiz tanks (copper tanks tinned inside with screw tapped covers and wooden containing chests). A. Taffe 5 June 1913 and VS to Desbarats, PAC RG42 vol. 475 file 84-2-29.
[85] 17 July 1913, PAC MG30 B66 vol. 1.
[86] VS to RMA, London, 1 March 1913, PAC MG30 B40 vol. 1.
[87] RMA, "The Canadian Arctic Expedition," typescript PAC MG30 B40 vol. 10 file 4.
[88] VS, "Solving the Problem of the Arctic: A Record of Five Years' Exploration," *Harper's Magazine* 138 (1919) 577 et seq. at 579.

The need for an international corps of scientists was probably the rule rather than the exception; the staff and publications of the *Challenger* expedition were by no means all British.[89]

We have already encountered several of the scientific staff. Among the others, the New Zealander, Diamond Jenness,[90] would turn out to be a real coup, for he became Canada's leading anthropologist. The Australian, George Wilkins, was the expedition's cinematographer, and went on to distinguish himself as a leading polar pilot and explorer. The Scots included James Murray, oceanographer, and Alistair Forbes Mackay, surgeon. Kenneth Chipman and John Cox were topographers and John O'Neill was seconded from the Geological Survey. George Malloch was added as a geologist for work with the northern party.[91]

From the outset, the appointees of the Geological Survey found themselves at odds with Stefansson's style. On 26 May Chipman sent a memorandum to the topographer in charge at the Survey: "I am convinced that the leader is unfamiliar with where scientific work needs to be done, methods of doing, equipment necessary[92] and facilities for carrying out. . . . This is especially true as concerning geography for the zoologist, anthropologist etc. have positive ideas as to their needs for maps on which to place their work. . . . "

Chipman as a topographer understandably viewed geography as the foundation of other scientific work; one needed to know exactly where a tribe lived or where a specimen had been collected: "I believe the leader's idea of travel living and work to be perhaps suitable for one or two man expedition in his own line of work, but not for organized work in the Coronation Gulf country. . . . "[93]

Here were complaints echoing those of Anderson from the previous expedition, pointing to a general incompatibility between Stefansson's buccaneering geographical and anthropological exploration, and the methodical demands of organized science – especially if the science was organized through a government department. No previously mounted Canadian expedition had been so organized, or given such careful instructions from Ottawa. On 29 May, Desbarats wrote to Stefansson, amplifying the instructions that had been set out in the Order-in-Council.[94]

89  See, e.g., T. H. Tizard, *Narrative of the Cruise of H.M.S. Challenger with a General Account of the Scientific Results of the Expedition* (Edinburgh, 1885).
90  Brief biographical accounts are W. E. Taylor Jr., "Diamond Jenness (1886–1969)," *Arctic* 23 (1970) 71–81, and "Foreword," D. Jenness, *The Indians of Canada*, 7th ed. (Toronto, 1977) pp. v–x.
91  Brock to RMA, 17 April 1913, PAC MG30 B40.
92  But see O'Neill's more favorable judgment above.          93  PAC MG30 B66 vol. 1.
94  PAC MG30 B57, file Correspondence 1903–12.

Figure 36. Scientific staff of CAE at Nome, Alaska, 13 July 1913, prior to sailing of the *Karluk*. Front row, l. to r.: F. Mackay, surgeon (died on the ice of Chukchi Sea); Capt. Robert Bartlett, skipper of C. G. S. *Karluk;* V. Stefansson; R. M. Anderson, zoologist and second in command; James Murray, marine biologist (died on the ice of Chukchi Sea); Fritz Johansen, marine biologist; Back row, l. to r.: B. Mamen, meteorologist (died on Wrangel Island); B. M. McConnell, secretary; K. G. Chipman, topographer; (behind Chipman) G. H. Wilkins, photographer; George Malloch, geologist (died on Wrangel Island); Henri Beuchat, anthropologist (died on ice of Chukchi Sea); J. J. O'Neill, geologist; D. Jenness, anthropologist; J. R. Cox, topographer; W. L. McKinlay, magnetician and meteorologist. Public Archives of Canada PA-74063.

The northern party under Stefansson would sail from Victoria on Vancouver Island, for "the exploration of unknown seas and lands" in the Beaufort Sea and neighboring waters. Every effort should be made to ascertain whether or not there was land there, and "Should any such lands be discovered, they should be taken possession of and annexed to His Majesty's Dominions." Given the uncertainties of arctic navigation and exploration, the work of the northern party could not be specified in detail, but was rather left to Stefansson. He was, however, given a clear statement of priorities, beginning with the search for new land:

*Exploration* - (1) Geographical. (2) oceanographical and Biological (marine). (3) Geological. (4) magnetical. (5) anthropological. (6) biological (Terrestrial). (Meteorological is not specified as that can always be carried on without interfering with other investigations.)

The southern party, in contrast, should operate from a fixed base "in the extreme northern land of Canada," perhaps in Coronation Gulf, and its priorities in order were: geological, geographical, anthropological, biological, and photographic. The geological work, their first priority, was directed especially to copper-bearing rocks along the coast. Anderson was the leader of the southern party, which had its own vessel, the schooner *Alaska*.

Anderson and Stefansson were thus made respective leaders of parties with different styles, different values, and different priorities, and the fact that Stefansson was overall expedition leader might conflict with but did not detract from Anderson's responsibility for ensuring that the work of the southern party went ahead. The relevant part of his instructions was emphatic:

> The work undertaken by these [southern] parties should be of a high order for this class of exploration and should make a distinct advance over previous work. To secure such results, the geological and topographical sub-parties should follow closely the regular scheme for field parties engaged on reconnaissance work adopted by the Geological Survey . . . [their] programme of work and freedom of movement should not be interfered with or hampered by biological or ethnological considerations.[95]

It was a prescient document in its anticipation of the conflicts that would emerge.

*The northern party* – Karluk. The northern party left Victoria on 20 July 1913. Their ship, *Karluk*, was a twenty-eight-year-old brigantine whaler,[96] with a bow strengthened for polar work, but as her captain, Robert Bartlett, later remarked, "she had neither the strength to sustain pressure, nor the engine power to force her way through loose ice."[97] Stefansson foresaw that miscalculation or accident could find the ship caught in the ice and launched upon a polar drift between the pole and the mainland of Asia, unless the ship was first crushed. "In the event of nothing being heard from us for a year or more, I want to urge that no great

95  Ibid.
96  K. G. Chipman, private diary no. 1, entry 17 June 1913, PAC MG30 B66 vol. 1.
97  R. Bartlett, *The Last Voyage of the Karluk: Flagship of Vilhjalmur Stefansson's Canadian Arctic Expedition of 1913–16* (Boston, 1916) quoted in McKinlay, *Karluk*, pp. 11–12. Bartlett was an experienced Arctic explorer; a native of Newfoundland, he commanded the *Roosevelt* and was "nearest the North Pole" with Peary in 1909.

fears for our safety need be entertained. We may, of course, all be gone; but the chances are that we shall all or most of us be safe."[98] Such breezy and irresponsible assurance was not what the scientists of the expedition wanted to hear from Stefansson, nor what one would expect from the leader of a national scientific expedition. If Bartlett and others were right, and *Karluk* could not withstand the pressure of the ice, then food and instruments[99] alike would be lost with the ship, so the members of the northern party promptly decided that they too needed a fixed base, on Prince Patrick Island or Banks Land. Stefansson's interest was in mobility, and opposed to the stasis of a permanent center of operations. The discussion was heated and convoluted, and never got round to resolving the question of such a base.[100]

*Karluk* steamed through Bering Strait and eastward along the coast of Alaska, until, in August, the ice seized her just to the west of Flaxman Island, where she began a fatal drift. In September, Stefansson and a small party left the ship for a caribou hunt, in line with his ideas of living off the land – and wind and currents swept the ship away from them.[101]

As the *Karluk* drifted west and north, the scientists left on board dredged and sounded. Early in October, they began to move into deeper water. Bartlett's journal records: "6 October 7:00 A.M. Soundings 28 fathoms. The dredge finds a different kind of sea-fauna – now that we are outside the 20 fathom curve . . . We are getting soft shell crabs, coral, &c, & losing the mud and silt . . . 10 October. Soundings 3 A.M. 180 fathoms – no bottom. We are now on the downward slope of the Continental shelf."[102] The depth increased rapidly to over 550 fathoms, and they switched from the Kelvin sounding machine to the Lucas Automatic Sounding Machine,

---

98 VS to J. D. Hazen, Minister of the Naval Service, 1 June 1913, en route to Victoria, PAC MG26H vol. 234 file RLB 2117.

99 VS to Mr. Reeves, W. Coast Amund Ringnes Island, 4 July 1916, DCL VS MSS 98 Folder IV-13: " . . . all the instruments I intended for my own use, as well as the instruments . . . of the *Karluk* part of our expedition went with her . . . and were lost with her when she sank. When I later on made connection with the *Alaska* branch of the expedition I was able to get only few instruments . . . as follows: one Nat. Phys. Lab, sextant by Heath, two 'astronomical' watches of the American Waltham Watch Company, a 'Battery' of three Waltham 'Chronometers' (really large watches), one prismatic compass, one maximum and one minimum Negretti and Zambra thermometers, one N. & Z. aneroid, one celluloid protractor, one pair of dividers and parallel rulers, and some pencils. I was able to buy from a Russian trapper of remarkable attainments for his station, a sextant and a set of Russian azimuth tables."

100 Copy of extract of letter from H. Beuchat to C. M. Barbeau, GSC, [1913], MS NMNS.

101 D. Jenness to Von Zedlitz, 16 Oct. 1913, PAC MG30 B89 vol. 2; McKinlay, *Karluk*, chaps. 4 and 5.

102 R. Bartlett, diary, PAC RG42 vol. 475.

operated through a hole in the ice at the ship's stern. At 995 fathoms we got a sample of the bottom, which was brown mud and sand. Winding in the wire was strenuous exercise, and working in relays, each relay winding in 100 fathoms, took us half an hour. The following day we had 1215 fathoms, about a mile-and-a-half deep, with the same bottom sample.[103]

To combat boredom, they organized themselves in teams, competing for the fastest time of wire retrieval. Murray meanwhile dredged at every opportunity. In the following month they recrossed the continental shelf. On 10 January 1914, the ship was crushed in the ice near Wrangel Island, on the western edge of the Chukchi Sea. They had time to unload the bulk of the stores on to the ice before the ship sank on the following day. Eight men died in the attempt to reach land, and the rest reached Wrangel Island. There followed a grim winter, a grim spring, and the rescue of the survivors late in the summer. They were taken back to Victoria.[104]

*The northern party. Ice expeditions. Tensions north and south.* Stefansson, with his hunting party, which included Jenness and Wilkins, returned safely to the mainland, rejoining the southern party at its winter base, Collinson Point, Alaska. Stefansson cabled from Barrow, Alaska to Ottawa in October,[105] explaining his separation from the ship, the probability that *Karluk* was still drifting, and his confident belief that all the ship's company were well – as indeed they still were in the fall of 1913.

Stefansson was always convinced that living off the land in the Arctic was a comfortable affair, a theme well worked in his expedition narrative, *The Friendly Arctic.* Anderson, in an interview following the publication of the book, described it as

a species of Arctic fiction couched in excellent literary style, pleasant to read while sitting in an easy chair, but precarious in the extreme to anyone who would seriously

[103]  McKinlay, *Karluk*, p. 38. They clearly did not have the steam-sounding machine that Johansen had requested.
[104]  A commemorative plaque erected in 1926 in the PAC is: "In memory of those who perished / Canadian Arctic Expedition / 1913–18. Alexander Anderson, Ships Officer; Charles Berker, Ships Officer; Peter Bernard, Ships Master; Henri Beuchat, Anthropologist; Daniel Wallace Blue, Ships Engineer; John Brady, Seaman; George Breddy, Ships Fireman; Edmund Lawrence Golightly, Seaman; John Jones, Ships Engineer; Alister Forbes Mackay, Surgeon: George Stewart Malloch, Geologist; Bjarne Mamen, Topographer; Thomas Stanley Morris, Seaman; James Murray, Oceanographer; André Norem, Steward; Charles Thompson, Seaman. / For Canada and for science / Pour la patrie et pour la science." Plenty of critics, then and now, have argued that the sacrifice was less for Canada and for science than for Stefansson's irresponsibility as a leader.
[105]  VS to Desbarats, 30 Oct. 1913, Barrow, DCL VS MSS 98 Folder IV-1.

Figure 37. *Karluk* crushed in ice. Public Archives of Canada C-24942.

Figure 38. Vilhjalmur Stefansson. Public Archives of Canada C-86406.

attempt to put the theories into practice . . . The exploitation of this theory by Mr. Stefansson . . . depends for its plausibility very largely on ignoring the heavy mortality among men who have tried it, including a considerable percentage of the men who have, unfortunately, been associated with Mr. Stefansson.[106]

[106]    PAC MG30 B40 vol. 10.

Stefansson did succeed in putting his theories into practice. Even if the *Karluk* were lost, there were other ships with the southern party. Altogether six ships were purchased during the expedition, of which only one, the C.G.S. *Alaska,* served from the start to the finish of the expedition.[107] Stefansson believed that even without ships, he could ski or sledge over the ice. The work of the southern party would not be affected, and he could carry on the work of the northern party somehow. He concluded his cable from Barrow: "Expedition's plans unchanged, merely delayed," a dramatic understatement of the fate of *Karluk* and those who had remained with the ship.

Stefansson was also underestimating the difficulties that he faced. Much essential gear for the northern party had been lost with the *Karluk,* and members of the southern party were unlikely to sacrifice their energies or their equipment for Stefansson. There was a division of loyalties, with the southern party determined to fulfill their responsibilities to the Geological Survey, and Stefansson determined that the exploratory work of the northern party would come first. Brock and Desbarats, in painfully delayed letters, backed up the southern party, and instructed Stefansson accordingly.[108] For Brock, the work of the southern party became "the paramount issue,"[109] and he roundly told Stefansson that "the disappearance of the KARLUK puts an end to the northern expedition, except what you may be able to accomplish yourself."[110]

Stefansson felt short-changed, and tried unsuccessfully to persuade Johansen to join the supporting party for the ice trip; in any event, he insisted that Johansen was to see that the proper sounding gear was made available, and while back at base was to carry out simultaneous tidal observations at various points along the coast, to help elucidate the character of "the unexplored polar sea."[111]

Stefansson tried to reassert the claims of his own part of the expedition, and according to Chipman he "made the remarkable statement that the Expedition was not essentially scientific, that scientific men were inclined to be narrow-minded and engrossed in their own lines;[112] that private

---

[107]   D. Gray, "C.G.S. *Alaska,*" *The Bulletin: Quarterly Journal of the Maritime Museum of British Columbia* (1979) no. 43 pp. 19–22, no. 44 pp. 2–4, 12.

[108]   Desbarats to VS, 30 April 1914, PAC MG30 B40 vol. 2.

[109]   Brock to RMA, 7 May 1914, PAC MG30 B40 vol. 2.

[110]   Brock to VS, 7 May 1914, PAC MG30 B40 vol. 2.

[111]   Johansen to VS, 6 Feb. 1914, VS to Johansen, 24 Jan. 1914 and 12 March 1914: DCL VS MSS 98 Folder IV-3.

[112]   The problem is an old one; cf. the troubles between Edmond Halley and his officers – see N. J. W. Thrower, ed., *The Three Voyages of Edmond Halley in the* Paramore *1698–1701,* 2 vols. (London, 1918).

individuals or a Government would not finance a scientific expedition on its own merits unless it had the spirit of adventure to catch the public interest."[113] Apart from ships, scientists, and sounding gear, Stefansson also wanted flags to plant on newly discovered lands. The flags were sent, with the advice that they were not really necessary, and that the mere act of planting them "has more of a sentimental than practical value."[114] Stefansson set off on his first ice journey delayed, frustrated, and under-equipped. He sent Wilkins back to base to ask for 5,000 meters of sounding wire:

I consider that the deep sea soundings we are making are of more scientific importance than any soundings that your party can make in your territory and for which they should conceivably need the wire, so even should the above quantity of wire leave Mr. Johansen short, that will not be a reason for not sending [it]. . . . We had only 1,386 metres of [braided] wire last year and could never get bottom except near land. This year we have about a 1,000 metres only left. It is of manifest importance for us to get bottom soundings if possible. [Send] any scientific equipment that you can spare – we have no thermometer except an advertising one – for instance.[115]

Soundings were important because a transition from shallow to deep water marked the edge of one continental shelf, and the reverse transition could mean the discovery of a new shelf, indicating a new continent. Stefansson's request, however, was not generously received, and his inventory of 27 April 1914 lists just one sounding machine and braided wire weighing 20 lbs.,[116] two leads of 3 and 6 lbs. respectively, and 4 lbs. of piano wire.[117] He also had two rifles and 305 rounds of ammunition to provide food. He took off across the ice, and by the late summer was assumed to be missing. Opinion was against sending out a rescue; certainly the Geological Survey would countenance no further loss of life from among their scientists, because, as LeRoy, Brock's lieutenant, put it, Stefansson's life was not worth it.[118]

Stefansson meanwhile had been demonstrating the validity of his own recipe for arctic travel, and by July had reached Banks Island. His party had "carried a line of soundings of over 4500 feet through 4 degrees of latitude and 19 degrees of longitude, most of it unexplored and all of it unsounded

---

113  Chipman to Brock, 15 May 1914, PAC MG30 B66 vol. 1.
114  Desbarats to VS, 31 March 1914, DCL VS MSS 98 Folder IV-2.
115  VS to RMA 6 April 1914, at sea, DCL VS MSS 98 Folder IV-4.
116  Ca. 1,000 meters.
117  Gauge unknown, so length unknown; in deep soundings (>1,000m) the wire broke under its own weight. We have seen that Johansen had received approval for his request for 6,000 fathoms of sounding wire.
118  Memorandum from O. E. Le Roy and W. H. Boyd to R. G. McConnell, Acting Deputy Minister, Department of Mines, 29 Oct. 1914, PAC MG30 B40 vol. 10.

territory [in the Beaufort Sea]: we have determined the 'continental shelf' off Alaska and off Banks Island, and have learnt something of the currents of the Beaufort Sea – most of this contrary to what men 'knew' before."[119] He wintered on Banks Island, meeting up with the expedition's second auxiliary schooner, the *Mary Sachs*, in September, to the surprise of its captain and crew, who had assumed that he was dead. Perhaps for this reason, or perhaps because of the jealousies and disagreements already indicated, none of his requests for sounding wire and related equipment had been met.

Stefansson remained on Banks Island as a base that winter, going north on to the ice in the spring of 1915, turning east to Prince Patrick Island, and discovering Brock Island. He discovered another landmass, only subsequently identifying it as a distinct island, Borden Island.[120] Stefansson wrote from the new land to the prime minister in Ottawa: "I feel strongly not only that Canada should explore the region to which she lays a claim as far as the pole; it is true also that by doing so she makes good her claim."[121]

In 1916, Stefansson sledded round most of the coasts of his newly discovered land, now known as Brock and Borden Islands, and went on to the northwest to discover Meighen Island to the north of the Ringnes Islands. Nineteen seventeen added Lougheed Island to the east of Borden Island. These sledge trips yielded not only new land, but also soundings of the ocean depths. Stefansson was taken ill in 1918, and while he recovered first from typhoid and then pneumonia, one of his crew, Storkerson, led an ice drift party out into the Beaufort Sea, making soundings and effectively proving the nonexistence of suppositious islands there.[122]

Borden was delighted with the results – thousands of square miles had been added to Canadian territory.[123] Desbarats subsequently listed the principal results of the northern party's work: definite information concerning Banks Island, Prince Patrick Island, and other islands formerly imperfectly known; the discovery of new land "which was claimed as part of the British Empire"; the investigation of a large area of the Beaufort Sea; the investigation into the life of the Eskimos frequenting the very northern islands; the investigation of animal life; and the proof that explorers could make long journeys in the Arctic without a mass of supplies.[124] Stefansson had discovered approximately 100,000 square miles of unknown sea and

[119]  VS CAE diary no. 1 f. 107, 29 June 1914, PAC MG30 B81 vol. 1.
[120]  Hunt, *Stef*, p. 128.
[121]  VS to Borden, 21 June 1915, PAC GM26H vol. 185 file RLB529.
[122]  Storker T. Storkerson in VS, *The Friendly Arctic*, pp. 689–703.
[123]  Borden in *The Friendly Arctic*, p. xxiv.
[124]  Desbarats to the Secretary to the High Commissioner for Canada, 3 May 1921, DCL VS MSS correspondence Box 10.

land. Anderson, whose differences with Stefansson had soared during the expedition, was predictably unconvinced. Stefansson for him was a fraud, a faker, a liar, and everything that was offensive. That included his mapping and sounding:

> I . . . know no[w] better than I ever did before that any geographical or mapping work (which is all that he is supposed to be doing) on his plans are mighty vague. He doesn't bother to carry surveying instruments to locate himself accurately. He couldn't use them if he had them, for want of knowledge of scientific geographical methods . . . Their method of rushing blindly ahead from one seal to the next seal they can kill does not allow the stops that must be made to get suitable weather for observations, and anything such parties can add to the map (if he can persuade Geographical Societies with sufficient eloquence to hoodwink them) will be as misleading as the old charts.[125]

Anderson's strictures were harsh. Stefansson had done valuable geographical work, had made some contribution to mapping the continental shelf, and had collected, although not digested, a good deal of ethnological data. In other respects the northern party accomplished very little. The official manuscript journal of the northern party from 1914 to 1918 shows more ambition than achievement.[126] Most extensive were the meteorological observations and the list of ethnological specimens, including thirty-nine "suits [of] native clothes" from Victoria Island. The zoological and botanical specimens were few. For the rest, there were section headings for ornithological, magnetical, tidal,[127] anthropological, and physiographical data, as well as for ice conditions . . . and each of these headings was followed by one or more blank pages. Stefansson's success as an explorer could not have been more clearly contrasted with his lack of scientific discipline, nor could the work of the northern party with that of the southern party.

*The southern party.* The southern party, under Anderson's leadership, had plenty to do apart from feuding with Stefansson. After the first winter

125   RMA to [MBA?], 17 Jan. 1916, PAC MG30 B40 vol. 2.
126   "Official Journal during the Canadian Arctic Expedition 1913–18," PAC RG42 vol. 345.
127   VS did accomplish more than this record suggests, although erratically. For example, at Cape Isachsen, Ellef Ringnes Island, from 7 a.m. on 2 June to 8 a.m. on 3 June he and two other members of the northern party made tidal observations at 15 min. intervals around the clock, and similar observations were made elsewhere on other occasions: PAC RG42 vol. 349. On 1 Feb. 1916, VS wrote to R. A. Harris at the U.S. Coast and Geodetic Survey, with data and arguments concerning Harris's theory of a new Arctic continent; Harris did not receive the letter before his death on 20 Jan. 1918 (R. L. Faris, Acting Superintendent, Dept. of Commerce, U.S. Coast and Geodetic Survey, Washington, to VS, 25 March 1918, DCL VS MSS 98 Folder IV-40: I have not seen VS's letter of 1 Feb. 1916).

at Collinson Point,[128] they continued eastward to Bernard Harbour in the Northwest Territories, where Dolphin and Union Strait open out into Coronation Gulf. There they established headquarters. That winter, Cox surveyed the coast eastward to Lockyer Point. Chipman in his surveying along the coast was able to appreciate the quality of Leffingwell's survey.[129] Jenness began to study the local Inuit and to "dig out some old village sites." Perhaps his most important work that winter was linguistic. He wrote to his mentor, Professor von Zedlitz in New Zealand:

I tried hard to learn Eskimo but it is frightfully difficult. The grammar is very complex and I had not a single book to help me. I tried to make a grammar for the dialect of the people I was with, but it is very incomplete, though as far as I have yet discovered comparatively accurate.[130]

Stefansson and Jenness were a study in contrasts. Jenness was painstaking, aware of the limits of his knowledge, scrupulous, and precise. What he achieved in the study of language had an enduring value.[131] Stefansson was more swashbuckling, boldly grasping new forms of language and building an extensive grammar and vocabulary. This linguistic knowledge served him well in the field, but was full of inaccuracies that greatly reduced the scholarly value (for him, the value to pedants) of his work; and besides, he never published anything approaching a comprehensive vocabulary or grammar of Inuktitut. The language *was* difficult in its strangeness. Jenness concluded his summary account, after a deal of grammatical information:

Finally if you are an expert Eskimo scholar you will find a multitude of alternate endings, long interminable suffixes, or you will pile up infix after infix in a single word till it reminds you of the pictures you sometimes see of John Bull – there is a head somewhere at one end and feet at the other, but they are swallowed up in the enormous body. / Are you bored with all this? Would you like to hear of the phonology? You know what an ignoramus I am about everything relating to that. Yet here I have to record Eskimo where scarcely a single sound coincides with any known to Europeans or to the rest of the world. Beuchat gave me some help on

---

128 RMA Journal 1 Jan. 1914 – 31 May 1914, CAE records vol. 2, NMNS.
129 Chipman to W. H. Boyd, GSC, 7 Sept. 1913, PAC MG30 B66 vol. 1 file 2: "At Flaxman Island he [Leffingwell] has established an astronomical station with a base line and azimuth mark. His longitude has been established from occultations and his latitude and azimuth from star observations. From a micrometer and later taped base he has carried a triang[ulation] net some 160 miles along the coast and has cut in the mountains as far as he could thus attempting to tie in his work to that from the interior of Alaska."
130 Jenness to Prof. von Zedlitz (in New Zealand), 29 June 1914, PAC MG30 B89 vol. 2.
131 See especially *Report of the Canadian Arctic Expedition 1913–18*, vol. xv, D. Jenness, *Eskimo Language and Technology Part A: Comparative Vocabulary of the Western Eskimo Dialects* (Ottawa, 1928), *Part B: Grammatical Notes on Some Western Eskimo Dialects* (Ottawa, 1944).

Figure 39. Canadian Arctic Expedition, southern party's permanent headquarters, Bernard Harbour, 20 July 1915. Public Archives of Canada C-57953.

Figure 40. R. M. Anderson. Public Archives of Canada C-23666.

Figure 41. Chipman and Cox taking observations. Public Archives of Canada PA-138631.

the Karluk. I started out, like any novice, with a sort of script, and have been changing it all the time. . . . Stefansson used what seemed to me a very poor script, very
inadequate.

But then Stefansson was an explorer first, an ethnologist second. For Jenness, exploration was merely incidental to ethnology.[132] In April, he headed
for Victoria Island, remaining with the Inuit there until November.

Geological and topographical survey parties that summer explored the
Mackenzie delta, part of the coast of Coronation Gulf, and inland along
Rae River. Anderson had been kept busy as leader of the southern party,
looking out for the men and ships, and was not satisfied with the extent of
his zoological collections.[133] He was a good leader, and won the loyalty of
his party: "it would be hard," wrote O'Neill, "to get a better man for his
job."[134] Anderson was determined that the loss of *Karluk* not prove detri-
mental to the work of the southern party.[135] With their support, Anderson
did his best to meet expectations.

Jenness headed for Victoria Island in April 1915, staying there with the
Inuit until November, and returning to them in 1916, when he undertook a

---

[132] A pointed exchange between Jenness and VS in 1929 underlines the contrast. VS wrote to
DJ, then Chief, Division of Anthropology, at the National Museum of Canada (DCL VS
MSS Correspondence Box 22): "Thanks for the copy of your important 'Comparative
Vocabulary of the Western Eskimo Dialects.' I feel a little guilty. Was I in any way to
blame for the fewness of your Eskimo words? I remember some conversation we had
about it years ago but the details have escaped me. Possibly I promised to send you some
material and forgot to do it." DJ replied to VS on 4 March (DCL VS MSS Correspondence Box 22): "You have no reason to reproach yourself for the smallness of my vocabulary. You promised me nothing and I did not know that you had any material in your
possession. I have always regretted that your dictionary of the Barrow dialect was lost on
the Karluk, because it represented months of work that no one is likely to do again. My
vocabulary was collected incidentally to other work, but since it contained ca. 60% of the
stems known in Greenland, I thought it probably worth publishing. After all a small vocabulary accurately transcribed is infinitely better than one twice or three times its size in
which the spelling is erratic and the phonetic distinctions inadequate, so perhaps this little
paper of mine will not be unworthy of a place in the Arctic Series."

[133] RMA to Desbarats, 18 May and 16 Aug. 1914, and RMA to Brock, 21 Aug. 1914, PAC
MG30 B40 vol. 2 file 2, and CAE records, NMNS; Cooke and Holland, *The Exploration
of Northern Canada* (1978) p. 337.

[134] O'Neill to Le Roy, 13 Oct. 1913, PAC MG30 B66 vol. 1 file 2. Other views of RMA at
this time include Leffingwell's, as reported by Chipman to Boyd on 15 Dec. 1913:
" 'Anderson is either a damn fool or one of the finest men on the face of the earth' – he
couldn't decide which"; and Chipman's own view that "Dr. Anderson is all right – sort
of harmless – and quite ready to do what is best from the survey standpoint": PAC MG30
B66 vol. 1.

[135] Anderson was following instructions from Ottawa. See Brock to MBA 22 June 1914,
CAE records, NMNS; cf. Brock to RMA, 7 May 1914, PAC MG30 B40 vol. 2: since
*Karluk*'s "disappearance the southern work becomes the paramount issue and upon it the
expedition must depend for success."

reconnaissance of all the tribes from the Coppermine River to Kent Peninsula. His aim in 1915 was to follow the Victoria Island people "through all their summer life, a thing which no one else, as far as I know, not even Stefansson, has ever done.[136] For this purpose I adopted, or was adopted into an Eskimo family," consisting of Ikpukkuak, his wife Taktu, and their children. It was a real adoption; Jenness later wrote of them as his parents, and of himself as their child.[137] He worked with them, first sealing and then, as the Inuit moved to the land, fishing and hunting with them. "I was an awful duffer at first, but became rather more skilful after a time, thanks to Ikpuks untiring coaching." The caribou moved from the Barren Grounds of the Canadian tundra to Victoria Island in the spring, then back in late summer, offering two occasions to hunt them during their northward and southward migrations:

> With a good Ross rifle & Ipkuk to train me in stalking I was more successful with the deer than the fish. In fact this was my one consolation. In every art of gaining a livelihood the Eskimos were immeasurably my superiors – they could fish better, stalk better, endure hunger & fatigue more easily, pack heavier loads in their skins, were less sensitive to cold – but when it came to straight shooting my Ross rifle generally did double the execution that Ipkuk's 44 Winchester did.[138]

Jenness did not endure as well as his family, and for a while was prostrate with illness from "bad food." His family carried his pack, fed and nursed him, and called in their " 'amyakok' or doctor [who] held a seance over me to drive away the 'evil spirits' who were trying to kill me. Really they were wonderful, I doubt if I should have pulled through but for them. We call them 'savages' but your savage is often 'whiter' than the majority of 'whites'."[139] Jenness learned not only the forms of Eskimo life, but also developed an understanding of its substance; he returned from the expedition with a personal as well as a scientific commitment to "the people of the twilight," and when he wrote his popular book of that title, he did so in hopes that publicity given to the needs of the Inuit might lead to constructive measures for their welfare.[140]

---

[136] VS to (Frank) Alexander Wetmore, Acting Secretary of the Smithsonian Institution, 21 Aug. 1928, DCL VS MSS Correspondence Box 21: "I have lived with the Eskimos as one of themselves for more than eleven years as against one for Boas, for instance, or none for Kroeber. If I say it myself as shouldn't (a saying I learned in Massachusetts), there are only two men now living who have a wide first hand knowledge of the Eskimos, Knud Rasmussen and myself."

[137] D. Jenness, *The People of the Twilight* (New York, 1928; reprinted Chicago, 1959; 7th impression, 1975).

[138] Jenness to von Zedlitz, Bernard Harbour, 11 Jan. 1916, PAC MG30 B89 vol. 2, file 1913–17.

[139] Ibid.          [140] Jenness to VS n.d., DCL VS MSS Correspondence Box 19.

Jenness's work was necessarily largely independent of that of the rest of the southern party, for whom geology, topography, and zoology were priorities. In March 1915, Chipman and O'Neill began a survey of the coast west from Bernard Harbour to Darnley Bay in Amundsen Gulf. In April and May, Cox and Sullivan surveyed stretches of the coast of Coronation gulf; later in the season, Cox embarked on a detailed survey of the coast eastward to Bathurst Inlet – an intricate piece of work.[141] Wilkins wrote to Leffingwell at his base on Flaxman Island that the southern party had carried out "a successful programme of work this spring covering the country geologically and topographically speaking from Cape Parry to Rae River finding many interesting outcrops but nothing of commercial value."[142]

Johansen's work had gone on through the year. He made an extensive collection of marine and freshwater specimens around Bernard Harbour, as well as a "practically complete" collection of the local flora, and he arranged the large collections made in Alaska and on Herschel Island in the Yukon Territory. He had managed to do a fair amount of dredging, not only in the harbor, but also in the waters of Dolphin and Union Strait, down to a depth of fifty fathoms, from which he obtained a good range of specimens.[143]

Johansen was not good as a member of a team, but was committed to his work: he wrote in frustration to Stefansson that his colleagues in the southern party did not share his sense of the importance of marine work, and Stefansson, glad to have an ally in the enemy's camp, agreed that it was a shame that so little oceanographic and marine biological work was done.[144]

After a slow start, Anderson's work had gone well, and by July 1915, he had collected about five hundred specimens of birds and animals, and made numerous photographs of the nests of arctic birds. He was frustrated by the paucity of the fauna around Bernard Harbour, "where we are perforce confined during the most interesting and fruitful part of the year for field collecting, i.e. the early part of the summer, on account of being unable to move either by sled or by boat."

Wilkins had contributed to the work of the expedition as well as providing future publicity for it. He had

---

141 Cooke and Holland, *The Exploration of Northern Canada*, p. 337. O'Neill to Le Roy, Port Epworth, Coronation Gulf, 17 July 1915, PAC MG30 B66 vol. 1 file 4(b).

142 Wilkins to E. deK. Leffingwell, 25 July 1915, PAC MG30 B40 vol. 2.

143 RMA report CAE, Bernard Harbour, 13 Jan. 1916, MSS CAE NMNS.

144 Johansen to VS, 16 Dec. 1916, DCL VS MSS 98 Folder IV-13. VS to Johansen, Barter Island, 18 Sept. 1917, PAC MG30 B66 vol. 1.

brought a cinematograph outfit with him from the northern party's base on Banks Island, and exposed about 2,000 feet of cinematograph film, principally views of the local Eskimos. He also obtained a small collection of Eskimo clothing, weapons, and instruments to send out for advertising purposes. Mr. Wilkins has made a very good series of portrait studies of most of the local Eskimos, men, women, and children, in full view and in profile, for Mr. Jenness' ethnological work.

Wilkins also made numerous botanical and zoological photographs, which Anderson welcomed for their scientific as well as their artistic value.[145]

In the spring and early summer of 1916, the main project was the completion of the survey of the Bathurst Inlet region, and filling in the gaps in the coastal survey west from Bathurst Inlet. Surveying and zoological collections were carried out "substantially as planned."[146] The southern party had worked hard from 1913, and effectively from 1914 to 1916, when they completed their program and headed for home and, in some cases, for the Great War, then two years old. The delays in the mail meant that from the end of May 1914 until November 1915, the expedition had had no news from the outside world; November's mail had therefore been intensely interesting to them, heightening their desire to finish their work.[147] That the Canadian government kept the expedition in the field in wartime suggests more than a commitment to science; hydrographic and territorial questions were both important for Canada as a maritime nation.

The ice broke up unusually early in 1916. On 13 July Anderson and his colleagues left Bernard Harbour, reaching Nome, Alaska, on 15 August.[148] Stefansson and his colleagues would remain in the Arctic for another two years, discovering new land and journeying over the ice.

*Results.* In November 1916, Anderson, back in Canada, could look back on effective leadership and fieldwork well performed, and look forward to the demanding business of preparing scientific reports. In his own field, and ignoring contributions still to come from the northern party, a preliminary account listed 616 specimens of 73 species of birds, and 422 specimens of 22 species of mammals. There were grounds for general satisfaction:

[145]  RMA, "Canadian Arctic Expedition 1915," GSC Summary Reports, Sessional Paper no. 26, 6 George V., A. 1916; the passage quoted is on p. 234. The reports are from RMA's formal letters to the director of the GSC.

[146]  [RMA] summary of work of the southern party, CAE 1913–16, PAC MG30 B40 vol. 10 file 4.

[147]  Jenness to von Zedlitz, 11 Jan. 1916.

[148]  RMA to MBA, cable 15 Aug. 1916, PAC MG30 B40 vol. 2.

All the members of the scientific staff of the Southern Party of the Expedition are here now working on their reports. To some extent I had covered some of the same ground zoologically while on the former expedition [i.e., the Stefansson-Anderson expedition of 1908–1912], but we brought back large collections in all branches of natural science – geology and mineralogy, terrestrial and marine biology, botany, entomology, ethnology and archaeology, and when the reports are all in we shall have made quite a showing. In addition to the collections our party made surveys of several hundred miles, scale 10 miles to the inch, of several hundred miles of the coast line of Yukon Territory and the Northwest Territories and adjacent islands, of the Mackenzie River delta, and of a number of other Arctic rivers.

Getting specimens identified and plates made meant conferring with experts at home and abroad. The Dominion Department of Agriculture was a great help with the entomology, and Anderson found he had to go to the United States to sort out the mammals and birds. He spent ten days at the end of April in Washington, "and finally succeeded in identifying my Arctic mammals, comparing them with the large series of Northern stuff in the collections of the U.S. Biological Survey and the U.S. National Museum. Also stayed three days in New York, studying specimens in the American Museum of Natural History."[149]

It is clear that Kennicott, Drexler, Preble, and other American collectors had done a splendid job in the American and Canadian Arctic,[150] and only now were comparable Canadian collections being made. By 1918 Anderson had his notes in shape, and his specimens determined. Thus prepared, he embarked on the real writing of his report on the animal life of the western Arctic.

In the midst of this sustained effort, Stefansson and his northern party reappeared, and returned to the world of newspapers, politicians, and professional jealousies. Stefansson gave press interviews, and headlines followed: "Great Potential Wealth is Lying in Arctic Lands," "Stefansson to make recommendations for reclamation of vast continent."[151] Stefansson's discoveries of new islands in the archipelago scarcely amounted to the discovery of a new continent, but he was convinced that his tidal and continental shelf determinations reinforced Harris's theory. Unfortunately, there was by now American evidence against it. The United States

---

[149]  RMA to J. H. Brownlee, 30 April 1919, PAC MG30 B40 vol. 2.
[150]  The Smithsonian collections in Arctic zoology predated those at the American Museum of Natural History. See, e.g., NARS RG. 401/1, Admiral R. E. Peary, letters received 1890: J. A. Allen, American Museum of Natural History, to Peary, 2 Sept. 1890: "I . . . would say that we have very little Arctic material in the museum, except in the Department of Ethnology. We have of course a few Arctic birds & mammals, but have thus far given no special attention to Arctic zoology."
[151]  Chipman to RMA, 2 Sept. 1918, quoting Vancouver paper of that date.

"Crocker Land" Expedition under Donald Baxter MacMillan had set out in the same year as the Canadian Arctic Expedition. The *Sunday Press* that summer had a banner headline, "NEW SEARCH FOR A NEW POLAR CONTINENT," above a magnificently fanciful picture of mountainous Crocker Land, and a photograph of Stefansson, elevated by the press on this occasion to the rank of captain.[152] When MacMillan returned in 1917, the question was resolved in the negative.[153] Stefansson was not convinced. He later wrote to George Sherwood, acting director of the American Museum of Natural History, which had cosponsored MacMillan's expedition, asking whether any soundings had been taken anywhere near the supposed location of Crocker Land:

I understand from Macmillan's popular book[154] that no soundings were taken, but I want to be quite sure, for it has an important bearing on the probability of land in that region. If deep soundings were taken, then land is almost certainly not to be found anywhere near; but if shallow soundings were taken, land may have been a few miles beyond the turning point. If there were no soundings, the conclusions to be drawn from the journey naturally become less definite.[155]

Sherwood replied that as far as he knew, there had been no adequate soundings[156] – so Stefansson was free to keep an open mind, or rather a bias toward the theory of an arctic continent. His own data were never subjected to public scientific scrutiny: the program of the American Philosophical Society for 24 April 1919 would list Stefansson's paper on his oceanographic results, but the paper was not given – Stefansson, who had not written the paper, explained that Canadian government work intervened.[157] If so, the Canadian government never saw the results. Similarly, the United States Coast and Geodetic Society waited in vain for Stefansson's tidal observations.[158]

Stefansson was not in the habit of writing scientific reports. Nevertheless, he had obtained government support for the Canadian Arctic Expedition as a scientific enterprise, and he needed to reassure Ottawa that the expedition had been justified by its scientific results. From Seattle, on his way back from the Arctic, he cabled Desbarats that the records and scien-

[152]  DCL VS MSS 98 Folder IV-15.
[153]  D. B. MacMillan, "Geographical Report of the Crocker Land Expedition, 1913–1917," *American Museum of Natural History: Bulletin 56* (1928) 379–435.
[154]  MacMillan, *Four Years in the White North* (New York, [1918]).
[155]  VS, 27 May 1926, DCL VS MSS 98 Folder VI-15.
[156]  Sherwood to VS, 11 June 1926, DCL VS MSS 98 Folder VI-15.
[157]  DCL VS Correspondence Box 3.
[158]  DCL VS MSS 98 Folder IV-40, R. L. Faris to VS, 25 March 1918.

tific specimens of the northern party were on their way to Ottawa: "COL-LECTIONS INCLUDE BIRDS AND EGGS SKINS AND BONES OF MAMMALS FISHES PRESERVED IN LIQUIDS AND GEOLOGICAL AND ETHNOLOGICAL SPECIMENS."[159] The specimens duly arrived, and the zoological items when listed occupied three pages[160] – a small haul for five years' work, but then zoology had been Anderson's responsibility, and was low in Stefansson's official and personal priorities. The ethnological items were more important to him, and he noted that the collection made by Harold Noice, who had been with him in 1915–16, might well be the last that would ever come out of the Copper Eskimo district: "Hereafter they will make no stone pots, copper knives or bows intended for actual use. Fashions in clothing are also rapidly changing. . . . "[161] Contact with whites meant irreversible change in northern cultures.

While Anderson worked on the detailed zoological reports of the expedition, and published brief preliminary accounts, which lacked much in the way of detailed zoological data,[162] Stefansson painted a larger picture with a broader brush, modestly[163] entitling his preliminary account "Solving the Problem of the Arctic."[164] In an interview with the *Christian Science Monitor*, Stefansson vented his irritation with the scientists of the southern party:

> Scientists who are in the civil service seem gradually to lose the scientific spirit, if they ever had it, and to become mere civil servants in outlook. . . . What one needs to be on an exploring expedition is a scientist of the Darwin type, who works day and night to the limits of his strength and whose mind is open to truth of every sort.[165]

[159]  VS to Desbarats, 5 Oct. 1918, DCL VS MSS 98 Folder IV-22.

[160]  PAC RG 45 vol. 67 folder 4078 B1.

[161]  VS to Dr. J. Walter Fewkes, Chief, Bureau of American Ethnology, Smithsonian Institution, 13 May 1922, DCL VS MSS Correspondence Box 9.

[162]  RMA, "Recent Zoological Explorations in the Western Arctic," *Journal of the Washington Academy of Science 9* (1919) 312–14; "Recent Explorations on the Canadian Arctic Coast," *Geographical Review 4* (1917) 241–66.

[163]  One of the most remarkable tributes to VS was a letter from Isaiah Bowman, Secretary of the Explorer's Club, to Sir John Scott Keltie, RGS, 11 Feb. 1920 (DCL VS MSS Correspondence Box 5), which paints a wholly unfamiliar portrait of its subject's character: "We think very highly of Stefansson's work in the Arctic, and have expressed our opinion in the form of a gold medal presented to him in December, 1918. He is far less sensational and pretentious than most American Arctic explorers, and in our opinion no one else has learned to live so successfully in the Arctic climate. He also has the merit of modesty, and if you have not met him before I should think you would enjoy a conversation with him."

[164]  VS, *Harper's Magazine 138* (1919) 577–90, 721–35; *139* 34–47, 193–203, 386–98, 709–20.

[165]  Interview 19 and 20 May 1919, clipping in PAC MG30 B40 vol. 10.

Stefansson may have seen himself as "a scientist of the Darwin type." Anderson saw him as a fraud, "a professional faker who was not even indirectly concerned with the scientific work."[166] Having developed a lively antagonism to Stefansson during their conflicts over the expedition's priorities, Anderson added to his contempt for him as a scientist, contempt for him as a man. By now, he had come to see Stefansson's separation from *Karluk* as "yellow and contemptible desertion . . . after getting her into a hole. / A commander *first to leave his ship and crew,* under such conditions, has forfeited the respect of every man who has been in the Arctic and every man who has any instinct of the duty of a seafaring man, and one should hope, of an explorer."[167] For Anderson, Stefansson was all of a piece, talking to journalists and politicians[168] for self-promotion, instead of getting down to the serious business of science. Anderson wrote in 1919 that Stefansson had thus far been scarcely involved in the preparation of scientific reports, although he should have something to contribute concerning tides, currents, geography, and some ethnology.[169] Anderson wanted to use material from Stefansson when he could get it. As he wrote to Bartlett, the commander of *Karluk* whom Stefansson blamed for the disaster, but whom Anderson admired:

> Whenever there is any material from the Northern Party we have been putting it into a joint report, but one really must have something in the way of specimens to base a biological report on. For example, in the Insect volume, Vol. III, Part D, Mr. Castel sent in a louse from a fox taken in 1918, so that gave the Northern Party a look-in. Similarly, one of the boys brought back a few broken butterflies from Victoria Island, some of them being identifiable as records, so that will give representation in Part I Lepidoptera. . . . One of Stefansson's Eskimos sent back some bumblebee larvae from Melville Island, so that adds a little to Part G, Hymenoptera. . . . We are giving full credit to any of the Northern Party, where it is due; though in most cases it would in justice be a very small footnote.[170]

At this stage, there were eight volumes already arranged in the expedition's biological series, with the work divided up among some fifty specialists in Canada, the United States, and Europe, with Anderson superintending the process, and writing a book on arctic mammals and part of one on arctic birds; neither of these volumes was published, nor, as far

[166]   RMA to Dr. W. B. Bell, Biological Survey, U.S. Department of Agriculture, Washington, D.C., 28 March 1921, CAE MSS NMNS.
[167]   RMA to Isaiah Bowman, 6 November 1918, DCL VS MSS 98 Folder IV-22.
[168]   Borden had VS to a private lunch at his house; see George W. Yates (Borden's secretary) to VS, 9 July 1919, and annotation by VS, DCL VS Correspondence Box 3.
[169]   [RMA] to Mme E. Beuchat (mother of Henri), 5 Oct. 1919, MSS CAE NMNS.
[170]   RMA to R. A. Bartlett, 1 Nov. 1919, PAC MG30 B40 vol. 3.

as I know, even completed. Anderson also planned to write a narrative account of the southern party.[171] Jenness was back from the war, in which he had served as a private, and was working on what would turn into five volumes of reports on *The Copper Eskimos, Eskimo Folk-Lore, Eskimo Songs, Eskimo Language and Technology,* and the *Material Culture of the Copper Eskimo.* Stefansson considered that Jenness's results were likely to be the most important scientific results of any single department of the expedition.[172] Two volumes of botanical reports were produced: part of one volume was written by James Macoun, curator of the National Herbarium and Anderson's best friend in Ottawa, who was planning an entire volume on arctic flora, but who died from cancer in January 1920.[173] There was a volume on fishes, one on crustacea, one on mollusks, echinoderms, and so forth, another on annelids, parasitic worms, and protozoans, as well as one on plankton, diatoms, and tides. Johansen did the preliminary sorting of marine invertebrates for shipment to specialists, and spent some time working up the expedition's results.[174] His work was erratic, and the Victoria Memorial Museum released him.[175] O'Neill wrote up the geology; Chipman and Cox undertook the geography. Both the quantity and the quality of the work were impressive, prompting a reviewer in *Science* to expect that the reports would be "the most extensive and comprehensive publication on

---

[171]   RMA to his nephew Horace E. Anderson, 1 Nov. 1919, PAC MG30 B40 vol. 3.
[172]   VS to Jenness, 25 May 1926, DCL VS MSS Correspondence Box 16.
[173]   RMA to J. E. Anderson, 14 Nov. 1919. For biographical information on James Macoun see W. A. Waiser, *The Field Naturalist: John Macoun, the Geological Survey, and Natural Science* (Toronto, 1989), numerous entries.
[174]   See, e.g., Johansen to Prof. Prince, 1 Dec. 1917, MSS CAE NMNS; FJ to VS, Aug. 1918, DCL MSS VS Correspondence Box 8.
[175]   See RMA to Dr. Albert Mann, Smithsonian Institution, 26 Nov. 1925, Smithsonian Institution Archives Record Unit 7292, Paul S. Conger Papers 1913–79, Box 4: "We have had three prominent Danish scientists in Ottawa within the past year or two and all of them agreed that Mr. J. whom they knew in the old country, was erratically abnormal and irresponsible at times. One of them went so far as to say that they feared for his sanity on the Denmark [*sic*] expedition to Greenland in 1906–08 and were relieved to get him back to Denmark. Then they shipped him off to the U.S. and Washington gave him a job and ultimately wished him off on the Canadian Arctic Expedition, 1913–18. It makes the Danes wonder about scientific institutions in Canada and the U.S. / He was only mildly erratic on the CAE. from 1913 to 1916, and collected a lot of miscellaneous stuff, most of which he did not know very much about, but on the whole I think his field labels were approximately correct. He had no axe to grind at that time. After we came back, he was given a temporary job in the Museum on my recommendation, to sort out specimens and do some cataloging. This lasted for a couple of years, until he got so impossible that the Department had to discharge him . . . / In 1913–1916 Mr. J. was really trying to work and make a creditable record, according to his lights, by the material he collected, while of late years he has had delusions of grandeur, indulges in . . . carping criticism to show that he is still doing something useful."

Canadian and Western Arctic biology" since Richardson's *Fauna* and Hooker's *Flora Boreali-Americana*.[176] They were right, although less was published than had been announced. Neither Stefansson nor Anderson would publish their official narratives, with Stefansson insisting on having an idiosyncratic last word, and Anderson opposing Stefansson while asserting the need to wait until all the scientific results were in.[177]

*Postscript: "The Friendly Arctic."* Stefansson's account of the Canadian Arctic Expedition was published in 1921, under the title, typical of its author, *The Friendly Arctic: The Story of Five Years in Polar Regions* (New York, 1921); the frontispiece showed Stefansson with harpoon and rifle, hauling home a seal across the ice. Borden wrote an approving introduction, and Greely and Peary, the latter in his last public appearance, paid tribute to Stefansson.

Stefansson wrote well, and was the hero of his own tale. In the spring of 1922 his book was among the ten best-sellers in America.[178] Anderson, who knew in advance that the book would have little to contribute to science, was predictably incensed when he saw it. Stefansson, "a professional entertainer and promoter," had written "a lie to a page."[179] Mrs. Anderson stirred the pot: "Steffy does not realize that he is fighting the Canadian Government when he fights Rudolph. . . . [T]here is no room at the present time in Canada for VS and Rudolph at the same time unless Steffy reforms. . . . He must be taught to play fair."[180] Personal differences apart, Stefansson's message was that, by emulating the Inuit, one could live off the land, and that the Arctic could thus be seen as a friendly, albeit demanding, place. By implication, those who died in the Arctic, including the sixteen lost during the Canadian Arctic Expedition, either rejected or did not know how to apply Stefansson's insight. Jenness, who had managed to learn a lot from the Inuit about survival, had a greater sensitivity to the environment, and to the impact on it of Stefansson's prescription. Living off the country

involves the destruction of entire herds of caribou and musk-oxen, males, females, and young. On Melville Island, one of the largest islands in the north where musk-oxen are still found, Mr. Stefansson and his companions killed, on their own estimate, about one tenth of the musk-oxen (400 out of an estimated four thousand). One can easily imagine how long the supply of game would last under these con-

---

176  *Science* N.S. *51* (1920) 167–9 at 168.
177  Zaslow, *Reading the Rocks* (Ottawa, 1975), pp. 324–5.
178  VS to Arthur R. Ford, 2 June 1922, PAC MG30 B81 vol. 2.
179  RMA to Fleming, 28 Feb. 1922, NMNS MSS CAE.
180  MBA to Mrs. A. Allstrand, 5 Jan. 1922, PAC RMA/15, quoted in Diubaldo, *Stefansson and the Canadian Arctic*, p. 196.

ditions. The musk-oxen have already been almost exterminated on the mainland of America and in Greenland; on Victoria and Banks Islands they were destroyed by the Eskimos prior to 1913, and the only places where they still remain in any numbers are Ellesmere Island and a few smaller islands adjacent to it . . . [181]

Jenness was right; but Stefansson had another message, one to which Canadian politicians were more sensitive than scientists, and that was about sovereignty. As Stefansson wrote, Canadians had the best chance of all nationalities to hold the North, but they had to act, "or the lands to the North of Barrow Strait and Lancaster Sound are likely to become foreign territory – perhaps American, perhaps Danish, more likely both. Under 'International Law' we cannot expect to hold our Northern possessions by merely coloring them red in Atlases published at Ottawa."[182] Effective occupation was needed; science as a state activity offered one form of such occupation. Stefansson's eagerness to help Canada toward sovereignty in the North extended to Wrangel Island, which, by the 1920s, he saw as a desirable station for transpolar aviation. The resulting tangle, involving Russia and other nations, was an embarrassment to the government in Ottawa, and helped undermine Stefansson's position in Canada.

Another route to sovereignty, unconsidered then, but important now, was through the recognition of the territorial rights of native peoples. Science, sovereignty, security, native rights, and environmental issues are seen today as interdependent. The seeds for that interdependence were already germinating in the aftermath of the Canadian Arctic Expedition under Stefansson and Anderson.

[181] Jenness, review of *The Friendly Arctic* in *Science* 56 (1922) offprint pp. 3–4. Jenness also quoted a number of documents contradicting VS's account of his clashes with the other members of the CAE, especially RMA and the southern party.

[182] VS to A. R. Ford, 2 June 1922, PAC MG30 B81 vol. 2.

# Afterword

In 1933, the Canadian ornithologist Percy Taverner wrote that there was still "a great deal to be done in arctic ornithology, as much to correct old errors as to make new discoveries. The trouble is that so much of the arctic work has been done by inexperienced and casual observers . . . "[1] We have seen that John Rae was not always accurate in recording the place where a specimen was taken, and that Feilden's gifts of arctic birds to the museum in Norwich sometimes failed to record place. For Charles Darwin, who was interested in the migration of species and in geographical distribution, such errors and omissions would have been serious, and so they were for later naturalists and zoologists. But if there were difficulties and gaps, there were also great gains in knowledge. The century from 1818 was one that, starting from almost total ignorance, witnessed the construction of a coherent flora and fauna, tied to biogeography; the variability and vulnerability of arctic systems, with their dramatic fluctuations in population and small number of species, were clear. Similar progress was made in geomagnetism, meteorology, geology, hydrography, and ethnology or anthropology. By 1918, the main landforms, currents, ice movements, geological features, human communities, and animal and plant populations had been recorded. These results could not be definitive, although they were sometimes final. Contact between isolated tribes and southern explorers led to rapid and irreversible changes: what the first arctic anthropologists and ethnologists found was often simply not there for later scientists to study. In spite, however, of occasional bouts of optimism or hubris by a few physicists, the sciences constitute an enterprise not susceptible of ultimate resolution:

---

[1] Taverner to W. B. Alexander, 6 Dec. 1933, quoted in J. Cranmer-Byng, *Birds and Men: A Life of Percy Taverner, Canadian Ornithologist, 1875–1947* (in preparation).

425

whether it be through refining or extending past observations, or moving to new questions requiring new data, scientists will always be engaged in an unfinished quest. That quest was energetically pursued in the Canadian Arctic in the nineteenth and early twentieth centuries, resulting in the acquisition of an impressive body of knowledge, much of it remarkably comprehensive.

There were inevitably major lacunas. Arctic geological work had been mainly coastal, so that the geology of the interior of the islands was mostly unknown. Studies of the sea bottom scarcely existed. Arctic oceanography was fragmentary, principally because of the enormous difficulties of studying oceans through thick ice, and for the same reason marine biology was extremely spotty. Glaciology was scarcely a coherent field until encouraged by the *Titanic*'s collision with an iceberg in 1912.[2] Meteorological and geomagnetic studies could be and were carried out throughout the year, but, especially in the case of the latter, mainly from fixed bases during the winter; summer observations from a moving ship or during a sledging trip were less frequent, and less regular. The reverse problem was true for the study of animals and birds. Most data in those fields were from spring and summer, when the land of the midnight sun was entirely visible. Even today, there are few year-round studies like David Gray's work on the musk-ox.[3] We lack extended studies; in so variable and extreme an environment as the Arctic, fluctuations often need to be studied over a term of years, not just seasons. Some of our concerns have shifted, to the human impact on the global environment, to holes in the ozone layer over the poles, to the smog from pollution that too often accompanies an arctic inversion, and to the use of seabirds, near the top of the food web, as environmental monitors.

We have seen the emergence of international work in the circumpolar North, with its first dramatic vindication in the International Polar Year of 1882–3. A second IPY followed fifty years later, and then, after only twenty-five years, came the International Geophysical Year.[4]

Alongside this internationalism was an emergent northern nationalism in Canada's concerns with arctic sovereignty. Political boundaries, drawn

---

[2] Even so, it was another 20 years before systematic glaciology and physical oceanography led to significant publication. See E. H. Smith, F. M. Soule, and O. Mosby, *The 'Marion' and 'General Green' Expeditions to Davis Strait and Labrador Sea, under direction of the United States Coast Guard, 1928–1931–1933–1934–1935: Part 2: Scientific Results* (Washington, D.C., 1937).

[3] D. Gray, *The Muskoxen of Polar Bear Pass* (Markham, Ontario, 1987).

[4] J. T. Wilson, *IGY: The Year of the New Moons* (Toronto, 1961).

up in the South, are of no obvious relevance in the North. Greenland Inuit used Ellesmere Island, whereupon Ottawa advised Denmark that hunting and trapping there were for Canadian natives only. The Danish response was that the contested territory was no-man's land; and in an attempt to enforce Canadian sovereignty, Ottawa ordered the Eastern Arctic Patrol to monitor the area on an annual basis.[5] Scientific work was also subject to territorial concerns. In 1925, the Northwest Territories Act was amended to require all scientists and explorers to obtain permits before entering the Canadian Arctic,[6] and more recently, in 1976, Canada drew up guidelines for scientific activities in the North that ensured a measure of control by Ottawa.[7] Such control does not rule out internationalism: in 1988, Canada, Denmark/Greenland,[8] Finland, Iceland, Norway, Sweden, the Soviet Union, and the United States agreed to set up an International Arctic Science Committee for cooperation and coordination in arctic scientific research.[9]

International politics have pushed in the same direction. Mikhail Gorbachev's Murmansk speech in September 1987 was a landmark, arguing for international cooperation and demilitarization in the Arctic. The Inuit have been advancing their own international organization, the Inuit Circumpolar Conference (ICC), which was set up in 1977 to further their shared concerns, including the protection of the arctic environment from oil spills, militarization, and other assaults.

Arctic science, from Ottawa's standpoint, has long emerged from its colonial phase. There is, however, a real sense in which northern regions around the pole have colonial economies in relation to southern governments and industries.[10] Northern development has generally been subordinated to southern economic goals; the ICC and national Inuit organizations have succeeded increasingly in getting their own concerns on the agenda, a difficult business when the scale of resource and energy developments in the North employs so many southerners. Apart from cultural and environmental differences, the Inuit are faced by the fact that their numbers are few, something that does not help them with elected governments. Nonetheless, they constitute a majority of the population over one-third of Canada: in

---

[5] S. Grant, *Sovereignty or Security?* (Vancouver, 1988) pp. 248, 14.     [6] Ibid., p. 17.
[7] Armstrong, Rogers, and Rowley, *The Circumpolar North* (London, 1978) pp. 268–9.
[8] Greenland is largely independent in domestic policy, but in foreign policy functions as Danish Greenland.
[9] VanderZwaag in VanderZwaag and Lamson, eds., *The Challenge of Arctic Shipping* (Montreal and Kingston, 1990) p. 229.
[10] Armstrong et al., *The Circumpolar North*, p. 265.

the current constitutional debates about Canada's future and Quebec separatism, we are at last hearing the proposal that native peoples should be asked to determine their own political future.[11]

This may have more than incidental importance for Canada. In 1969 and 1970 the United States sent the *Manhattan* through the Northwest Passage without first seeking Canadian permission, and repeated the stunt in 1985 with the *Polar Sea*. These were clear challenges to Canadian sovereignty over the waters of the arctic archipelago. It has been suggested that Inuit hunters living on the ice may provide Canada's best evidence of sovereignty.[12] Evidence is one thing; exercising sovereignty is another, requiring at the very least the ability to navigate the archipelago, and to assist ships of other nations in making the Northwest Passage, just as Soviet icebreakers operate in the Northeast Passage. The legal, technological, environmental, political, and military problems are all great, but if Canada is to take the lead in the passage, it will need, besides cooperation with the circumpolar Inuit and the resolution of defense issues, excellence in arctic science.[13] There are still plenty of challenges.

[11]  S. D. Grant, "Northern Identity: Barometer or Convector for National Unity," in J. L. Granatstein and K. McNaught, eds., *"English Canada" Speaks Out* (Toronto, 1991) pp. 150–61 at 161.

[12]  P. Jull, "Inuit Concerns and Environmental Assessment," in VanderZwaag and Lamson, *Arctic Shipping* p. 139. For a full discussion of the legal issues, see D. Pharand, *Canada's Arctic Waters in International Law* (Cambridge, 1988).

[13]  Cf. F. Griffiths, *Politics of the Northwest Passage* (Kingston and Montreal, 1987) p. 21. An exciting account of a wide range of current scientific activity is given in C. R. Harington, ed., *Canada's Missing Dimension: Science and History in the Canadian Arctic Islands,* 2 vols. (Ottawa, 1990): in spite of its title, this work emphasizes the current state of knowledge, and most of it looks forward rather than backward.

# Index

Sciences in the nineteenth century often had boundaries different from modern sciences with the same name; there was, for example, a substantial overlap between hydrography and oceanography, and, rather than seek to draw uncertain distinctions, I have placed entries for both under the joint heading "oceanography and hydrography." Similarly, anthropology and ethnology are grouped together under the former heading. Scientific entries are for the most part grouped under major arctic sciences, such as geomagnetism, zoology, and meteorology. Expeditions are listed either under their commanding officer or their ship. Individuals are listed by name, not by title. In some instances, cross-references are given, for example, "Melville, Lord, *see* Dundas, Robert Saunders." I have followed the spelling of British names given in the *Dictionary of National Biography* (e.g., McClure, McCormick).

432    *Index*

Forster, Johann Reinhold, 34, 177
Fort Conger, 318–20
Fort Rae, 315, 324–7, 348 n43, 353
Fort Yukon, 348, 352
Foster, Henry, 80, 142
Fox, Robert Were, 4, 152–6
Foxe, Luke, 23
*Fram*, 362–4, 379, 389
Franklin, Jane, 203, 209–10, 218–19, 225, 228, 251
Franklin, John, 78, 129–30, 136, 140–1, 218; 1819–22 expedition, 103–10; 1825–7 expedition, 110–25, 343; last expedition, 1845–, 198–202; fate, discovery of, 224–5, 230–1; and Fox, Robert Were, 4, 154–5; and geomagnetism, 4, 161, 199–201, 215–16; on North Pole expedition (1818), 44; Royal Society of London, election to Council, 143; *see also* Franklin search expeditions
Franklin search expeditions: and the Arctic Council, 239; Austin, 210; Belcher, 216; cost, financial, 218–19; Hall, 248–51; Inglefield, 211; Kane, 219–23; Kellett, 208; Maguire, 215; McClintock, 212–13, 228–31; McClure, 213–15; Osborn, 217–18; Penny, 210–11; Rae, 224–5; Richardson and Rae, 1847–9 expedition, 203–7; Ross, James, 208; Ross, John, 202–3, 208–10; *see also* Franklin, Jane; botany; geology; Haughton, Samuel; natural history; phytogeography; Sutherland, Peter Cormack
Fritz, Herman, 326
Frobisher, Martin, 21, 248
Frobisher Bay, 21, 248
fur trade, *see* Hudson's Bay Company

Gauss, Carl Friedrich, 9, 148, 150, 156
Gellibrand, Henry, 30
geodesy: earth, ellipticity of, measurement by seconds pendulum, 35, 54, 75, 80, 83, 246, 252; *see also* Sabine, Edward
geography: and Canadian Arctic Expedition, 393, 398, 409, 415, 421; and Franklin search expeditions, 207; a science, 3–4, 27, 60, 143, 167–9, 265, 312, 338–9; *see also* exploration, conflict with science
Geological Society of London, 117, 256
Geological Survey of Canada, 172–3, 323, 342, 344, 356–61, 371–6; and Canadian Arctic Expedition, 390, 392–3, 398, 400
geology, 117, 139, 426; in *Admiralty Manual*, 170–6; and British Arctic Expedition, 269–71; and Canadian Arctic Expedition, 392–3, 415, 421; Cape Raw-

son Beds, 284; fossil mammals, 208; and Franklin's 1819–22 expedition, 105, 109; Franklin's 1825–7 expedition, 120–1, 125; Franklin search expeditions, 211–15, 223, 229–30, 232–4; and *Neptune* 1903–4 expedition, 372–5; and Northwest Passage expeditions, 83, 92, 192; and Sverdrup expedition, 368–71; utility of, 5, 358–60, 372, 375; *see also* arctic archipelago, Canadian; Fitton, William; Geological Survey of Canada; Jameson, Robert; Lyell, Charles; mineralogy; Murchison, Roderick Impey
geomagnetism, 28, 48, 119–20, 125, 198–201, 247, 307; and British Arctic Expedition 1875–6, 278–82, 292; and Canadian Arctic Expedition, 393; and Carnegie Institution, 381–2; circle, Barrow, 222, 267, 281–2; circle, Fox, 152–6, 159–60, 201, 267, 282, 292; compasses, 81–2; compasses, azimuth, 49, 108, 195, 267; compasses, effects of rock on, 92; compasses, effects of ship's iron on, 70, 82–3, 161, 201; compasses in high latitudes, 65, 70, 103; declinometer, 150, 206, 215, 221; dip, 28, 48, 125, 149–50, 158, 193, 282; dip circle, Casella, 331; dip circle, Nairne and Blunt, 35; and Franklin's 1819–22 expedition, 105, 107–8; and Franklin's last expedition, 4, 198–201, 216; and Franklin search expeditions, 215, 221–2; German leadership in, 157; Gilbert, William, 23, 25; inductor, earth, Edelmann, 331; instruments, Smithsonian's, 249; intensity, 50, 81, 151–2, 282; and International Polar Year, 317–19, 321, 324–6, 331–3; international studies, 9, 148, 199, 307; isodynamic lines, 149–58; observatories, arctic, 197, 215, 230, 278–81, 318, 331; observatories, colonial, 156–8, 199; observatory, Dublin, 156, 340; observatories, Gauss's model for, 150, 156, 200–1; pole(s), magnetic, 65, 70, 149, 158, 231, 324, 326, 373; pole, magnetic dip, discovery of, 91, 158; *magnetische Verein*, 148; and, Parry 65–6; term days, 160, 222, 307, 333; variation, 23, 28, 30, 48–9, 120, 149, 193; variation, secular, 29–30, 81; *see also* Fox, Robert Were; Lefroy, John Henry; Sabine, Edward
geophysics, 313; earth temperatures, 325; *see also* geomagnetism
German expeditions, 260; *see also* International Polar Year
Giese, W., 328
Gilbert, William, 23, 25, 48

Index 435

Printed in the United States
By Bookmasters